Selected Titles in This Series

AF699249

282 **Arlan Ramsay and Jean Renault, Editors,** Groupoids in analysis, geometry, and physics, 2001

281 **Vadim Olshevsky, Editor,** Structured matrices in mathematics, computer science, and engineering II, 2001

280 **Vadim Olshevsky, Editor,** Structured matrices in mathematics, computer science, and engineering I, 2001

279 **Alejandro Adem, Gunnar Carlsson, and Ralph Cohen, Editors,** Topology, geometry, and algebra: Interactions and new directions, 2001

278 **Eric Todd Quinto, Leon Ehrenpreis, Adel Faridani, Fulton Gonzalez, and Eric Grinberg, Editors,** Radon transforms and tomography, 2001

277 **Luca Capogna and Loredana Lanzani, Editors,** Harmonic analysis and boundary value problems, 2001

276 **Emma Previato, Editor,** Advances in algebraic geometry motivated by physics, 2001

275 **Alfred G. Noël, Earl Barnes, and Sonya A. F. Stephens, Editors,** Council for African American researchers in the mathematical sciences: Volume III, 2001

274 **Ken-ichi Maruyama and John W. Rutter, Editors,** Groups of homotopy self-equivalences and related topics, 2001

273 **A. V. Kelarev, R. Göbel, K. M. Rangaswamy, P. Schultz, and C. Vinsonhaler, Editors,** Abelian groups, rings and modules, 2001

272 **Eva Bayer-Fluckiger, David Lewis, and Andrew Ranicki, Editors,** Quadratic forms and their applications, 2000

271 **J. P. C. Greenlees, Robert R. Bruner, and Nicholas Kuhn, Editors,** Homotopy methods in algebraic topology, 2001

270 **Jan Denef, Leonard Lipschitz, Thanases Pheidas, and Jan Van Geel, Editors,** Hilbert's tenth problem: Relations with arithmetic and algebraic geometry, 2000

269 **Mikhail Lyubich, John W. Milnor, and Yair N. Minsky, Editors,** Laminations and foliations in dynamics, geometry and topology, 2001

268 **Robert Gulliver, Walter Littman, and Roberto Triggiani, Editors,** Differential geometric methods in the control of partial differential equations, 2000

267 **Nicolás Andruskiewitsch, Walter Ricardo Ferrer Santos, and Hans-Jürgen Schneider, Editors,** New trends in Hopf algebra theory, 2000

266 **Caroline Grant Melles and Ruth I. Michler, Editors,** Singularities in algebraic and analytic geometry, 2000

265 **Dominique Arlettaz and Kathryn Hess, Editors,** Une dégustation topologique: Homotopy theory in the Swiss Alps, 2000

264 **Kai Yuen Chan, Alexander A. Mikhalev, Man-Keung Siu, Jie-Tai Yu, and Efim I. Zelmanov, Editors,** Combinatorial and computational algebra, 2000

263 **Yan Guo, Editor,** Nonlinear wave equations, 2000

262 **Paul Igodt, Herbert Abels, Yves Félix, and Fritz Grunewald, Editors,** Crystallographic groups and their generalizations, 2000

261 **Gregory Budzban, Philip Feinsilver, and Arun Mukherjea, Editors,** Probability on algebraic structures, 2000

260 **Salvador Pérez-Esteva and Carlos Villegas-Blas, Editors,** First summer school in analysis and mathematical physics: Quantization, the Segal-Bargmann transform and semiclassical analysis, 2000

259 **D. V. Huynh, S. K. Jain, and S. R. López-Permouth, Editors,** Algebra and its applications, 2000

258 **Karsten Grove, Ib Henning Madsen, and Erik Kjær Pedersen, Editors,** Geometry and topology: Aarhus, 2000

(Continued in the back of this publication)

Groupoids in Analysis, Geometry, and Physics

CONTEMPORARY MATHEMATICS

282

Groupoids in Analysis, Geometry, and Physics

AMS-IMS-SIAM Joint Summer Research Conference
on Groupoids in Analysis, Geometry, and Physics
June 20–24, 1999
University of Colorado, Boulder

Arlan Ramsay
Jean Renault
Editors

American Mathematical Society
Providence, Rhode Island

The 1999 AMS-IMS-SIAM Joint Summer Research Conference in the Mathematical Sciences on "Groupoids in Analysis, Geometry, and Physics" was held at the University of Colorado, Boulder on June 20–24, 1999, with support from the National Science Foundation, Grant DMS-9618514.

Any opinions, findings, and conclusions or recommendations expressed in this material are those of the authors and do not necessarily reflect the views of the National Science Foundation.

2000 *Mathematics Subject Classification.* Primary 22A22, 35S05, 46Lxx, 53C12, 53D50, 57R30, 58H05.

Library of Congress Cataloging-in-Publication Data

AMS-IMS-SIAM Joint Summer Research Conference on Groupoids in Analysis, Geometry, and Physics (1999 : University of Colorado)

Groupoids in analysis, geometry, and physics : AMS-IMS-SIAM Joint Summer Research Conference on Groupoids in Analysis, Geometry, and Physics, June 20–24, 1999, University of Colorado, Boulder / Arlan Ramsay, Jean Renault, editors.

p. cm. — (Contemporary mathematics, ISSN 0271-4132 ; 282)

Includes bibliographical references.

ISBN 0-8218-2042-7 (alk. paper)

1. Groupoids—Congresses. I. Ramsay, Arlan. II. Renault, Jean, 1949– III. Title. IV. Contemporary mathematics (American Mathematical Society) ; 282.

QA181.A67 1999
512′.2–dc21

2001046076
CIP

∞ The paper used in this book is acid-free and falls within the guidelines
established to ensure permanence and durability.
Visit the AMS home page at URL: http://www.ams.org/

10 9 8 7 6 5 4 3 2 1   06 05 04 03 02 01

Contents

Contemporary Mathematics
Volume **282**, 2001

Introduction

Arlan Ramsay and Jean Renault

Groupoids were first noticed by Brandt in his study of quadratic forms. The reader can find a reference to his paper in the article by Alan Weinstein in the present volume. Since then they have been used in other areas, and we simply suggest some writers Weinstein mentions in his bibliography: algebraic geometry (Grothendieck), differential geometry (Ehresmann, Mackenzie), algebraic topology (Brown), group representations (Mackey), mathematical physics (Karasev, Weinstein), non-commutative geometry (Connes), and operator algebras (Renault and Connes). Of course there are many other writers.

We hope that the readers of this volume will go beyond it to look at journal articles on groupoids and their uses, as well as other books already in print, and that some of them will join in. We have found groupoids both useful and interesting in their own right.

While the material in these Proceedings is not comprehensive on the topic of groupoids, even in the three areas mentioned, we believe it will still be useful to a researcher or student interested in acquiring insight and information about the uses of these group-like structures. In some cases, there is symmetry of a nature not quite expressible in terms of groups. Other uses of groupoids can involve something of a dynamical nature. Indeed, some of the main examples come from group actions.

It should be noted that in many situations where groupoids have been used, the main emphasis has not been on symmetry or dynamics issues. For example, a foliation is an equivalence relation and has another groupoid associated with it, called the holonomy groupoid. While the implicit symmetry and dynamics are relevant, the groupoid records mostly the structure of the space of leaves and the holonomy. More generally, the use of groupoids is very much related to various notions of orbit equivalence. The point of view that groupoids describe "singular spaces" can be found in work of A. Grothendieck and is prevalent in the non-commutative geometry of A. Connes.

As with groups, measure theoretic, topological and differentiable structures can be added to the algebraic one, according to the context. The purely topological addition can be suitable for the topological context, but many situations in analysis, geometry or physics demand at least the addition of measures, or an awareness of measures that are implicit. Some facts about groupoids in analysis can be explained using only a measure theoretic structure, but that minimum is more often accompanied by a locally compact topology or a manifold structure. Some readers may be surprised that spaces that are only locally Hausdorff are mentioned, but that latitude is needed in the context of foliations and pseudo-groups of transformations.

Adding another structure requires more care for groupoids than it does for groups, mainly because of the fact that only certain pairs of elements can be multiplied. This feature can be expressed in terms of two maps from the groupoid to the space of units (identities), one selecting the right unit and the other the left unit. These maps are usually referred to as *source* and *range* maps, respectively, and denoted by s and r. (Sometimes, they are s and t, for source and target.) The convention is that the source of the left factor must equal the range of the right factor.

It is to be expected that the level sets of these two maps, often called *fibers for the range and source maps,* enter into the proper formulation of how an additional structure relates to the algebraic structure.

For example, a groupoid does not have a single left Haar measure, but instead a family of measures, each of which gives measure zero to the complement of one of the range fibers. A given groupoid element 'translates' the range fiber over its source to the range fiber over its range, by groupoid multiplication. (If $\gamma' \in r^{-1}(s(\gamma))$, then $\gamma\gamma' \in r^{-1}(r(\gamma))$.) Left invariance just means that the measure in the family that is concentrated on the fiber over its source is carried by this translation to the measure in the family that is concentrated on the fiber over its range. Of course, the family of measures must vary in a Borel fashion in the purely measure theoretic setting and continuously in the locally compact setting.

Beyond the difference in formulation, which is expected for groupoids, we have the fact that the existence of a left Haar system is not automatic. Also, a given left Haar system can be multiplied by a well-behaved function of the form $g \circ s$ to obtain another left Haar system, so the uniqueness result for groups must be weakened considerably for groupoids. On the other hand, many processes can be formulated in terms of the class of the measure, even for groups (two measures are in the same class iff they have the same sets of measure 0). Thus the non-uniqueness is not a great hindrance.

For topological groupoids, the range and source maps need to be open as well as continuous. Without this condition, the set of products of elements of two open sets need not be open, for example.

The notion of smooth or Lie groupoid is particularly interesting. For the groupoid structure to be compatible with the manifold structure, the range and source maps need to be submersions, which is analogous to the requirement that they be open in the topological case. In the smooth case, having the range and source maps be submersions guarantees that the set of pairs in $G \times G$ that can be multiplied is a closed submanifold. We refer the reader to articles in these Proceedings but also to the original articles by J. Pradines and to various books (Landsman, MacKenzie, Paterson, and Reinhart) and to papers mentioned there. Let us point out that Lie groupoids and Lie algebroids occur naturally in various contexts of pseudo-differential calculus (e.g. foliated manifolds, manifolds with boundaries, manifolds with corners).

To finish this introduction, we include a few remarks on the content of the articles for the assistance of the reader.

The article by A. Weinstein appeared first in the *Notices of the American Mathematical Society.* It is a good introduction to the way groupoids provide a more comprehensive context for thinking about symmetry than groups do. One of his examples is that of a finite part of a tiling of the plane.

The other articles are more specifically about one of the three topics of the conference, or relations between two or all of them.

According to the introduction of the book by K. Mackenzie mentioned in the article by A. Weinstein, Ehresmann was the first person to use groupoids in differential geometry. We refer to Mackenzie's book for much further information on groupoids in differential geometry.

The article by A. Haefliger is one of the two that are primarily about geometry. That of Haefliger is about foliations. He explains several of the groupoids that can be associated to a foliation and points out results for which groupoids are important. He puts an emphasis on the ones that are étale. The techniques include sheaves of groupoids, cohomology of groupoids, etc.

The article of I. Moerdijk shows how étale groupoids can be used to give convenient generalizations of ordinary topological spaces and manifolds. He explains G-sheaves, cohomology and other constructions for étale groupoids from the viewpoint of derived categories.

The article by G. Della Rocca and M. Takesaki is primarily in the area of analysis. They are specifically interested in understanding equivalence relations and equivalence classes in the context of classification problems. They expand the usual discussion of classification of representations or operator algebras by including the ways that equivalences can occur as well as the fact of equivalence. That is, they want to have a functor defined on the groupoid that produces the equivalence relation on the concretely constructed objects of interest. They make it clear what the advantages of this approach are by working out the case of UHF algebras in detail.

P. Muhly introduces a new perspective for thinking about Fell bundles over groupoids. The approach is to think of them as subcategories of the category of C*-algebras in which the mappings are C*-correspondences. The discussion centers around bundles based on groupoids, as might be expected from the title.

The article by D. Williams is primarily in the area of groupoids themselves, explaining a general tool for obtaining information about groupoids. He defines the Brauer group of a groupoid, and explains how it fits into various perspectives on groupoids. He describes the behavior of the Brauer group relative to equivalence and extensions, and provides cohomological meanings.

A. Paterson has written about index theory for a well-behaved class of groupoid actions. Thus his article stands at the intersection of geometry and analysis, and has groupoids involved in the formulation of the result.

Since G. Mackey introduced the virtual group philosophy, groupoids have been present in group representation theory and in operator algebras. Four contributions to this volume are directly related to operator algebras.

The article by C. Anantharaman and J. Renault is a short survey on the amenability of groupoids, a notion directly prompted by the group case. One needs to distinguish the case of measured groupoids and the case of topological groupoids. The main problem considered is how amenability is reflected on the associated operator algebras. In the measured case, it is known since the work of R. Zimmer that it is related to the hyperfiniteness of the von Neumann algebras. In the topological context, it is related to the nuclearity of the C*-algebras and has interesting applications to the exactness of group C*-algebras.

As stated earlier, groupoids often appear as symmetries of a system or of an algebra. P.-Y. Le Gall gives a concise survey of the Kasparov KK-theory equivariant with respect to a groupoid. This requires a precise definition, in the C*-algebraic setting, of a G-algebra, when G is a groupoid. The corresponding KK-functor, denoted by $KK_G(A, B)$, has now three variables, namely the groupoid G and the G-algebras A, B. Besides its usual properties in A and B, in particular the existence of the Kasparov product, its functoriality in G is useful in computing K- groups and has applications to the Baum-Connes conjecture.

The two other contributions are concerned with Lie groupoids and their Lie algebroids. B. Monthubert shows, in the case of manifolds with corners, how a pseudo-differential calculus is related to the construction of a Lie groupoid G and how to get an index theorem from it. The smoothing operators have a kernel defined on G and the pseudo-differential operators have a symbol defined on the dual $\mathcal{G}^*$ of the Lie algebroid of G. The tangent groupoid, a notion introduced by A. Connes for that purpose, is a key ingredient in the proof of the index theorem. N. Landsman and B. Ramazan show that it also provides a deformation quantization of the Poisson bracket of $\mathcal{G}^*$, in a C*-algebraic context, but with a definition different from that introduced by M. Rieffel.

DEPARTMENT OF MATHEMATICS, 395 UCB, UNIVERSITY OF COLORADO, BOULDER, COLORADO 80309, U.S.A.

E-mail address: ramsay@boulder.colorado.edu

DÉPARTEMENT DE MATHÉMATIQUES, UNIVERSITÉ D'ORLÉANS, B.P.6759, 45068 ORLEANS CEDEX 2, FRANCE

E-mail address: Jean.Renault@labomath.univ-orleans.fr

Contemporary Mathematics
Volume **282**, 2001

Groupoids: Unifying Internal and External Symmetry

A Tour Through Some Examples

Alan Weinstein*
Department of Mathematics
University of California
Berkeley, CA 94720 USA
(alanw@math.berkeley.edu)

1 Introduction

Mathematicians tend to think of the notion of symmetry as being virtually synonymous with the theory of *groups* and their actions, perhaps largely because of the well known Erlanger program of F. Klein and the related work of S. Lie, which virtually defined geometric structures by their groups of automorphisms. (See, for example, Yaglom's account in [27].) In fact, though groups are indeed sufficient to characterize homogeneous structures,

*Research partially supported by NSF Grant DMS-93-09653. This article is based on a talk given first at UCB (the University of California at Berkeley) in April 1995, during Mathematics Awareness Week, the theme of which was Mathematics and Symmetry. I would like to thank the Department of Mathematics of another UCB (the University of Colorado at Boulder) for an invitation to Boulder in November 1995, where a similar talk was most recently given, and where the first draft of this manuscript was largely written in the downtown cafes. Thanks should also be given to a third UCB (Université Claude-Bernard, Lyon), where I started my intensive work on groupoids in the summer of 1986. George Bergman, Ronnie Brown, Pierre Dazord, Jack Feldman, Kirill Mackenzie, George Mackey, Saunders MacLane, Jean Pradines, Arlan Ramsay, Sasha Voronov, and the referees made helpful comments on the original manuscript. Finally, I would like to thank Paul Brown for preparing the figures.

there are plenty of objects which exhibit what we clearly recognize as symmetry, but which admit few or no nontrivial automorphisms. It turns out that the symmetry, and hence much of the structure, of such objects can be characterized algebraically if we use *groupoids* and not just groups.

The aim of this paper is to explain, mostly through examples, what groupoids are and how they describe symmetry. We will begin with elementary examples, with discrete symmetry, and end with examples in the differentiable setting which involve *Lie groupoids* and their corresponding infinitesimal objects, *Lie algebroids*. These objects play a role in the study of general differentiable manifolds and partial differential equations which is, in a sense, an extension of the role which Lie groups play in the geometry and analysis of highly symmetric manifolds.

1.1 Some history

The following historical remarks are not intended to be complete but merely to indicate the breadth of areas where groupoids have been used. An extensive survey of groupoids as of 1986 can be found in R. Brown's article [2].

Groupoids were introduced (and named) by H. Brandt [1] in a 1926 paper on the composition of quadratic forms in four variables. C. Ehresmann [6] added further structures (topological and differentiable as well as algebraic) to groupoids, thereby introducing them as a tool in differentiable topology and geometry.

In algebraic geometry, A. Grothendieck used groupoids extensively and, in particular, showed how they could be used to tame the unruly equivalence relations which arise in the construction of moduli spaces. (See [8], with further details in [7].) As the role of moduli spaces expands in physics as well as mathematics, groupoids continue to play an essential role.

In analysis, G. Mackey [16] used groupoids under the name of *virtual groups* to allow the treatment of ergodic actions of groups, "as if" they were transitive, while it was observed that the convolution operation, extended from groups to groupoids, made possible the construction of a multitude of interesting noncommutative algebras. (See [21] for a survey.) In this context, the use of groupoid convolution algebras as a substitute for the algebras of functions on badly behaved quotient spaces is a central theme in the noncommutative geometry of A. Connes [4]. The book just cited also shows the extent to which groupoids provide a framework for a unified study of operator algebras, foliations, and index theory.

In algebraic topology, the *fundamental groupoid* of a topological space has been exploited by P. Higgins, R. Brown, and others (see Brown's textbook [3], or Higgins' book [9], which is a good general introduction to groupoids as well) in situations where the use of a fixed base point as imposed by the usual fundamental group would be too restrictive. Groupoid methods are thus well adapted to disconnected spaces (which arise frequently when connected spaces are cut into pieces, as in the van Kampen theorem) and to spaces with fixed-point-free group actions.

The extension of Lie theory from differentiable groups to groupoids was carried out for the most part by J. Pradines, as described in a series of notes ending with [20]. An exposition of this work, together with a detailed study of the role of groupoids and Lie algebroids in differential geometry, can be found in the book of K. Mackenzie [13].[1] There is also a close relation between Lie algebroids and Lie groupoids and the better-known pseudogroups of differentiable transformations. (See the brief remark at the end of the following section, or [11] for more details.)

Finally, the author's own interest in groupoids arose from the discovery [24][25], made independently at around the same time by Karasev [10] and Zakrzewski [28], that a groupoid with compatible *symplectic* structure is the appropriate geometric model for a family of noncommutative algebras obtained by deformation from the algebra of functions on a manifold with a Poisson bracket structure.

2 Global and local symmetry groupoids

We begin our exposition with a simple example which will lead to the definition of a groupoid. Consider a tiling of the euclidean plane $\mathbb{R}^2$ by identical 2×1 rectangles, specified by the set X (idealized as 1-dimensional) where the grout between the tiles lies: $X = H \cup V$, where $H = \mathbb{R}\times\mathbb{Z}$ and $V = 2\mathbb{Z}\times\mathbb{R}$. (See Figure 1.) We will call each connected component of $\mathbb{R}^2 \setminus X$ a *tile*.

The symmetry of this tiling is traditionally described by the group Γ consisting of those rigid motions of $\mathbb{R}^2$ which leave X invariant. It consists of the normal subgroup of translations by elements of the lattice $\Lambda = H\cap V = 2\mathbb{Z}\times\mathbb{Z}$ (corresponding to corner points of the tiles), together with reflections

[1]We note that, before the appearance of [5], Lie groupoids were called differentiable groupoids, the term "Lie" being reserved for those satisfying a certain transitivity property.

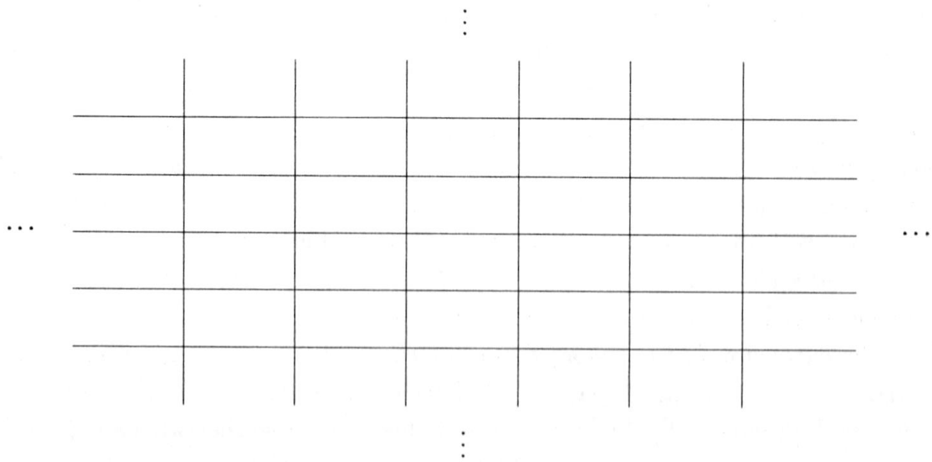

Figure 1: Tiling of the plane by 2×1 rectangles.

through each of the points in $\frac{1}{2}\Lambda = \mathbb{Z} \times \frac{1}{2}\mathbb{Z}$ and across the horizontal and vertical lines through those points.

Now consider what is lost when we pass from X to Γ.

- The same symmetry group would arise if we had replaced X by the lattice Λ of corner points, even though Λ looks quite different from X. (Of course, there are reasons why this loss of detail in passing from X to Γ is welcome, as is the case for many mathematical abstractions, but read on.)

- The group Γ retains no information about the local structure of the tiled plane, such as the fact that neighborhoods of all the points inside the tiles look identical if the tiles are uniform, while they may look different if the tiles are painted with a design (which could still be invariant under Γ).

- If, as is the case on real bathroom floors, the tiling is restricted to a finite part of the plane, such as $B = [0, 2m] \times [0, n]$, the symmetry group shrinks drastically. The subgroup of Γ leaving $X \cap B$ invariant contains just 4 elements, even though a repetitive pattern on the bathroom floor is clearly visible.

Our first groupoid will enable us to describe the symmetry of the finite tiled rectangle. We first define the *transformation groupoid* of the action of

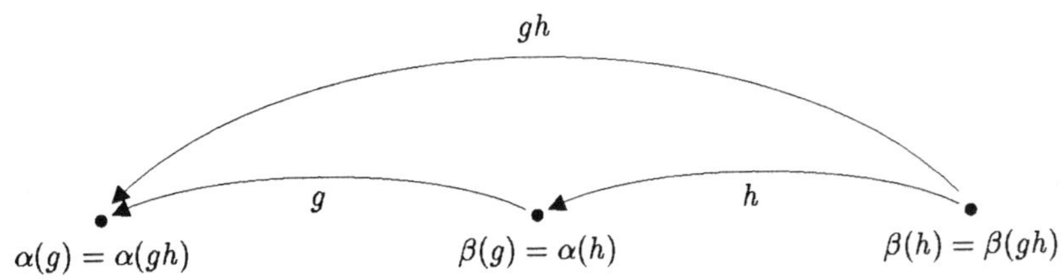

Figure 2: Multiplication in a groupoid.

Γ on $\mathbb{R}^2$ to be the set[2]

$$G(\Gamma, \mathbb{R}^2) = \{(x, \gamma, y) | x \in \mathbb{R}^2, y \in \mathbb{R}^2, \gamma \in \Gamma, \text{ and } x = \gamma y\}$$

with the partially defined binary operation

$$(x, \gamma, y)(y, \nu, z) = (x, \gamma\nu, z).$$

This operation on $G = G(\Gamma, \mathbb{R}^2)$ has the following properties.

1. It is defined only for certain pairs of elements: gh is defined only when $\beta(g) = \alpha(h)$ for certain maps α and β from G onto $\mathbb{R}^2$; here $\alpha : (x, \gamma, y) \mapsto x$ and $\beta : (x, \gamma, y) \mapsto y$.

2. It is associative: if either of the products $(gh)k$ or $g(hk)$ is defined, then so is the other, and they are equal.

3. For each g in G, there are left and right identity elements λ_g and ρ_g such that $\lambda_g g = g = g\rho_g$.

4. Each g in G has an inverse g^{-1} for which $gg^{-1} = \lambda_g$ and $g^{-1}g = \rho_g$.

More generally, a *groupoid* with base B is a set G with mappings α and β from G onto B and a partially defined binary operation $(g, h) \mapsto gh$ satisfying the conditions 1-4 above. We may think of each element g of G as an arrow pointing from $\beta(g)$ to $\alpha(g)$ in B; arrows are multiplied by placing them head to tail, as in Figure 2.

Many properties of groupoids suggested by this picture are easily proven. for instance, $\alpha(g^{-1}) = \beta(g)$ (since gg^{-1} is defined), and $\alpha(gh) = \alpha(g)$ (since $(\lambda_g g)h = \lambda_g(gh)$ implies that $\lambda_g(gh)$ is defined). One also shows easily

[2]As a set, $G(\Gamma, \mathbb{R}^2)$ is isomorphic to $\Gamma \times \mathbb{R}^2$, but we prefer the following more symmetric description, which makes the composition law more transparent.

that $\alpha(g)$ and λ_g (or $\beta(g)$ and ρ_g) determine one another, thus producing a bijective mapping ι from the base B to the subset $G^{(0)} \subseteq G$ consisting of the identity elements. The reader familiar with categories will probably have realized by now that *a groupoid is just a category in which every morphism has an inverse.*[3]

Now we return to our transformation groupoid $G(\Gamma, \mathbb{R}^2)$ and form its *restriction* to $B = [0, 2m] \times [0, n]$ (or any other subset B of $\mathbb{R}^2$) by defining

$$G(\Gamma, \mathbb{R}^2)|_B = \{g \in G(\Gamma, \mathbb{R}^2) | \alpha(g) \text{ and } \beta(g) \text{ belong to } B\}.$$

The following concepts from groupoid theory, applied to this example, now show that $G(\Gamma, \mathbb{R}^2)|_B$ indeed captures the symmetry which we see in the tiled floor.

- An *orbit* of the groupoid G over B is an equivalence class for the relation

  $x \sim_G y$ if and only if there is a groupoid element g with $\alpha(g) = x$ and $\beta(g) = y$.

  In the example, two points are in the same orbit if they are similarly placed within their tiles or within the grout pattern. (We are allowed to turn the tiles over, as well as to translate them or rotate them by 180°.)

- The *isotropy group* of $x \in B$ consists of those g in G with $\alpha(g) = x = \beta(g)$. In the example, the isotropy group is trivial for every point except those in $\frac{1}{2}\Lambda \cap B$, for which it is $\mathbb{Z}_2 \times \mathbb{Z}_2$.

We can describe local symmetries of our tiling by introducing another groupoid. Consider the plane $\mathbb{R}^2$ as the disjoint union of $P_1 = B \cap X$ (the grout), $P_2 = B \setminus P_1$ (the tiles), and $P_3 = \mathbb{R}^2 \setminus B$ (the exterior of the tiled room). Let E be the group of all euclidean motions of the plane, and define the local symmetry groupoid G_{loc} as the set of triples (x, γ, y) in $B \times E \times B$ for which $x = \gamma y$, and for which y has a neighborhood $\mathcal{U}$ in $\mathbb{R}^2$ such that $\gamma(\mathcal{U} \cap P_i) \subseteq P_i$ for $i = 1, 2, 3$. The composition is given by the same formula as for $G(\Gamma, \mathbb{R}^2)$.

[3]Actually, not quite. The elements of a category can constitute a *class* rather than a set, so a groupoid is a *small* category with every morphism invertible. This is just a question of terminology, though; many authors allow their groupoids to be classes.

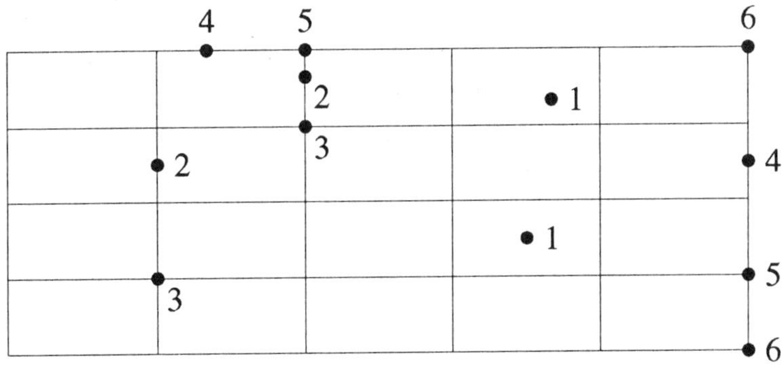

Figure 3: The point labeled by j belongs to the orbit $\mathcal{O}_j$ of the groupoid $G(\Gamma, \mathbb{R}^2)|_B$.

For this groupoid, there are only a finite number of orbits (see Figure 3).

$\mathcal{O}_1 =$ interior points of tiles
$\mathcal{O}_2 =$ interior edge points
$\mathcal{O}_3 =$ interior crossing points
$\mathcal{O}_4 =$ boundary edge points
$\mathcal{O}_5 =$ boundary "T" points
$\mathcal{O}_6 =$ boundary corner points

The isotropy group of a point in $\mathcal{O}_1$ is now isomorphic to the entire rotation group $O(2)$; it is $\mathbb{Z}_2 \times \mathbb{Z}_2$ for $\mathcal{O}_2$, the 8 element dihedral group D_4 for $\mathcal{O}_3$, and $\mathbb{Z}_2$ for $\mathcal{O}_4$, $\mathcal{O}_5$, and $\mathcal{O}_6$.

Now we can create a completely finite object by restricting G_{loc} to the finite set $L = \mathcal{O}_3 \cup \mathcal{O}_5 \cup \mathcal{O}_6$ of tile corner points. $G_{\text{loc}}|_L$ now has just three orbits, with isotropy groups isomorphic to D_4, $\mathbb{Z}_2$, and $\mathbb{Z}_2$ as before. (See Figure 4.) At this point, we have something like the smile which remains after the Cheshire Cat has disappeared. The set L has no structure at all except for that provided by the groupoid $G_{\text{loc}}|_L$, which is a sort of relic of the symmetry structure which was given by the original decomposition into grout, tiles, and and exterior. We could say that the points of L have "internal symmetry" specified by the isotropy groups of $G_{\text{loc}}|_L$, and that the "types" of these points are classified by the orbit structure of the groupoid. This internal symmetry and type classification can be viewed either as a relic of the original tiling, or simply as a "geometric structure" on L.

The situation here is similar to that in gauge theory, where the points of

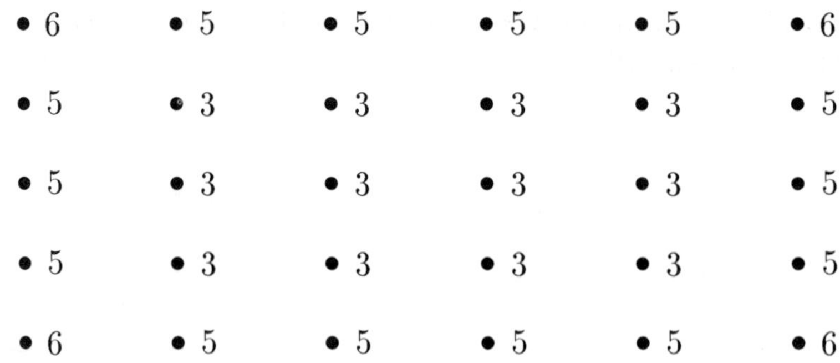

Figure 4: Orbit decomposition of the finite groupoid $G_{\mathrm{loc}}|_L$.

space-time are equipped with internal symmetry groups whose representations correspond to physical particles. But gauge groups and gauge transformations are applicable only because all the points of space time look alike. To describe non-homogeneous structures like the tiled floor requires groupoids rather than groups.[4]

To close this section, we mention that another approach to local symmetry is through the theory of **pseudogroups**, whose elements typically consist of selected homeomorphisms between open subsets of a space of interest, with the composition of two transformations defined on the largest possible domain. The "germs" of elements of a pseudogroup form a groupoid, while in the other direction certain subsets of a groupoid admit a multiplication which makes them the abstract counterpart of a pseudogroup (in the same sense that a group is the abstract counterpart of a set of invertible mappings closed under composition), called an **inverse semigroup**. We refer to [21] for an exposition on inverse semigroups and their relation to groupoids.

3 Groupoids and equivalence relations

If B is any set, the product $B \times B$ is a groupoid over B with $\alpha(x, y) = x$, $\beta(x, y) = y$, and $(x, y)(y, z) = (x, z)$. The identities are the (x, x), and $(x, y)^{-1} = (y, x)$. We call this the *pair groupoid* of B.[5] Note that a subgroupoid of $B \times B$, i.e. a subset closed under product and inversion, and

[4]Actually, groupoids are useful even in the realm of gauge theory. (See [14].)

[5]Other authors use the terms *coarse groupoid* or *banal groupoid* for this object.

containing all the identity elements,[6] is nothing but an equivalence relation on B. The orbits are the equivalence classes, and the isotropy subgroups are all trivial.

If G is any groupoid over B, the map $(\alpha, \beta) : G \to B \times B$ is a morphism from G to the pair groupoid of B. (A groupoid morphism from G over B to G' over B' is a pair of maps $G \to G'$ and $B \to B'$ compatible with the multiplication, "head," and "tail" maps of the two groupoids.) The image of (α, β) is the orbit equivalence relation $\sim_G$, and the kernel is the union of the isotropy groups.

The morphism (α, β) suggests another way of looking at groupoids. Until now, we have been describing them as generalized groups, but now we want to think of groupoids as generalized equivalence relations as well. From this point of view, a groupoid over B tells us not only *which* elements of B are equivalent to one another (or "isomorphic"), but it also parametrizes the *different ways* ("isomorphisms") in which two elements can be equivalent. This leads us to the following guiding principle of Grothendieck, Mackey, Connes, Deligne, ...

> Almost every interesting equivalence relation on a space B arises in a natural way as the orbit equivalence relation of some groupoid G over B. Instead of dealing directly with the orbit space B/G as an object in the category $\mathcal{S}_{\text{map}}$ of sets and mappings, one should consider instead the groupoid G itself as an object in the category $\mathcal{G}_{\text{htp}}$ of groupoids and homotopy classes of morphisms.

Here is the definition. If

$$\begin{array}{ccc} G & \xrightarrow{\;f^i_G\;} & G' \\ \alpha \Big\downarrow \Big\downarrow \beta & & \alpha' \Big\downarrow \Big\downarrow \beta' \\ B & \xrightarrow{\;f^i_B\;} & B' \end{array}$$

are morphisms of groupoids for $i = 1, 2$, a *homotopy* (or *natural transformation* in the language of categories) from f^1 to f^2 is a map $h : B \to G'$ with the following properties.

[6] One often omits this last condition from the definition of a subgroupoid. To emphasize it, we can refer to a *wide subgroupoid*.

- For each $x \in B$, $\alpha'(h(x)) = f_B^1(x)$ and $\beta'(h(x)) = f_B^2(x)$.
- For each $g \in G$, $h(\alpha(g))f_G^2(g) = f_G^1(g)h(\beta(g))$.

An isomorphism in $\mathcal{G}_{\mathrm{htp}}$ between the groupoids G and G' (sometimes called an *equivalence of groupoids*) gives a bijection between the orbit spaces B/G and B/G', as well as group isomorphisms between corresponding isotropy subgroups in G and G'. In the absence of further structure on the groupoids, an equivalence is essentially no more than this, but when the groupoids (and their bases) have measurable, topological, differentiable, algebraic, or symplectic (and Poisson) structures, the quotient spaces can be rather nasty objects, outside the category originally under consideration, and it is essential to focus attention on the groupoids themselves. One tool which makes this focus effective is an algebra associated to the groupoid G over B which is a useful substitute for the space of functions on B/G when the quotient is a "bad" space. We begin to describe this algebra in the next section.

4 Groupoid convolution and matrices

The *convolution* of two complex valued functions on a group G is defined as the sum

$$(a * b)(g) = \sum_{k \in G} a(k)b(k^{-1}g),$$

at least when the support of each function (i.e. the set where it is not zero) is finite. When the supports are infinite, the sum may still be defined if the functions are absolutely summable. One obtains in this way, for instance, the rules for multiplying Fourier series as functions on the additive group $\mathbb{Z}$.

To extend the definition of convolution from groups to groupoids, one replaces the range of summation G by the α-*fibre* $G_g = \alpha^{-1}(\alpha(g))$. Alternatively, one may use the more symmetric formula for convolution:

$$(a * b)(g) = \sum_{\{(k,\ell) \mid k\ell = g\}} a(k)b(\ell).$$

For instance, if $G = B \times B$ is the pair groupoid over $B = \{1, 2, \dots, n\}$, the convolution operation (defined on all functions) is

$$(a * b)(i,j) = \sum_{(i,k)(k,j)=(i,j)} a(i,k)b(i,j),$$

and if we write a_{ij} for $a(i,j)$ we get

$$(a * b)_{ij} = \sum_{k=1}^{n} a_{ik} b_{kj},$$

which exhibits $\mathcal{F}(\{1, 2, \ldots, n\} \times \{1, 2, \ldots, n\})$ as the algebra of $n \times n$ matrices.

This may look like a peculiar way of viewing the algebra of matrices, but Connes ([4] pp. 33-39, and elsewhere) has made the point that it was precisely as a groupoid algebra that Heisenberg constructed his original formulation of quantum mechanics. (The elements of the groupoid were transitions between excited states of an atom. Only later were the elements of the groupoid algebra identified as matrices.) The noncommutativity of the algebra of observables in quantum mechanics is then seen to be a direct consequence of the noncommutativity of the pair groupoid $B \times B$, as compared with the commutativity of the group $\mathbb{Z}$ underlying the Fourier series representation of oscillatory motion in classical mechanics.

5 Topological groupoids and groupoid algebras

For many interesting groupoids, such as the pair groupoid of the real numbers, the functions with finite or even countable support are rather unnatural objects, and it is more common to consider continuous functions. To define the convolution of such functions, it is necessary to replace the sum in the definition by an integral. An appropriate general setting for this extended notion of convolution is that of topological groupoids.

A groupoid G is a *topological groupoid* over B if G and B are topological spaces, and α, β, and multiplication are continuous maps. It turns out that this notion is useful for many purposes beyond the definition of convolution. A criticism sometimes applied to the theory of groupoids is that their classification up to isomorphism is nothing other than the classification of equivalence relations (via the orbit equivalence relation) and groups (via the isotropy groups). The imposition of a compatible topological structure produces a nontrivial interaction between the two structures. (The theory of principal bundles is just a special case this.)[7]

[7]Of course, a completely different retort could be made to the criticism above. By the same criterion, finite-dimensional vector spaces should be completely uninteresting, but the study of their morphisms is a very rich subject.

For instance, the groupoid $G_{\mathrm{loc}}|_L$ of local symmetries introduced in Section 2 can be entirely decomposed into its restrictions to the orbits $\mathcal{O}_3$, $\mathcal{O}_5$, and $\mathcal{O}_6$. In other words, the groupoid contains no information about the relation of the orbits to one another. This relation can be encoded, however, in a suitable topology on the groupoid.

To return to our discussion of convolution on groupoids, we consider an example. If $G = \mathbb{R}^n \times \mathbb{R}^n$, we can define a convolution operation on the space $C_c(G)$ of continuous functions with *proper* support (i.e. for each compact subset K of $\mathbb{R}^n$, the intersections of $K \times \mathbb{R}^n$ and $\mathbb{R}^n \times K$ with $\{(x,y)|a(x,y) \neq 0\}$ have compact closure). Multiplication in $C_c(G)$ is given by the integral formula

$$(a * b)(x,y) = \int_{\mathbb{R}^n} a(x,z)b(z,y)\, dz, \tag{1}$$

where dz is Lebesgue measure.[8]

This "continuous matrix multiplication" is precisely the composition law which we obtain by thinking of each $a \in C_c(G)$ as the " kernel" (here the term is used in the sense of Schwartz [23]) of an operator $\tilde{a}$ on $L^2(\mathbb{R}^n)$:

$$(\tilde{a}\psi)(x) = \int_{\mathbb{R}^n} a(x,y)\psi(y)dy.$$

Not every operator (not even the identity) is realized by a continuous kernel, but all operators (including unbounded ones) satisfying reasonable continuity properties can be realized by kernels which are *distributions*, or generalized functions on $\mathbb{R}^n \times \mathbb{R}^n$. (See [23].)

To define convolution operations on functions on a general locally compact topological groupoid, we need to start with a family $\{\mu_x\}$ of measures along the α-fibres, which will fill the role played by the measure dz in (1) above. Convolution is then defined[9] by

$$(a * b)(g) = \int_{G_g} a(k)b(k^{-1}g)d\mu_g(k).$$

This operation is associative if the measures μ_g satisfy a suitable left-invariance property.

[8]Dirac's notation $\langle x|a|y\rangle$ for $a(x,y)$ $a(x,y)$ emphasizes the idea that the values of a are "transition amplitudes" between "states" and so fits nicely with the groupoid interpretation of (1).

[9]This definition seems to be due to J. Westman [26].

From continuous functions on G, on can pass by various processes of completion and localization to larger *groupoid algebras*, which play the role of continuous functions, measurable functions, or even distributions on the orbit space of G. We refer to the books [4], [19], and [21] for more details.

A more intrinsic construction of groupoid algebras, free of the choice of measures, can also be given. It includes as special cases the convolution of measures on a group G and the multiplication of kernels of operators on Mackey's "intrinsic Hilbert spaces" [15]. The definition can be found on p. 101 of [4] for the case of Lie groupoids (see Section 6 below). A construction modeled on the definition of the intrinsic Hilbert spaces should extend this idea to a general locally compact groupoid with a prescribed left-invariant family of measure *classes* along the α-fibres.

6 Lie groupoids and Lie algebroids

A groupoid G over B is a *Lie groupoid* if G and B are differentiable manifolds, and α, β, and multiplication are differentiable maps.[10] For example, the pair groupoid $M \times M$ over any differentiable manifold M is a Lie groupoid, and any Lie group is a Lie groupoid over a single point.

The infinitesimal object associated with a Lie groupoid is a *Lie algebroid.* A Lie algebroid over a manifold B is defined to be a vector bundle A over B with a Lie algebra structure $[\ ,\ ]$ on its space of smooth sections, together with a bundle map ρ (called the *anchor* of the Lie algebroid) from A to the tangent bundle TB satisfying the conditions:

1. $[\rho(X), \rho(Y)] = \rho([X,Y])$;
2. $[X, \varphi Y] = \varphi[X,Y] + (\rho(X)\cdot\varphi)Y$.

Here, X and Y are smooth sections of A, φ is a smooth function on B, and the bracket on the left hand side of the first condition is the usual Jacobi-Lie bracket of vector fields. For example, the tangent bundle TB with the usual bracket, and the identity map as anchor, is a Lie algebroid over B, while any Lie algebra is a Lie algebroid over a single point.

The Lie algebroid of a Lie groupoid is defined via left-invariant vector fields, just as it is in the case of Lie groups. Since the left translation by

[10] A further technical condition is included in the definition to insure that the domain of multiplication is a smooth submanifold of $G \times G$; the derivatives of α and β are required to have maximal rank everywhere.

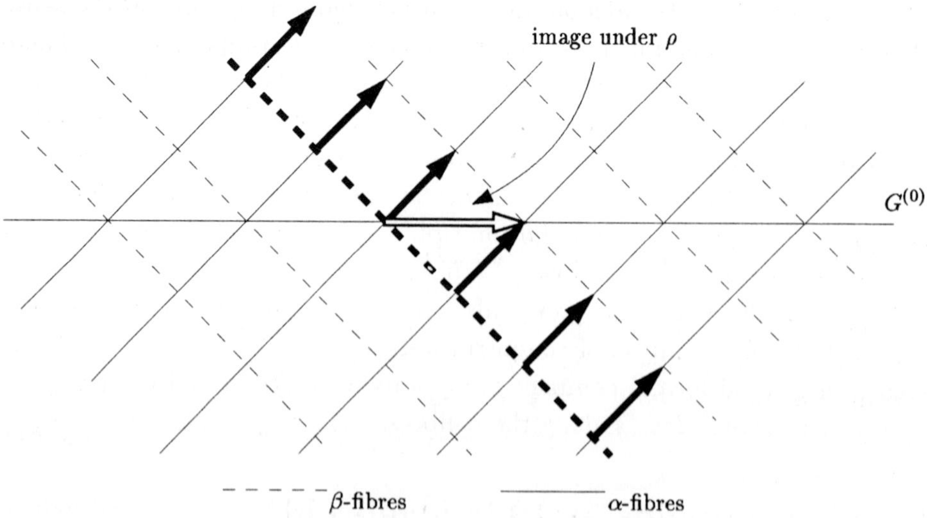

Figure 5: The groupoid G with an element of $\mathcal{A}$ and its image under ρ.

a groupoid element g maps only from $G_{\beta(g)}$ to $G_{\alpha(g)}$, we must first restrict the values of our vector fields to lie in the subbundle $T^\alpha G = \ker(T\alpha)$ of TG consisting of vectors tangent to the α-fibres. (Here, $T\alpha$ is the derivative of α.) For every x in B and every g in the β-fibre $\beta^{-1}(x)$, the fibre $T^\alpha_g G$ can be identified with $T^\alpha_{\iota(x)}G$ via left translation by g. Thus, the space of left-invariant sections of $T^\alpha G$ along $\beta^{-1}(x)$ can be identified with the single space $\mathcal{A}_x G = T^\alpha_{\iota(x)}G$, and $T\beta$ gives a well defined map $\rho_x : \mathcal{A}_x G \to T_x B$. The (disjoint) union of these objects for all x in B forms a vector bundle $\mathcal{A}G$ over B with a bundle map $\rho : \mathcal{A}G \to TB$. (See Figure 5.) Identifying the sections of $\mathcal{A}G$ with left-invariant sections of $T^\alpha G \subseteq TG$ and using the usual bracket on sections of TG defines a bracket of sections on $\mathcal{A}G$ making it into a Lie algebroid with anchor ρ.

For instance, the Lie algebroid of $B \times B$ is TB, and by extension of a well-known idea from Lie group theory, one may think of the Jacobi-Lie bracket on sections of TB as being the infinitesimal remnant of the noncommutativity of the composition law $(x, y)(y, z) = (x, z)$.

The local versions of Lie's three fundamental theorems all extend from groups to groupoids, but there are some interesting differences at the global level. Although every subalgebroid A' of $\mathcal{A}G$ is indeed $\mathcal{A}G'$ for a Lie groupoid G', there may be no such G' for which the induced morphism $G' \to G$ is globally one-to-one. More seriously (and contrary to an incorrect assertion in [20]), there exist Lie algebroids which are "non-integrable" in the sense that they are not the Lie algebroid of any global Lie groupoid at all. (There is a *local* Lie groupoid corresponding to any Lie algebroid, though.) We refer to Chapter V of [13] for a discussion of all this, together with further references.

7 Boundary Lie algebroids

In this section, we will explain how Lie algebroids provide a natural setting for understanding some recent work by R. Melrose [17] [18] on the analysis of (pseudo)differential operators on manifolds with boundary. (Some prior knowledge about such operators on manifolds *without* boundary will be assumed here.)

The space $\mathcal{X}(M, \partial M)$ of vector fields which are tangent to the boundary ∂M of a manifold M forms a Lie algebra over $\mathbb{R}$ and a module over $C^\infty(M)$, and the condition $[X, \varphi Y] = \varphi[X, Y] + (X \cdot \varphi)Y$ is satisfied (since it is satisfied for all vector fields and functions). Remarkably, there is a vector bundle ${}^bTM$ over M whose sections are in one-to-one correspondence with the elements of $\mathcal{X}(M, \partial M)$ via a bundle map $\rho : {}^bTM \to TM$ which is an isomorphism over the interior of M, and whose image over a boundary point is the tangent space to ∂M at that point. These structures make ${}^bTM$ into a Lie algebroid. We refer to [17] for a precise definition of ${}^bTM$, mentioning here only that, when $(y_1, \cdots, y_{n-1}, x)$ are local coordinates for M defined on an open subset of the half space $x \geq 0$, a local basis over $C^\infty(M)$ for the sections of ${}^bTM$ is given by $(\frac{\partial}{\partial y_1}, \cdots, \frac{\partial}{\partial y_{n-1}}, x\frac{\partial}{\partial x})$.

Melrose develops analysis on M by regarding the *b*-tangent bundle ${}^bTM$ as the "correct" tangent bundle for this manifold with boundary. Thus, the algebra of *b*-differential operators is the algebra of operators on $C^\infty(M)$ generated by the sections of ${}^bTM$ (acting on functions via ρ) and $C^\infty(M)$ (acting on itself by multiplication).[11] The principal symbols of these operators are homogeneous functions on the dual, or *b*-cotangent bundle ${}^bT^*M$. Inverting (modulo smoothing operators) elliptic *b*-differential operators leads to the notion of *b*-pseudodifferential operators, whose symbols are again functions on the *b*-cotangent bundle.

Since the *b*-pseudodifferential operators act on $C^\infty(M)$, their Schwartz

[11] Such a "universal enveloping algebra" was already defined for general Lie algebroids by Rinehart [22].

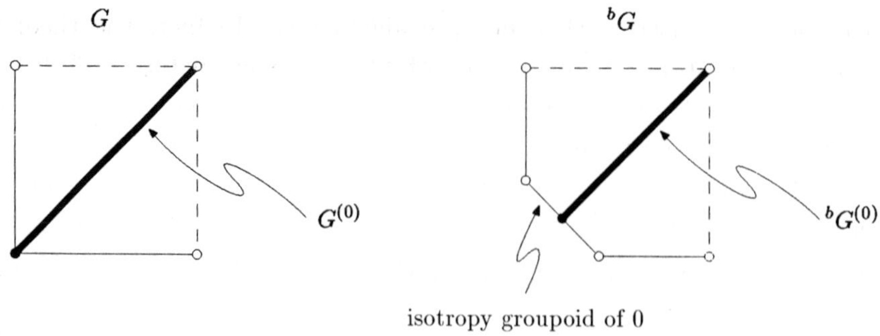

Figure 6: The groupoids $G = M \times M$ and ${}^bG = {}^b(M \times M)$, where $M = [0, 1)$.

kernels are distributions on $M \times M$, which is a manifold with corners. Analysis of these kernels is facilitated, though, by lifting them to a certain manifold with boundary, ${}^b(M \times M)$, obtained from $M \times M$ by a blowing-up operation along the corners. The case where M is a half-open interval is illustrated in Figure 6. The general case is locally the product of this one with a manifold without boundary.

It turns out that ${}^b(M \times M)$ is a groupoid whose Lie algebroid is ${}^bTM$, and that the composition law for b-kernels ([17], Equation (5.66)) is precisely the convolution operation in the groupoid algebra of ${}^b(M \times M)$. Thus, we can think of the groupoid ${}^b(M \times M)$ and its Lie algebroid ${}^bTM$ as providing a new "internal symmetry structure" appropriate to our manifold with boundary.

It is therefore interesting to understand the structure of the groupoid ${}^b(M \times M)$. In fact, it decomposes algebraically (though not topologically) as the disjoint union of the pair groupoid over the interior of M and the groupoid G over ∂M for which $(\alpha, \beta)^{-1}(x, y)$ consists of the orientation preserving linear isomorphisms between the normal lines $T_yM/T_y(\partial M)$ and $T_xM/T_x(\partial M)$. Thus, with the "geometric structure" on M provided by the enlarged pair groupoid ${}^b(M \times M)$, each point of the boundary ∂M has the internal symmetries are those of an oriented 1-dimensional vector space. We should think of the vectors in this space as being points "infinitesimally far" from the boundary, so that the boundary of M becomes an "infinitesimal boundary layer." The section $x\frac{\partial}{\partial x}$ of ${}^bTM$ can now be seen as pointing into this layer in a non trivial way along ∂M; that is why it is not zero there.

At this point, our use of Lie algebroids and Lie groupoids adds only a geometric picture to Melrose's analysis. It does, however, suggest the possibility of extending the theory of pseudodifferential operators to algebras of operators associated with other Lie algebroids. In fact, the theory of longitudinal operators on foliated manifolds (see Section I.5.γ) of [4] is an example of such an extension, but the class of Lie algebroids which can be treated in this way remains to be determined. An interesting test case should be the Lie algebroids associated with the Bruhat-Poisson structures on flag manifolds [12], since the corresponding orbit decomposition on a flag manifold is the Bruhat decomposition, which plays a central role in representation theory.

8 Conclusion

This article has presented a small sample of the applications of groupoids in analysis, algebra, and topology. R. Brown has reported in [2] the suggestion of F.W. Lawvere that groupoids should perhaps be renamed "groups", and those special groupoids with just one base point given a new name to reflect their singular nature. Even if this is too far to go, I hope to have convinced the reader that groupoids are worth knowing about, and worth looking out for.

References

[1] Brandt, W., Über eine Verallgemeinerung des Gruppenbegriffes, *Math. Ann.* **96** (1926), 360-366.

[2] Brown, R., From groups to groupoids: a brief survey, *Bull. London Math. Soc.* **19** (1987), 113-134.

[3] Brown, R., *Topology: a geometric account of general topology, homotopy types, and the fundamental groupoid*, Halsted Press, New York, 1988.

[4] Connes, A., *Noncommutative Geometry*, Academic Press, San Diego, 1994.

[5] Coste, A., Dazord, P., et Weinstein, A., Groupoïdes symplectiques, *Publications du Département de Mathématiques, Université Claude Bernard-Lyon I* **2A** (1987), 1-62.

[6] Ehresmann, C., Oeuvres complètes et commentées, ed. A.C. Ehresmann, Suppl. *Cahiers Top. Géom. Diff.*, Amiens, 1980-1984.

[7] Gabriel, P., Construction de préschémas quotient, *Schémas en Groupes, Sém. Géométrie Algébrique, Inst. Hautes Études Scientifiques, 1963/64*, Fasc. 2a, Exposé 5.

[8] Grothendieck, A., Techniques de construction et théorèmes d'existence en géométrie algébrique III: préschemas quotients, *Séminaire Bourbaki* 13e année 1960/61, no. 212 (1961).

[9] Higgins, P.J., *Notes on Categories and Groupoids*, Van Nostrand Reinhold, London, 1971.

[10] Karasev, M.V., Analogues of objects of Lie group theory for nonlinear Poisson brackets, *Math. USSR Izvestia* **28**, (1987), 497-527.

[11] Kumpera, A., and Spencer, D., *Lie Equations*, Princeton, N.J. Princeton University Press, 1972.

[12] Lu, J.-H., and Weinstein, A., Poisson Lie groups, dressing transformations, and the Bruhat decomposition, *J. Diff. Geom.* **31** (1990), 501-526.

[13] Mackenzie, K., *Lie Groupoids and Lie Algebroids in Differential Geometry*, LMS Lecture Notes Series, **124**, Cambridge Univ. Press, 1987.

[14] Mackenzie, K., Classification of principal bundles and Lie groupoids with prescribed gauge group bundle, *J. Pure Appl. Algebra* **58** (1989), 181-208.

[15] Mackey, G.W., *The Mathematical Foundations of Quantum Mechanics*, W.A. Benjamin, New York, 1963.

[16] Mackey, G.W., Ergodic theory and virtual groups, *Math. Ann.* **166** (1966), 187-207.

[17] Melrose, R.B., *The Atiyah-Patodi-Singer Index Theorem*, A.K. Peters, Wellesley, 1993.

[18] Melrose, R.B., *Geometric Scattering Theory*, Cambridge Univ. Press, Cambridge, 1995.

[19] Moore, C.C. and Schochet, C., *Global Analysis on Foliated Spaces*, MSRI Publications **9**, Springer-Verlag, New York (1988).

[20] Pradines, J., Troisième théorème de Lie sur les groupoïdes différentiables, *C. R. Acad. Sc. Paris,* **267** (1968), 21 - 23.

[21] Renault, J., A groupoid approach to C^* algebras, *Lecture Notes in Math.* **793** (1980).

[22] Rinehart, G.S., Differential forms on general commutative algebras, *Trans. Amer. Math. Soc.* **108** (1963), 195-222.

[23] Schwartz, L., *Théorie des Distributions*, Nouvelle edition, Hermann, Paris (1966).

[24] Weinstein, A., Symplectic groupoids and Poisson manifolds, *Bull. Amer. Math. Soc.* **16**, (1987), 101-104.

[25] Weinstein, A., Noncommutative geometry and geometric quantization, *Symplectic Geometry and Mathematical Physics: actes du colloque en l'honneur de Jean-Marie Souriau*, P. Donato et al eds., Birkhäuser, Boston, 1991, pp. 446-461.

[26] Westman, J., Harmonic analysis on groupoids, *Pacific J. Math.* **27** (1968), 621-632.

[27] Yaglom, I. M., Felix Klein and Sophus Lie: evolution of the idea of symmetry in the Nineteenth Century, Birkhäuser, Boston, 1988.

[28] Zakrzewski, S., Quantum and classical pseudogroups, I and II, *Comm. Math. Phys.* **134** (1990), 347-370, 371-395.

Contemporary Mathematics
Volume **282**, 2001

A Primer for the Brauer Group of a Groupoid

Dana P. Williams

ABSTRACT. This is an expository article outlining the definition and basic properties of the Brauer Group of a locally compact groupoid.

As the title suggests, this article is meant to be a elementary introduction and guide to the theory of the Brauer group of a locally compact groupoid. This material is primarily based on the article [**18**] by Alex Kumjian, Paul Muhly, Jean Renault and the author. Throughout, "groupoid" will mean a second countable locally compact groupoid with open range and source maps. In fact, all topological spaces should be assumed to be second countable, and virtually all Banach algebras which appear — with the exception of the algebra of bounded operators on Hilbert space and other multiplier algebras — are assumed to be separable. I will refer to [**18**] for the full details, complete statements of results, and further references.

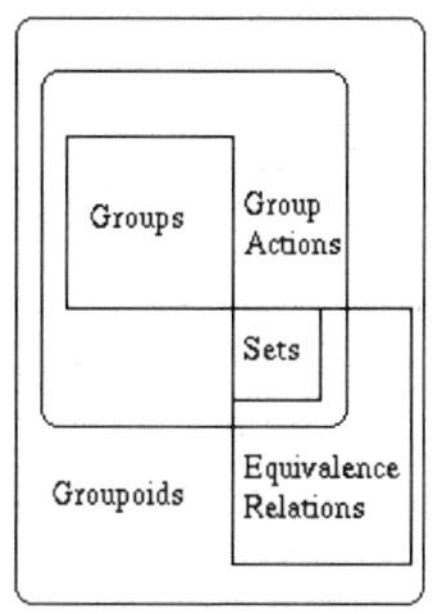

A groupoid G can mean many different things to different audiences. This is due in part to the multifaceted nature of groupoids illustrated by the diagram at right. Therefore if we want to associate an invariant — in the spirit, say, of a cohomology group — to a groupoid, then ideally this invariant should reflect all the various characteristics of groupoids: topological space, group, transformation group, etc. Our Brauer group $\operatorname{Br}(G)$ is meant to be such an object. For motivation, I'll mention what happens in some simple cases. First, suppose G is the cotrivial or diagonal equivalence relation in $T \times T$; that is, G is simply a topological space T. In [**12**], Green defined the Brauer group $\operatorname{Br}(T)$ of a space T to be the set of Morita equivalence classes of continuous-trace C^*-algebras with spectrum T. He proved that with multiplication defined via a balanced tensor product, $\operatorname{Br}(T)$ is a group isomorphic to the second sheaf cohomology group $H^2(T, \mathcal{S})$ [**28**, Theorem 6.3]. Thus $\operatorname{Br}(T)$, which is apparently an analytical object, is in fact a well-understood topological invariant associated to T. (In fact, $H^2(T, \mathcal{S})$ is naturally isomorphic to the integral cohomology group $H^3(T; \mathbf{Z})$. Although integral cohomology groups are more familiar objects, it is more natural for our purposes to think of $\operatorname{Br}(T)$ as a 2-cohomology group, and I will use the sheaf group $H^2(T, \mathcal{S})$ here.) In [**3**] we defined the equivariant Brauer

1991 *Mathematics Subject Classification.* Primary 46L55; Secondary 46M20, 22A30.
Key words and phrases. Brauer group, groupoid, continuous-trace C^*-algebra.

group $\mathrm{Br}_H(T)$ associated to a transformation groupoid $G = T \times H$ where H is a locally compact group acting on T by homeomorphisms. The structure of this group is very interesting and has important consequences for actions of groups on continuous trace C^*-algebras [**24, 25, 20, 10, 9**]. I will have more to say about this group shortly, but at the moment I want to specialize to the case T is a single point (and G is simply the group H). Then as described in [**28**, §7.3], $\mathrm{Br}(H)$ is Moore's group 2-cohomology group $H^2(H, \mathbf{T})$ which describes the locally compact extensions of H by the circle $\mathbf{T}$. (See [**28**, §7.4] for a discussion of Moore's theory.) Since our Brauer group $\mathrm{Br}(G)$ will generalize the above constructions, it will reflect both the "space-like" and "group-like" characteristics of G. Since our definition of $\mathrm{Br}(G)$ goes via continuous-trace C^*-algebras, it is, in view of Green's result, most directly tied to the space-like nature of G. To show that $\mathrm{Br}(G)$ also captures the group-like nature, we want to tie $\mathrm{Br}(G)$ with extensions. This is a more subtle business for groupoids than in the case of groups. We do this, as in [**18**], by introducing a group $\mathrm{Ext}(G, \mathbf{T})$ which organizes the groupoid extensions of trivial $\mathbf{T}$-bundles by groupoids equivalent to G. The punch line is that $\mathrm{Ext}(G, \mathbf{T})$ is naturally isomorphic to $\mathrm{Br}(G)$.

1. The definition of $\mathrm{Br}(G)$

The notation used here will be consistent with that used in [**21**]. The range and source maps of a groupoid will be denoted by r and s, respectively. The unit space of a groupoid G is $G^{(0)}$.

Groupoids act on fibred objects. In particular, G acts on a space X if there is a continuous, open surjection $r : X \to G^{(0)}$ and a continuous map $(\gamma, x) \mapsto \gamma \cdot x$ from $G * X := \{ (\gamma, x) \in G \times X : s(\gamma) = r(x) \}$ to X such that

(a) $r(\gamma \cdot x) = r(\gamma)$,
(b) $(\alpha\gamma) \cdot x = \alpha \cdot (\gamma \cdot x)$ if $(\alpha, \gamma) \in G^{(2)}$ and
(c) $r(x) \cdot x = x$ for all $x \in X$.

In this case, X is called a left G-space. The definition for a right G-space is similar, but then the structure map from X to $G^{(0)}$ is usually denoted by s.

The action is *free* if $\gamma \cdot x = x$ implies $\gamma = r(x)$, and *proper* if $(\gamma, x) \mapsto (\gamma \cdot x, x)$ is a proper map from $G * X$ to $X * X$. Free and proper actions are called *principal.*

If G is to act on a Banach space, that space will have to be fibred over $G^{(0)}$ and we will want that fibering to respect the vector space structure. The needed formalism is spelled out in §§13–14 of Chapter II of [**11**].

DEFINITION 1.1. Suppose that T is a locally compact space. A *Banach bundle over* T is a topological space $\mathcal{A}$ together with a continuous, open surjection $p = p_{\mathcal{A}} : \mathcal{A} \to T$ and Banach space structures on each fibre $p^{-1}(\{ t \})$ $(t \in T)$ satisfying the following axioms.

(a) The map $a \mapsto \|a\|$ is continuous from $\mathcal{A}$ to $\mathbf{R}^+$.
(b) The map $(a, b) \mapsto a + b$ is continuous from
$$\mathcal{A} * \mathcal{A} := \{ (a, b) \in \mathcal{A} \times \mathcal{A} : p(a) = p(b) \}$$
to $\mathcal{A}$.
(c) For each $\lambda \in \mathbf{C}$, the map $b \mapsto \lambda b$ is continuous from $\mathcal{A}$ to $\mathcal{A}$.
(d) If $\{ a_i \}$ is a net in $\mathcal{A}$ such that $p(a_i) \to t$ in T and such that $\|a_i\| \to 0$ in $\mathbf{R}$, then $a_i \to 0_t$ in $\mathcal{A}$. (Of course, 0_t denotes the zero element in the Banach space $p^{-1}(\{ t \})$.)

A *C^*-bundle over T* is a Banach bundle $p : \mathcal{A} \to T$ such that each fibre is a C^*-algebra satisfying, in addition to axioms (a)–(d) above, the following axioms.

(e) The map $(a, b) \mapsto ab$ is continuous from $\mathcal{A} * \mathcal{A}$ to $\mathcal{A}$.

(f) The map $a \mapsto a^*$ is continuous from $\mathcal{A}$ to $\mathcal{A}$.

A continuous function $f : T \to \mathcal{A}$ such that $p \circ f = \mathrm{id}_T$ is called a *section* of $\mathcal{A}$. The collection of sections f for which $t \mapsto \|f(t)\|$ is in $C_0(T)$ is denoted by $A := C_0(T; \mathcal{A})$.

The point is that if $\mathcal{A}$ is a Banach bundle (resp. C^*-bundle) over T, then $A = C_0(T; \mathcal{A})$ is a Banach space (resp. C^*-algebra) with respect to the obvious pointwise operations and norm $\|f\| := \sup_{t \in T} \|f(t)\|$. We will use the notation $A(t)$ for the fibre $p^{-1}(\{t\})$.

Now suppose that G is a locally compact groupoid and that $p : \mathcal{A} \to G^{(0)}$ is a Banach bundle (resp. C^*-bundle). A collection $\{\alpha_\gamma\}_{\gamma \in G}$ of isometric isomorphisms (resp. $*$-isomorphisms) $\alpha_\gamma : A\big(s(\gamma)\big) \to A\big(r(\gamma)\big)$ is called a *G-action by isomorphisms* or simply a G-action on $\mathcal{A}$ if $\gamma \cdot a := \alpha_\gamma(a)$ makes $(\mathcal{A}, p)$ into a left G-space as above. In this case, we say that $\mathcal{A}$ is a Banach G-bundle (resp. C^*-G-bundle). As the next example illustrates, a C^*-G-bundle is the natural generalization of a C^*-dynamcial system.

EXAMPLE 1.2. Let G be the transformation groupoid $T \times H$ and let $p : \mathcal{A} \to T$ be a C^*-bundle with $A = C_0(T; \mathcal{A})$. Suppose that $\beta : H \to \mathrm{Aut}(A)$ is a dynamical system which induces the given action on T: that is, $\beta_h(f \cdot a)(t) = f(h^{-1} \cdot t)\beta_h(a)(t)$ for all $f \in C_0(T)$ and $a \in A$. Then, if $\alpha_{(t,h)} : A(t) \to A(h^{-1} \cdot t)$ is given by

$$\alpha_{(t,h)}\big(a(t)\big) = \beta_h(a)(h^{-1} \cdot t),$$

it follows that $(\mathcal{A}, \alpha)$ is a C^*-G-bundle over T.

The basics of Morita equivalence of C^*-algebras and continuous-trace C^*-algebras are covered in [**28**]. Although the Brown-Green-Rieffel Theorem implies that A and B are Morita equivalent if and only if they are stably isomorphic [**28**, Theorem 5.55] (as everything in sight is separable), it is usually more useful to recall that by definition A and B are Morita equivalent if and only if there is an A–B-imprimitivity bimodule X; that is, X is an A–B-bimodule which is both a full left Hilbert A-module and a full right Hilbert B-module such that $\langle a \cdot x \,,\, y\rangle_B = \langle x \,,\, a^* \cdot y\rangle_B$, ${}_A\langle x \cdot b \,,\, y\rangle = {}_A\langle x \,,\, y \cdot b^*\rangle$, and ${}_A\langle x \,,\, y\rangle \cdot z = z \cdot {}_A\langle y \,,\, z\rangle$ for all $x, y, z \in \mathsf{X}$, $a \in A$, and $b \in B$. A C^*-algebra A with Hausforff spectrum has continuous trace if and only if A is locally Morita equivalent to an abelian C^*-algebra [**28**, Proposition 5.15]. Although this is the characterization I prefer, it can be shown, with some hard work, that A has continuous trace if and only if it is "stably locally trivial" in the sense that $A \otimes K$ is locally isomorphic to $C_0(\hat{A}) \otimes K$ [**28**, Proposition 5.59]. In this note, we want to specialize to Morita equivalence between two fibred algebras.

DEFINITION 1.3. Suppose that $p_{\mathcal{A}} : \mathcal{A} \to T$ and $p_{\mathcal{B}} : \mathcal{B} \to T$ are C^*-bundles with section algebras A and B, respectively. Then a Banach bundle $q : \mathcal{X} \to T$ is called an *$\mathcal{A}$–$\mathcal{B}$-imprimitivity bimodule bundle* if each fibre $\mathsf{X}(t)$ is an $A(t)$–$B(t)$-imprimitivity bimodule in such a way that the natural maps $(a, x) \mapsto a \cdot x$ and $(x, b) \mapsto x \cdot b$ define continuous maps from $\mathcal{A} * \mathcal{X}$ to $\mathcal{X}$ and from $\mathcal{X} * \mathcal{B}$ to $\mathcal{X}$, respectively.

If $A = C_0(T;\mathcal{A})$ and $B = C_0(T;\mathcal{B})$ are the section algebras, then the continuity conditions in Definition 1.3 are exactly what is required to define A and B valued inner products on $\mathsf{X} := C_0(T;\mathcal{X})$:

$$ {}_A\langle x\,,\,y\rangle(t) = {}_{A(t)}\langle x(t)\,,\,y(t)\rangle \quad\text{and}\quad \langle x\,,\,y\rangle_B = \langle x(t)\,,\,y(t)\rangle_{B(t)}. \tag{1.1}$$

Then it is easy to see that X is an A–B-imprimitivity bimodule, and A and B are Morita equivalent. In fact, although Definition 1.3 might seem a bit restrictive at first glance, the existence of an $\mathcal{A}$–$\mathcal{B}$-imprimitivity bimodule bundle is equivalent to a Morita equivalence of the section algebras which is compatible with the identification of the spectrums of A and B with T [**18**, Proposition 2.18]. Anyway, with Definition 1.3 and the inner products (1.1) in hand, a definition of Morita equivalence for two C^*-G-bundles which generalizes that for dynamical systems (see [**2**] and [**6**]) is straightforward.

DEFINITION 1.4. Two C^*-G-bundles $(\mathcal{A},\alpha)$ and $(\mathcal{B},\beta)$ are *Morita equivalent* if there is a $\mathcal{A}$–$\mathcal{B}$-imprimitivity bimodule bundle $\mathcal{X}$ which admits an action V of G by isomorphisms such that

$$ {}_{\mathcal{A}}\langle V_\gamma(x)\,,\,V_\gamma(y)\rangle = \alpha_\gamma\big({}_{\mathcal{A}}\langle x\,,\,y\rangle\big) \quad\text{and}\quad \langle V_\gamma(x)\,,\,V_\gamma(y)\rangle_{\mathcal{B}} = \beta_\gamma\big(\langle x\,,\,y\rangle_{\mathcal{B}}\big).$$

It is not hard to see that Definition 1.4 defines an equivalence relation on the set of C^*-G-bundles [**18**, Lemma 3.2].

As in Green's definition of $\mathrm{Br}(T)$, it will be necessary to work with continuous-trace C^*-algebras, or more precisely, the associated C^*-bundles. But the section algebra of a C^*-bundle $\mathcal{A}$ has continuous-trace if and only if each fibre is isomorphic to the compact operators on some Hilbert space (which can vary from fibre to fibre) and $\mathcal{A}$ has local rank-one projections. Such bundles are called *elementary C^*-bundles satisfying Fell's condition.* The *Brauer group* $\mathrm{Br}(G)$ of G is defined to be the set of Morita equivalence classes of C^*-G-bundles $(\mathcal{A},\alpha)$ for which $\mathcal{A}$ is an elementary C^*-bundle satisfying Fell's condition. There is a natural binary operation on $\mathrm{Br}(G)$ defined as follows. Given G-bundles $(\mathcal{A},\alpha)$ and $(\mathcal{B},\beta)$, then there is unique C^*-bundle $\mathcal{A}\otimes\mathcal{B}$ over $G^{(0)}\times G^{(0)}$ such that the fibre over (u,v) is $A(u)\otimes B(v)$ and $(u,v)\mapsto f(u)\otimes g(v)$ is a section whenever f and g are sections of $\mathcal{A}$ and $\mathcal{B}$, respectively. The restriction $\mathcal{A}\otimes_{G^{(0)}}\mathcal{B}$ of $\mathcal{A}\otimes\mathcal{B}$ to the diagonal in $G^{(0)}\times G^{(0)}$ is another elementary C^*-bundle over $G^{(0)}$ satisfying Fell's condition. The diagonal action $\alpha\otimes\beta$ of G makes $(\mathcal{A}\otimes_{G^{(0)}}\mathcal{B},\alpha\otimes\beta)$ into a C^*-G-bundle: $(\alpha\otimes\beta)_\gamma$ is the isomorphism $\alpha_\gamma\otimes\beta_\gamma$ from $A\big(s(\gamma)\big)\otimes B\big(s(\gamma)\big)$ to $A\big(r(\gamma)\big)\otimes A\big(r(\gamma)\big)$.

Of course, we have to prove that the Brauer group is a group.

THEOREM 1.5. *The operation* $[\mathcal{A},\alpha][\mathcal{B},\beta] := [\mathcal{A}\otimes_{G^{(0)}}\mathcal{B},\alpha\otimes\beta]$ *is a well-defined commutative and associative operation on* $\mathrm{Br}(G)$. *With respect to this operation,* $\mathrm{Br}(G)$ *is a group. The neutral element is the class of* $(\mathcal{T},\tau)$, *where* $\mathcal{T} = G^{(0)}\times\mathbf{C}$ *is the trivial line bundle with* τ *the action of* G *on* $\mathcal{T}$ *given by* $\gamma\cdot\big(s(\gamma),\lambda\big) := \big(r(\gamma),\lambda\big)$. *The inverse of* $[A,\alpha]$ *is the conjugate bundle* $[\overline{A},\bar\alpha]$.

The proof of Theorem 1.5 is nontrivial. Here I will confine myself to a rather inadequate sketch. The details can be found in [**18**, Theorem 3.7] and [**3**, §3].

SKETCH OF PROOF FOR THEOREM 1.5. The main issue is to see that $[\mathcal{A},\alpha]^{-1}$ exists and equals $[\overline{\mathcal{A}},\bar\alpha]$. Let $A = C_0(G^{(0)};\mathcal{A})$. Since every continuous-trace C^*-algebra is locally Morita equivalent to an abelian C^*-algebra, we can, with a bit of work, localize and reduce to the case were there is an A–$C_0(G^{(0)})$-imprimitivity bimodule X. Now $\mathsf{X} \otimes_{G^{(0)}} \overline{\mathsf{X}}$ is a $A \otimes_{G^{(0)}} \overline{A}- C_0(G^{(0)})$-imprimitivity bimodule, but it carries no obvious G-action. But it is possible to show that it is isomorphic to

$$\mathsf{N} = \{\, a \in A : u \mapsto \operatorname{tr}\bigl(a^* a(u)\bigr) \in C_0(G^{(0)}) \,\}.$$

Furthermore N is isomorphic to the section algebra $C_0(G^{(0)}, \mathcal{N})$, where

$$\mathcal{N} := \{\, a \in \mathcal{A} : \operatorname{tr}(a^* a) < \infty \,\}.$$

Now very much unlike X, $\mathcal{N}$ admits a natural G-action: $V_\gamma(a) := \alpha_\gamma(a)$. Therefore, it is possible to show that $(\mathcal{N}, V)$ implements an equivalence between $(\mathcal{A} \otimes_{G^{(0)}} \overline{\mathcal{A}}, \beta)$ and $(\mathcal{T}, \tau)$ for some action β. Some untangling is required to verify that $\beta = \alpha \otimes \bar{\alpha}$. □

REMARK 1.6. The restriction to continuous-trace algebras is essential. The multiplication in $\operatorname{Br}(G)$ could easily be extended to Morita equivalence classes of C^*-G-bundles over $G^{(0)}$ — as in [**14**, §2] — but a result of Green shows that the invertible elements of the resulting semigroup are exactly $\operatorname{Br}(G)$ [**14**, Theorem B.1].

The structure of $\operatorname{Br}(G)$ is quite rich and intriguing. As would be expected, most of the detailed information known is for the case of a transformation groupoid.

EXAMPLE 1.7. Let $G = T \times H$ be a transformation groupoid as in Example 1.2. Then $\operatorname{Br}(G)$ is the equivariant Brauer group $\operatorname{Br}_H(T)$ of [**3**] consisting of Morita equivalence classes of C^*-dynamical systems (A, H, β) in which A is a continuous-trace C^*-algebra with spectrum T and β induces the given action on T.

There is a deep structure theorem for $\operatorname{Br}_H(T)$ [**3**, Theorem 5.1], but it would take us too far a field to detail that here. Instead let me mention a result due to Packer and generalized by a number of authors which involves the calculation of $\operatorname{Br}_H(T)$ when the action of H on T is *trivial.* The answer is surprisingly complex.

THEOREM 1.8 ([**24, 25, 10, 9**]). *Suppose that H acts trivially on T and that H is either compact, discrete, or compactly generated abelian. If G is the transformation groupoid $T \times H$, then*

$$\operatorname{Br}(G) = \operatorname{Br}_H(T) \cong H^1(T, \widehat{\mathcal{H}}_{\mathrm{ab}}) \oplus C\bigl(T, H^2(H, \mathbf{T})\bigr) \oplus H^2(T, \mathcal{S}).$$

The formidable looking $H^1(T, \widehat{\mathcal{H}}_{\mathrm{ab}})$ is merely a sheaf cohomology group which is isomorphic to the isomorphism classes of principle $\widehat{H}_{\mathrm{ab}}$-bundles over T. ($\widehat{H}_{\mathrm{ab}}$ is the Pontryagin dual of the abelian group $H_{\mathrm{ab}} = H/\overline{[H,H]}$.) Thus, in this case, the Brauer group is made up of three groups which clearly reflect both the group-like and space-like structure of G.

2. The rôle of equivalence

It is generally accepted that homomorphism and isomorphism are too rigid (at least in some cases) for the study of groupoids. A particularly useful replacement for isomorphism is Renault's notion of groupoid equivalence.

DEFINITION 2.1. Two groupoids H and G are *equivalent* if there is a space Z which is a left principal H-space and a right principal G-space, with structure maps $r : Z \to H^{(0)}$ and $s : Z \to G^{(0)}$, such that

(a) The actions commute.
(b) The map r induces a homeomorphism $Z/G \cong H^{(0)}$.

(c) The map s induces a homeomorphism $H\backslash Z \cong G^{(0)}$.

The space Z is called a (H, G)-*equivalence.*

In particular, groupoid equivalence implies Morita equivalence of the corresponding groupoid C^*-algebras (Theorem 2.2), and as we shall see, the Brauer group is invariant under equivalence of groupoids (Theorem 2.3).

THEOREM 2.2 ([**21**, Theorem 2.2]). *If G and H are equivalent and if λ and β are Haar systems on G and H, respectively, then $C^*(G,\lambda)$ and $C^*(H,\beta)$ are Morita equivalent.*

To connect with our Brauer group data, we use what I call the "basic construction". Let Z be a (H, G)-equivalence.[1] If $[\mathcal{A}, \alpha] \in \operatorname{Br}(G)$, then

$$Z * \mathcal{A} := \{\, (z,a) \in Z \times \mathcal{A} : s(z) = p(a)\,\}$$

is a right principal G-space: $(z,a)\cdot\gamma := \big(z\cdot\gamma, \alpha_\gamma^{-1}(a)\big)$. The quotient $\mathcal{A}^Z = Z*\mathcal{A}/G$ is a elementary C^*-bundle over $H^{(0)}$ satisfying Fell's condition, and

$$\alpha_h^Z\big([z,a]\big) := [h\cdot z, a]$$

defines a left H-action on $\mathcal{A}^Z$ making $(\mathcal{A}^Z, \alpha^Z)$ a C^*-H-bundle representing a class in $\operatorname{Br}(H)$.

This construction provides a natural isomorphism between the Brauer groups of equivalent groupoids.

THEOREM 2.3 ([**18**, Theorem 4.1]). *If H and G are equivalent groupoids via a (H,G)-equivalence Z, then $[\mathcal{A},\alpha] \mapsto [\mathcal{A}^Z, \alpha^Z]$ is a an isomorphism φ^Z of* $\operatorname{Br}(G)$ *onto* $\operatorname{Br}(H)$.

SKETCH OF PROOF. First one verifies that $\varphi^Z\big([\mathcal{A},\alpha]\big) := [\mathcal{A}^Z, \alpha^Z]$ is well defined. Now if Z is a (H,G)-equivalence and Y is a (G,K)-equivalence, then the quotient $Z \mathbin{\underset{G}{*}} Y$ by the diagonal G-action is a (H,K)-equivalence, and

$$\big[x, [y,a]\big] \mapsto \big[[x,y], a\big] \tag{2.1}$$

gives an equivariant Banach bundle map from $(\mathcal{A}^Y)^Z$ onto $\mathcal{A}^{Z \overset{*}{_G} Y}$. Therefore,

$$\varphi^{Z \overset{*}{_G} Y} = \varphi^Z \circ \varphi^Y. \tag{2.2}$$

If Z^{op} is the opposite space (a (G,H)-equivalence), then $Z^{\mathrm{op}} \mathbin{\underset{G}{*}} Z \cong G$. Since $\varphi^G = \mathrm{id}$, it follows that $\varphi^{Z^{\mathrm{op}}}$ is an inverse for φ^Z. □

Theorem 2.3 has a number of interesting corollaries. One is a fair bit of the main result from [**20**].

COROLLARY 2.4 ([**20**]). *Suppose that H and K are* groups *acting freely and properly on T and that these actions commute. Then*

$$\operatorname{Br}_H(T/K) \cong \operatorname{Br}_K(T/H).$$

PROOF. The transformation groupoids $(T/K, H)$ and $(T/H, K)$ are equivalent via T. □

Other corollaries of Theorem 2.3 and further examples are given in Section 4.

[1]For the basic construction, it would suffice for Z to be a (H,G)-*correspondence*; that is, a principle right G-space, and a left H-space such that the actions commute and $r : Z \to H^{(0)}$ induces a homeomorphism of Z/G with $H^{(0)}$.

3. The Brauer group and extensions

To see how extensions are related to the Brauer group, it seems best to consider first the special case where G is a topological space T. The theory of continuous-trace C^*-algebras allows us to assign a class in the sheaf cohomology group $H^2(T, \mathcal{S})$ to each continuous-trace C^*-algebra with spectrum T (or to each elementary C^*-bundle $\mathcal{A}$ over T) [**28**, Proposition 5.24]. I'll abuse terminology a bit, and call this element $\delta(A)$ the *Dixmier-Douady* class of the C^*-algebra A (or of the bundle $\mathcal{A}$).[2] The Dixmier-Douady theory (cf., [**28**, Theorem 5.29]) implies that $\delta(A)$ depends only on the Morita equivalence class of A and Green's theorem gives a group isomorphism $\delta : \mathrm{Br}(T) \to H^2(T, \mathcal{S})$. A profitable way to think of $\delta(A)$ is as the obstruction to A being globally Morita equivalent to an abelian C^*-algebra. To go from a class $\delta(A) \in H^2(T, \mathcal{S})$ to an extension, we need to recall a bit about the group $H^2(T, \mathcal{S})$. Following [**28**, Chap. 4], the set of locally finite open covers $\boldsymbol{U}$ of T is directed by refinement and $H^2(T, \mathcal{S})$ is the direct limit

$$\varinjlim_{\boldsymbol{U}} H^2(\boldsymbol{U}, \mathcal{S}). \tag{3.1}$$

Thus if $\delta(A) \in H^2(T, \mathcal{S})$, then there is a cover $\boldsymbol{U} = \{ U_i \}_{i \in I}$ and a element of $H^2(\boldsymbol{U}, \mathcal{S})$ represented by an alternating 2-cocycle $\nu = \{ \nu_{ijk} \}_{i,j,k \in I}$ consisting of functions $\nu_{ijk} : U_i \cap U_j \cap U_k \to \mathbf{T}$ such that $\nu_{ijk}(t)\nu_{ikl}(t) = \nu_{jkl}(t)\nu_{ijl}(t)$ for all $i, j, k, l \in I$ and $t \in U_i \cap U_j \cap U_k \cap U_l$. The cover $\boldsymbol{U}$ determines a groupoid $G(\boldsymbol{U})$ equivalent to $G = T$ called the *localization of* G:

$$G(\boldsymbol{U}) := \{ (i, x, j) : x \in U_i \cap U_j \}$$

with unit space the disjoint union $\coprod_{i \in I} U_i = \{ (i, x) : x \in U_i \}$. The groupoid structure on $G(\boldsymbol{U})$ comes by identifying $G(\boldsymbol{U})$ with the equivalence relation on $\coprod U_i$ in which $(i, x) \sim (j, y)$ if and only if $x = y$. (It is not hard to see that $\coprod U_i$ implements an equivalence between $G(\boldsymbol{U})$ and G.) Then we can define a multiplication on $E_\nu := G(\boldsymbol{U}) \times \mathbf{T} = \{ (i, x, j, \zeta) : x \in U_i \cap U_j \text{ and } \zeta \in \mathbf{T} \}$ as follows:

$$(i, x, j, \zeta)(j, x, k, \omega) := (i, x, k, \nu_{ijk}(x)\zeta\omega).$$

Therefore we get a groupoid extension

$$G^{(0)} \times \mathbf{T} \hookrightarrow E_\nu \twoheadrightarrow G(\boldsymbol{U}), \tag{3.2}$$

determined by $\delta(A) = [\nu]$. A groupoid extension of $G^{(0)} \times \mathbf{T}$ such as (3.2) is called a $\mathbf{T}$-*groupoid* or *twist* over $G(\boldsymbol{U})$, and play an important rôle in the theory. Notice that in general, a twist over a groupoid G is just a principal circle bundle over G with a compatible groupoid structure.

REMARK 3.1. Although it will not be needed in what follows, note that given a twist E over G (and a Haar system λ on G), one can form the twisted groupoid C^*-algebra $C^*(G; E, \lambda)$ as in [**30, 22, 23**]. For a localization of a topological space, there is a natural Haar system λ on $G(\boldsymbol{U})$ and it can be shown that $C^*(G(\boldsymbol{U}), E_\nu, \lambda)$ has continuous-trace and Dixmier-Douady class coinciding with that of ν in $H^2(T, \mathcal{S})$ [**27**, Remark 3]. The groupoid E_ν is known to some as the Raeburn-Taylor groupoid, and by still others as the third jewel in the Raeburn trifecta.

[2] Formally, the Dixmier-Douady class lives in $H^3(T; \mathbf{Z})$. It is the image of what we have defined to be $\delta(A)$ in $H^3(T; \mathbf{Z})$ by the natural isomorphism of $H^2(T, \mathcal{S})$ with $H^3(T; \mathbf{Z})$.

The bit I wish to distill from the the case $G = T$, is that $\mathrm{Br}(G)$ corresponds to certain twists over groupoids equivalent to G. Furthermore one ought to be able to recover $\mathrm{Br}(G)$ from a direct limit of groups these twists with the direct limit taken over groupoids H equivalent to G directed in some fashion. To accomplish this for general G, we'll first have to find some twists related to a groupoid H equivalent to G. Notice that each element in the subgroup

$$\mathrm{Br}_0(H) := \{\, [\mathcal{A}, \alpha] \in \mathrm{Br}(H) : \delta(\mathcal{A}) = 0 \,\}$$

has a representative with $\mathcal{A}$ equal to the trivial bundle $H^{(0)} \times K(\ell^2)$ [**28**, Theorem 4.85 and Proposition 5.59]. With some more work, one can prove the following.

LEMMA 3.2 ([**18**, Lemma 8.2]). *If $[\mathcal{A}, \alpha] \in \mathrm{Br}_0(H)$, then there is a strongly continuous homomorphism $\pi : H \to \mathrm{Aut}\, K(\ell^2)$ such that $(\mathcal{A}, \alpha)$ is Morita equivalent to $\big(H^{(0)} \times K(\ell^2), \bar{\pi}\big)$, where for all $\gamma \in G$ and $T \in K(\ell^2)$*

$$\bar{\pi}_\gamma\big(s(\gamma), T\big) := \big(r(\gamma), \pi_\gamma(T)\big).$$

Lemma 3.2 allows us to associate a *twist* to each element in $\mathrm{Br}_0(H)$:

$$H^{(0)} \times \mathbf{T} \hookrightarrow E(\pi) \twoheadrightarrow H,$$

where $E(\pi) := \{\, (\gamma, U) \in H \times U(\ell^2) : \pi_\gamma = \mathrm{Ad}\, U \,\}$. It should be noted that $E(\pi)$ may depend on our choice of stabilization in Lemma 3.2. However, two twists E and F are called *equivalent* if there is a groupoid isomorphism φ such that

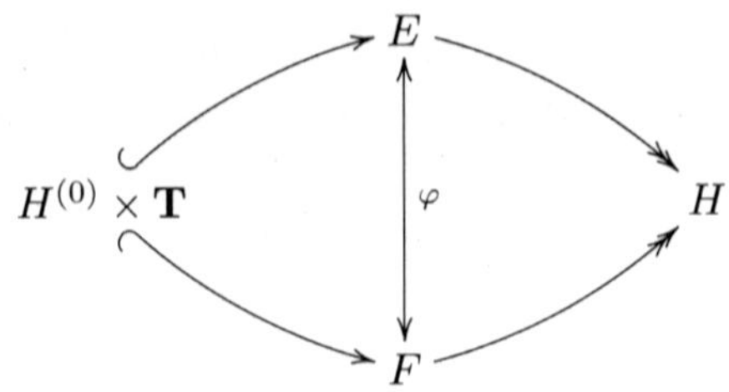

commutes. The set $\mathrm{Tw}(H)$ of equivalence classes of twists over H is group with respect to Baer sum. Furthermore, if E is a twist over H and if H has a Haar system, then there is a homomorphism $\pi : H \to \mathrm{Aut}\, K(\ell^2)$ such that $E(\pi)$ is equivalent to E [**18**, Proposition 8.7]. If $E(\pi)$ is equivalent to $E(\eta)$, then π and η are *exterior equivalent.* That is, there is a continuous map $u : H \to U(\ell^2)$ such that

(a) $\pi_\gamma = (\mathrm{Ad}\, u_\gamma) \circ \eta_\gamma$, and
(b) $u_{\gamma\beta} = u_\gamma \eta_\gamma(u_\beta)$ for all $(\gamma, \beta) \in H^{(2)}$.

It is straightforward to check that π exterior equivalent to η implies that $[H^{(0)} \times K(\ell^2), \bar{\pi}] = [H^{(0)} \times K(\ell^2), \bar{\eta}]$ in $\mathrm{Br}_0(H)$. This, however, is not the whole story. It is possible to construct nontrivial twists as follows. (Keep in mind that a twist over H is principal circle bundle over H with a compatible groupoid structure.) Let $q : \Lambda \to H^{(0)}$ be a circle bundle viewed as a right $\mathbf{T}$-space. Let $\Lambda \overset{*}{\underset{\mathbf{T}}{}} \Lambda$ be the circle bundle over $H^{(0)} \times H^{(0)}$ which is the quotient of $\{\, (\lambda_1, \lambda_2) \in \Lambda \times \Lambda : q(\lambda_1) = q(\lambda_2) \,\}$ by the diagonal $\mathbf{T}$-action: $(\lambda_1, \lambda_2) \cdot \zeta := (\lambda_1 \cdot \zeta, \lambda_2 \cdot \zeta)$. If $\epsilon(\Lambda)$ is the pull back of $\Lambda \overset{*}{\underset{\mathbf{T}}{}} \Lambda$ by the map $d : H \to H^{(0)} \times H^{(0)}$ sending γ to $\big(r(\gamma), s(\gamma)\big)$, then $\epsilon(\Lambda)$ is a twist over H. To connect this construction up with $\mathrm{Br}_0(H)$, suppose that $\varphi : H^{(0)} \to \mathrm{Aut}\, K(\ell^2)$ is continuous. Then

$$\Lambda := \{\, (u, W) \in H^{(0)} \times U(\ell^2) : \varphi_u = \mathrm{Ad}\, W \,\}$$

is a circle bundle over $H^{(0)}$ and if $\pi_\gamma := \varphi^*_{r(\gamma)}\varphi_{s(\gamma)}$, then $E(\bar{\pi})$ is equivalent to $\epsilon(\Lambda)$ and potentially nontrivial. But $[H^{(0)} \times K(\ell^2), \bar{\pi}]$ is covariantly isomorphic to the trivial element $[H^{(0)} \times K(\ell^2), \tau]$. Thus the correspondence between twists over H and $\mathrm{Br}_0(H)$ need not be one-to-one. Fortunately, twists of the form $\epsilon(\Lambda)$ are essentially the only complication.

THEOREM 3.3 ([**18**, Theorem 8.3]). *Let $\mathcal{E}(H)$ be the quotient of* $\mathrm{Tw}(H)$ *by the subgroup of twists of the form $\epsilon(\Lambda)$ where Λ is a principal circle bundle over $G^{(0)}$. Let the class of $E \in \mathrm{Tw}(H)$ in $\mathcal{E}(H)$ be denoted by $[\![E]\!]$. Then the map*

$$[H^{(0)} \times K(\ell^2), \bar{\pi}] \mapsto [\![E(\pi)]\!]$$

defines an isomorphism of $\mathrm{Br}_0(H)$ *onto $\mathcal{E}(H)$.*

Now we have to see how to organize the groupoids equivalent to G in such a way that we can form a direct limit generalizing (3.1). To do this, we use the *imprimitivity groupoid* as defined in, for example, [**23**]. Let X be a principal right G-space and let $X * X := \{\, (x, y) \in X \times X : s(x) = s(y) \,\}$. Then the quotient $G^X := X * X/G$ by the diagonal action admits the structure of a groupoid, and G^X is equivalent to G via X. Conversely, if X is a (H, G)-equivalence, then $\eta \mapsto [\eta \cdot x, x]$ is an isomorphism of H onto G^X. Although it need not greatly concern us here, we will have to ensure that our imprimitivity groupoids G^X have a Haar system. For this, it is necessary and sufficient that X have a *full equivariant s-system* [**18**, Proposition 5.2]. That is, measures $\{\, \beta_u \,\}_{u \in G^{(0)}}$ such that $\operatorname{supp} \beta_u = s^{-1}(u)$ and

$$\int_X f(x \cdot \gamma)\, d\beta_{r(\gamma)}(x) = \int_X f(x)\, d\beta_{s(\gamma)}(x) \quad \text{for all } \gamma \in G.$$

Fortunately, this condition is trivially satisfied for all the G^X we need to consider.

Anyway, we let $\mathcal{P}(G)$ be the collection of principal right G-spaces with full equivariant s-systems. We say that $Y \preceq X$ in $\mathcal{P}(G)$ if there a continuous open G-equivariant surjection $\varphi : X \to Y$. Then $\mathcal{P}(G)$ is directed by $\preceq$. This just means that given X and Y, with s-systems α and β, we just have to find $Z \in \mathcal{P}(G)$ such that $X \preceq Z$ and $Y \preceq Z$. But $Z := X * Y = \{\, (x, y) : s(x) = s(y) \,\}$ together with s-system

$$\mu_u(f) = \int_X \int_Y f(x, y)\, d\beta_u(y)\, d\alpha_u(x)$$

does the job.

Suppose $Y \preceq_\varphi X$; that is, $\varphi : X \to Y$ is a continuous open G-equivariant surjection and $X, Y \in \mathcal{P}(G)$. Then $[x, y] \mapsto [\varphi(x), \varphi(y)]$ is a homomorphism $\tilde{\varphi}$ from G^X onto G^Y. If E is a twist over G^Y, then the pull-back $\tilde{\varphi}^*(E)$ is a twist over G^X. However, if $Y \preceq_\psi X$, there is no reason to believe that $\tilde{\varphi}^*(E)$ is equivalent to $\tilde{\psi}^*(E)$. But the importance of the quotients $\mathcal{E}(G^X)$ becomes evident again since it is true — but not so obvious — that $[\![\tilde{\varphi}^*(E)]\!] = [\![\tilde{\psi}^*(E)]\!]$, and we obtain homomorphisms $\Phi_{Y,X} : \mathcal{E}(G^Y) \to \mathcal{E}(G^X)$ depending only on X and Y [**18**, Theorem 9.1]. It is now an easy matter to see that $\{\, \mathcal{E}(G^X), \Phi_{Y,X} \,\}_{X,Y \in \mathcal{P}(G)}$ is a directed system of groups. Therefore we can define

$$\operatorname{Ext}(G, \mathbf{T}) := \varinjlim_{X \in \mathcal{P}(G)} \mathcal{E}(G^X).$$

Thus $\operatorname{Ext}(G, \mathbf{T})$ effectively organizes the union of all the extension groups $\mathcal{E}(H)$ with H equivalent to G.

The next step is to see that $\mathrm{Br}(G)$ is (isomorphic to) $\mathrm{Ext}(G, \mathbf{T})$ by showing that it has the appropriate universal properties for the direct limit. The tool we need for this is the following lemma.

LEMMA 3.4 ([**18**, Corollary 6.6]). *If $[\mathcal{A}, \alpha] \in \mathrm{Br}(G)$, then there is a $X \in \mathcal{P}(G)$ such that*

$$[\mathcal{A}^X, \alpha^X] \in \mathrm{Br}_0(G^X).$$

SKETCH OF PROOF. Since $\mathcal{A}$ satisfies Fell's condition, we can find a locally finite open cover $\boldsymbol{U} = \{ U_i \}$ of $G^{(0)}$ such that $\delta(\mathcal{A}|_{U_i}) = 0$ for all i. Then it is not hard to see that

$$X := \coprod G^{U_i} = \coprod \{ (i, \gamma) : r(\gamma) \in U_i \}$$

belongs to $\mathcal{P}(G)$. Then

$$G^X \cong \{ (i, \gamma, j) : r(\gamma) \in U_i \text{ and } s(\gamma) \in U_j \},$$

and is commonly referred to as a local trivalization of G. If $\psi : \coprod U_i \to G^{(0)}$ is the natural projection, then it is easy to see that the pull-back $\psi^* \mathcal{A} \cong \bigoplus \mathcal{A}|_{U_i}$. Hence $\delta(\psi^* \mathcal{A}) = 0$. But

$$[i, \gamma, a] \mapsto [i, r(\gamma), \alpha_\gamma(a)]$$

is a bundle map, and therefore an isomorphism, from $\mathcal{A}^X$ to $\psi^* \mathcal{A}$. Thus $\delta(\mathcal{A}^X) = 0$ as well. □

If $X \in \mathcal{P}(G)$, then X is a (G^X, G)-equivalence and Theorem 2.3 implies that $\varphi^{X^{\mathrm{op}}} : \mathrm{Br}(G^X) \to \mathrm{Br}(G)$ is an isomorphism. Theorem 3.3 implies that we have an isomorphism $\Theta_{G^X} : \mathrm{Br}_0(G^X) \to \mathcal{E}(G^X)$. Thus we have monomorphisms

$$\tau_X = \varphi^{X^{\mathrm{op}}} \circ \Theta_{G^X}^{-1} : \mathcal{E}(G^X) \to \mathrm{Br}(G).$$

The maps φ^X enjoy a nice calculus. For example, $\varphi^{X \ast_G Y} = \varphi^X \circ \varphi^Y$ and $\varphi^{X^{\mathrm{op}}} \circ \varphi^X = \mathrm{id}_{\mathrm{Br}(G)}$ [**18**, Lemma 4.5]. These sorts of considerations allow us to show that the τ_X are compatible with the $\Phi_{Y,X}$; that is,

$$\begin{array}{ccc} \mathcal{E}(G^Y) & \xrightarrow{\Phi_{Y,X}} & \mathcal{E}(G^X) \\ & {\scriptstyle \tau_Y} \searrow \quad \swarrow {\scriptstyle \tau_X} & \\ & \mathrm{Br}(G) & \end{array}$$

commutes for all $Y \preceq X$. Thus if $h_X : \mathcal{E}(G^X) \to \mathrm{Ext}(G, \mathbf{T})$ are the canonical maps, then the universal property of the direct limit implies that there is a monomorphism $\tau : \mathrm{Ext}(G, \mathbf{T}) \to \mathrm{Br}(G)$ such that $\tau \circ h_X = \tau_X$. But if $[\mathcal{A}, \alpha] \in \mathrm{Br}(G)$, Lemma 3.4 implies there is an $X \in \mathcal{P}(G)$ such that $[\mathcal{A}^X, \alpha^X] = \varphi^X([\mathcal{A}, \alpha]) \in \mathrm{Br}_0(G^X)$. Then $\tau_X\big(\Theta_{G^X}([\mathcal{A}^X, \alpha^X])\big) = [\mathcal{A}, \alpha]$. It follows that τ is surjective and therefore an isomorphism. Thus we have sketched the proof of the following.

THEOREM 3.5 ([**18**, Theorem 10.1]). *The Brauer group $\mathrm{Br}(G)$ together with the homomorphisms $\{ \tau_X \}_{X \in \mathcal{P}(G)}$ form a direct limit for the directed system $\{ \mathcal{E}(G^X), \Phi_{Y,X} \}$. Therefore, $\mathrm{Br}(G)$ is naturally isomorphic to $\mathrm{Ext}(G, \mathbf{T})$.*

4. Examples

As I detailed above (e.g., Theorem 1.8 and Corollary 2.4), we've made quite a good start in understanding the structure of $\mathrm{Br}(G)$ when G is a transformation groupoid. Naturally, the situation for general G is a bit more complicated, and a general structure theorem along the lines of [**3**, Theorem 5.1] is out of reach. (This is at least partly because a satisfactory analogue of Moore group cohomology, which is used in the structure theory, has yet to be developed for groupoids.) However, as I stressed in the introduction, the Brauer group is meant to specialize to the known theories when G is a space or a group. With the help of Theorem 2.3, we can say a bit more. For example, a groupoid is equivalent to a group precisely when it is transitive. In particular, if G is transitive and $u \in G^{(0)}$, then $G^u := r^{-1}(u)$ implements an equivalence between G and the isotropy group $G|_u := r^{-1}(u) \cap s^{-1}(u)$, Thus Theorem 2.3 gives us:

PROPOSITION 4.1. *Let G be a transitive groupoid, $u \in G^{(0)}$, and $G|_u$ the isotropy group at u. Then*

$$\mathrm{Br}(G) \cong \mathrm{Br}(G|_u) \cong H^2(G|_u, \mathbf{T}).$$

Another extreme is the class of groupoids which are totally space-like in that they are equivalent to a topological space. This happens exactly when G is principal and proper in the sense that $\gamma \mapsto \big(r(\gamma), s(\gamma)\big)$ is a proper injection from G to $G^{(0)} \times G^{(0)}$. In particular, a proper principal groupoid is equivalent to $G^{(0)}/G$, and Theorem 2.3 implies the following.

PROPOSITION 4.2. *If G is a proper principal groupoid, then*

$$\mathrm{Br}(G) \cong H^2(G^{(0)}/G, \mathcal{S}) \cong \mathrm{Br}(G^{(0)}/G).$$

A groupoid Γ is called *r-discrete* if its range map r is a local homeomorphism. This definition is at slight odds with [**29**], but now has been universally adopted. The term *étale* groupoid is also used — especially in applications to geometry. In [**17**], r-discrete groupoids in the sense we are using the term are called sheaf groupoids. In that paper, Kumjian developed a cohomology theory for r-discrete groupoids acting on sheaves of abelian groups for the purpose of identifying Brauer groups. Fix an r-discrete groupoid Γ, let $\mathcal{S}$ denote the sheaf of germs of continuous $\mathbf{T}$-valued functions on $\Gamma^{(0)}$, and let Γ act on $\mathcal{S}$ in the usual way. Then Kumjian shows how to build a cohomology theory $H^*(\Gamma, \mathcal{S})$, which reduces to Grothendieck's equivariant cohomology theory [**13**], $H^*_G(X, \mathcal{S})$, when Γ is the transformation group groupoid $X \times G$ determined by the action of a discrete group G on a locally compact space X. It may also be worthwhile to point out that certain equivariant sheaf cohomology groups have been "computed" in [**13**] (using spectral sequences) and in [**17**] (using an analog of Milnor's $\lim^1$ sequence). Theorem 4.19 of [**17**] may be phrased as

PROPOSITION 4.3. *If Γ is an r-discrete groupoid, then* $\mathrm{Br}(\Gamma) \cong H^2(\Gamma, \mathcal{S})$.

Of course, we can combine Theorem 2.3 and Proposition 4.3 to compute the Brauer group of any groupoid equivalent to an r-discrete groupoid. The range of applicability of this observation is quite large as many groupoids of interest are known to be equivalent to r-discrete groupoids. In particular, the holonomy groupoid of a foliation has this property. See [**21**] for a discussion of this. Also, the groupoids arising in the theory of hyperbolic dynamical systems and Smale spaces have this property (see [**26**] and [**31**]). The exact range of applicability of this proposition is not known. There are groupoids that are not equivalent to r-discrete groupoids. In fact, a locally compact *group* is equivalent to an r-discrete groupoid if and only if the group is itself discrete. As far as I know, the situation for *principal* groupoids is unknown.

Given a directed graph E, there is an associated C^*-algebra $C^*(E)$ [**19, 16**], and this construction is a natural generalization of that of the Cuntz-Krieger aleg-bras $\mathcal{O}_A$ associated to an integer matrix A [**5, 4**]. These algebras have enjoyed considerable attention recently; Kumjian has written a nice survey in [**15**], and [**1**] is a particular nice overall treatment. If the graph E has no sinks, then $C^*(E)$ is the groupoid C^*-algebra $C^*(G_E)$ of an associated groupoid G_E. Since the unit space of G_E is 0-dimensional, it follows that $\operatorname{Br}(G_E) = \operatorname{Br}_0(G_E)$ and is therefore isomorphic to $\mathcal{E}(G_E)$ by Theorem 3.3. However, as we show in [**18**], one actually has $\operatorname{Tw}(G_E) = \{0\}$, so obtain the following result.

PROPOSITION 4.4 ([**18**, Proposition 11.8]). *If E is a locally finite directed graph with no sinks and G_E the associated groupoid, then* $\operatorname{Br}(G_E) = \{0\}$.

Proposition 4.4 is also a special case of a nice computation of $H^2(\Gamma, \mathcal{S})$ for an interesting class of r-discrete groupoids Γ due to Deaconu, Kumjian and Muhly [**8**]. Let σ be a local homeomorpism of a locally compact space X. Then Deaconu [**7**] associated a r-discrete groupoid to (X, σ):

$$\Gamma := \{ (x, m, y) \in X \times \mathbf{Z} \times X : \\ \sigma^k(x) = \sigma^l(x) \text{ for some nonnegative intergers } k - l = m \}.$$

The multiplication is given by $(x, m, y)(y, n, z) := (x, m + n, z)$ and $(x, m, y)^{-1} := (y, -m, x)$. This class of groupoids includes the graph groupoids G_E above as well as transformation groupoids associated to homeomorphisms. Using Proposition 4.3, they prove the following.

PROPOSITION 4.5 ([**8**, Corollary 4]). *Let $\Gamma = \Gamma(X, \sigma)$ be as above. Then there is an exact sequence*

$$H^2(X; \mathbf{Z}) \xrightarrow{1-\sigma^*} H^2(X; \mathbf{Z}) \longrightarrow \operatorname{Br}(\Gamma) \longrightarrow H^3(X; \mathbf{Z}) \xrightarrow{1-\sigma^*} H^3(X; \mathbf{Z}).$$

5. Speculation

There are a number of reasons to believe that the Brauer group is an important and fundamental invariant associated to a groupoid G. Amoung other things, it gives good information about C^*-G-bundles, as well as insight into the structure of the groupoid G itself. But as I have alluded to above, I want to argue that $\operatorname{Br}(G)$ is a 2-cohomology group. In general, given a groupoid G and an group bundle $\mathcal{C}$ of abelian groups over $G^{(0)}$ on which G acts on the left, we would like to associate a family $\{ \mathbb{H}^n(G, \mathcal{C}) \}$ of abelian groups with the usual sorts of properties enjoyed by cohomology groups. Renault defined one such family in [**29**], which I'll denote by $\{ H^n_{\text{cont}}(G, \mathcal{C}) \}$ here, but these groups, based on continuous cocylces modeled closely after the theory for groups, do not seem to capture the full spectrum of groupoid properties. If $\mathcal{C}$ is the trivial bundle $\mathcal{S} := G^{(0)} \times \mathbf{T}$ and G is a r-discrete groupoid Γ, then Kumjian's $H^*(\Gamma, \mathcal{S})$ is the sort of thing one wants. In view of Proposition 4.3, our $\operatorname{Br}(G)$ is a reasonable candidate for $\mathbb{H}^2(G, \mathcal{S})$ for general G. There are also quite respectable candidates for $\mathbb{H}^0(G, \mathcal{C})$ and $\mathbb{H}^1(G, \mathcal{C})$ for any $\mathcal{C}$.

The group $\mathbb{H}^0(G, \mathcal{C})$ ought to be the group of equivariant sections of $r : \mathcal{C} \to G^{(0)}$, and $\mathbb{H}^1(G, \mathcal{C})$ the set of isomorphism classes of $(G, \mathcal{C})$-correspondences. Recall that a $(G, \mathcal{C})$-correspondence X is a right principal $\mathcal{C}$-space equipped with a commuting left G-action such that $r : X \to G^{(0)}$ induces a homeomorphism of $X/\mathcal{C}$ with $G^{(0)}$. I will omit the details here, but it can be shown that, in analogy with Theorem 3.5,

$$\mathbb{H}^1(G, \mathcal{C}) \cong \varinjlim_{X \in \mathcal{P}(G)} H^1_{\text{cont}}(G^X, \mathcal{C}^X)$$

(where $\mathcal{C}^X := G\backslash X * \mathcal{C}$ as in our basic construction preceeding Theorem 2.3). Thus, $\mathbb{H}^1(G, \mathcal{S})$ is an invariant in the spirit of $\operatorname{Br}(G)$. However, as of this writing, a complete theory is still out of reach.

References

1. Teresa Bates, David Pask, Iain Raeburn, and Wojciech Szymański, *The C^*-algebras of row-finite graphs*, preprint, February 1999.
2. François Combes, *Crossed products and Morita equivalence*, Proc. London Math. Soc. **49** (1984), 289–306.
3. David Crocker, Alexander Kumjian, Iain Raeburn, and Dana P. Williams, *An equivariant Brauer group and actions of groups on C^*-algebras*, J. Funct. Anal. **146** (1997), 151–184.
4. Joachim Cuntz, *A class of C^*-algebras and topological Markov chains. II. Reducible chains and the Ext-functor for C^*-algebras*, Invent. Math. **63** (1981), 25–40.
5. Joachim Cuntz and Wolfgang Krieger, *A class of C^*-algebras and topological Markov chains*, Invent. Math. **56** (1980), 251–268.
6. Raul E. Curto, Paul S. Muhly, and Dana P. Williams, *Crossed products of strongly Morita equivalent C^*-algebras*, Proc. Amer. Math. Soc. **90** (1984), 528–530.
7. Valentin Deaconu, *Groupoids associated with endomorphisms*, Trans. Amer. Math. Soc. **347** (1995), 1779–1786.
8. Valentin Deaconu, Alex Kumjian, and Paul S. Muhly, *Cohomology of topological graphs and Cuntz-Pimsner algebras*, J. Operator Theory (in press).
9. Siegfried Echterhoff and Ryszard Nest, *The structure of the Brauer group and crossed products of $C_0(X)$-linear group actions on $C_0(X, \mathcal{K})$*, preprint, 1999.
10. Siegfried Echterhoff and Dana P. Williams, *Locally inner actions on $C_0(X)$-algebras*, preprint, June 1997.
11. J. M. G. Fell and Robert Doran, *Representations of $*$-algebras, locally compact groups, and Banach $*$-algebraic bundles*, vol. I & II, Academic Press, New York, 1988.
12. Philip Green, *The Brauer group of a commutative C^*-algebra*, unpublished seminar notes, University of Pennsylvania, 1978.
13. Alexander Grothendieck, *Sur quelques points d'algèbre homologique*, Tǒhoku Math. J. **9** (1957), 119–221.
14. Astrid an Huef, Iain Raeburn, and Dana P. Williams, *The equivariant Brauer semigroup and the symmetric imprimitivity theorem*, Trans. Amer. Math. Soc. **352** (2000), 4789–4813.
15. A. Kumjian, *Notes on C^*-algebras of graphs*, Operator algebras and operator theory (Shanghai, 1997), Amer. Math. Soc., Providence, RI, 1998, pp. 189–200.
16. Alex Kumjian, David Pask, and Iain Raeburn, *Cuntz-Krieger algebras of directed graphs*, Pacific J. Math. **184** (1998), 161–174.
17. Alexander Kumjian, *On equivariant sheaf cohomology and elementary C^*-bundles*, J. Operator Theory **20** (1988), 207–240.
18. Alexander Kumjian, Paul S. Muhly, Jean N. Renault, and Dana P. Williams, *The Brauer group of a locally compact groupoid*, Amer. J. Math. **120** (1998), 901–954.
19. Alexander Kumjian, David Pask, Iain Raeburn, and Jean N. Renault, *Graphs, groupoids and Cuntz-Krieger algebras*, J. Funct. Anal. (1997), 505–541.
20. Alexander Kumjian, Iain Raeburn, and Dana P. Williams, *The equivariant Brauer groups of commuting free and proper actions are isomorphic*, Proc. Amer. Math. Soc. **124** (1996), 809–817.
21. Paul S. Muhly, Jean Renault, and Dana P. Williams, *Equivalence and isomorphism for groupoid C^*-algebras*, J. Operator Theory **17** (1987), 3–22.

22. Paul S. Muhly and Dana P. Williams, *Continuous trace groupoid C^*-algebras. II*, Math. Scand. **70** (1992), 127–145.
23. ———, *Groupoid cohomology and the Dixmier-Douady class*, Proc. London Math. Soc. (3) (1995), 109–134.
24. Judith A. Packer, *Moore cohomology and central twisted crossed product C^*-algebras*, Canad. J. Math. **48** (1996), 159–174.
25. Judith A. Packer, Iain Raeburn, and Dana P. Williams, *The equivariant Brauer groups of principal bundles*, J. Operator Theory **36** (1996), 73–105.
26. Ian F. Putnam, *Hyperbolic systems and generalized Cuntz-Krieger algebras*, Lecture notes from the Summer School in Operator Algebras, Odense, Denmark, 1996.
27. Iain Raeburn and Joseph L. Taylor, *Continuous-trace C^*-algebras with given Dixmier-Douady class*, J. Austral. Math. Soc. (Series A) **38** (1985), 394–407.
28. Iain Raeburn and Dana P. Williams, *Morita equivalence and continuous-trace C^*-algebras*, Mathematical Surveys and Monographs, vol. 60, American Mathematical Society, Providence, RI, 1998.
29. Jean Renault, *A groupoid approach to C^*-algebras*, Lecture Notes in Mathematics, vol. 793, Springer-Verlag, New York, 1980.
30. ———, *Représentations des produits croisés d'algèbres de groupoïdes*, J. Operator Theory **18** (1987), 67–97.
31. David Ruelle, *Themodynamic formalism*, Encyclopedia of Mathematics and its Applications, vol. 5, Addison-Wesley, Reading, 1978.

6188 Bradley Hall, Dartmouth College, Hanover, NH 03755
E-mail address: Dana.Williams@Dartmouth.EDU

Contemporary Mathematics
Volume **282**, 2001

Amenable Groupoids

Claire Anantharaman and Jean Renault

ABSTRACT. We give a survey of the notion of amenability for groupoids, both in the measured and in the topological framework. We discuss some examples and some recent applications to the theory of C*-algebras.

1. Introduction

The amenability of a measured groupoid was defined around 1977 by R. Zimmer in [**44**], [**45**],[**46**]. A strong motivation then was the theory of hyperfinite factors. Zimmer showed in particular that, in the case of the groupoid of a measured action of a discrete group or of a discrete measured equivalence relation, amenability is a necessary and sufficient condition for the injectivity of the von Neumann algebra of the measured groupoid. A direct connection between amenability and hyperfiniteness was given some four years later by Connes-Feldman-Weiss in [**14**] who proved that a discrete measured equivalence relation is amenable if and only if it is hyperfinite. A notion of topological amenability for locally compact groupoids with a Haar system was defined in [**36**]. It implies measurewise amenability, i.e. the amenability with respect to every quasi-invariant measure. It was shown in [**5**] that the converse holds for discrete group actions. Amenability for topological groupoids plays a role in recent developments of the theory of C*-algebras. It provides a convenient way to prove the nuclearity of some C*-algebras, in particular those purely infinite algebras arising from graphs (see [**30**]) and from boundary actions (see [**6**]). Moreover, the existence of an amenable action on a compact space, is a convenient sufficient condition for the C*-exactness (see [**29**]) of locally compact groups. There are several connections between topological amenability and the Baum-Connes conjecture; they are given in the article [**42**] by J.-L. Tu. First, amenable locally compact groupoids with Haar systems admit proper affine actions on Hilbert spaces; in particular, they satisfy Baum-Connes conjecture [**41**]. Second, it was observed by N. Higson and J. Roe [**26**] that, for a finitely generated discrete group, the existence of an amenable action on a compact space is exactly the property A introduced by G. Yu in [**43**]; these authors show that the groups which have this property satisfy the strong Novikov conjecture (i.e. the injectivity

1991 *Mathematics Subject Classification.* Primary 46L55, 22A22, 22D25; Secondary 28C10, 43A05, 58H05.

Key words and phrases. Amenable groupoids. C*-algebras. Boundary actions. Exact C*-algebras. Baum-Connes conjecture.

of the Baum-Connes homomorphism). We also refer the reader to the forthcoming article [**40**] for the use of groupoids in the coarse Baum-Connes conjecture.

In this report, which is based on the monograph [**7**], we shall give some equivalent definitions and properties of the amenability of a measured groupoid (section 2), the definition and properties of the amenability of a topological groupoid and its relation to measured amenability (section 3). Applications to C*-algebras will be given in section 4. The reader is referred to [**7**] for proofs and a more complete exposition. The primary goal of this survey is a quick access and a convenient reference to the various notions of amenability and some of their applications.

We shall use the same definitions and notations for groupoids as in [**33**]. Given a groupoid G, $G^{(0)}$ will denote its unit space and $G^{(2)}$ the groupoid of composable pairs. The range and source maps from G to $G^{(0)}$ will be denoted by r, s respectively. Elements of G will usually be denoted by γ, γ'; elements of $G^{(0)}$ will usually be denoted by x, y. We shall write $G^x = r^{-1}(x), G_x = s^{-1}(x)$ and $G_x^y = G_x \cap G^y$.

2. Amenable measured groupoids

2.1. Measured groupoids. Here is a short review of the notion of a measured groupoid. Detailed expositions can be found in [**23, 35**]. We shall only consider Borel spaces which are analytic. A *Borel groupoid* G is a set endowed with compatible groupoid and Borel space structures. This means that the range, source, inverse and product maps are Borel, where $G^{(2)}$ has the Borel structure induced by $G \times G$ and $G^{(0)}$ has the Borel structure induced by G. A *Borel Haar system* on a Borel groupoid G is a family $\lambda = \{\lambda^x\}$ of measures indexed by $x \in G^{(0)}$; we assume that λ^x is a σ-finite measure on $G^x = r^{-1}(x)$, that λ is Borel in the sense that for every non-negative Borel function f on G, the function $\lambda(f) : x \mapsto \lambda^x(f)$ is Borel and proper in the sense that there exists a positive Borel function f on G such that $\lambda(f)$ is identically one, and invariant in the sense that the equality $\gamma\lambda^{s(\gamma)} = \lambda^{r(\gamma)}$ holds for every $\gamma \in G$. A σ-finite measure μ on $G^{(0)}$ is called quasi-invariant (with respect to G and a Borel Haar system λ) if the measure $\mu \circ \lambda$ on G is equivalent to its image $\mu \circ \lambda^{-1}$ by the inverse map. Its module is the Radon-Nikodym derivative $\delta = d(\mu \circ \lambda)/d(\mu \circ \lambda^{-1})$.

DEFINITION 2.1. A *measured groupoid* is a triple (G, λ, μ), where G is a Borel groupoid, λ is a Borel Haar system and μ is a quasi-invariant measure.

Some remarks should be made about this definition. What will matter for us are measure classes rather than measures; two pairs (λ, μ) and (λ', μ') giving equivalent measures $\mu \circ \lambda$ and $\mu' \circ \lambda'$ define the same measured groupoid. Second, some conditions in the definition of the Haar system can be weakened; it suffices to have λ^x defined for a.e. $x \in G^{(0)}$ and the invariance condition satisfied for a.e. $\gamma \in G$. Third, the proper framework for measure theory for groupoids is A. Connes transverse measure theory (see [**11**] and [**7**], Appendix A.1); it explains the role played by the pair (λ, μ) and leads to the notion of equivalent measured groupoids.

Here are three basic examples of measured groupoids.

EXAMPLE 2.2. *Group actions.* Let a second countable locally compact group G with left Haar measure λ, an analytic Borel space X, a Borel right action $X \times G \to X$ and a measure μ on X quasi-invariant by every $g \in G$ be given. The semi-direct product of the action is the measured groupoid $(X \times G, \tilde{\lambda}, \mu)$, where $X \times G$ has the groupoid structure such that $r(x, g) = x, s(x, g) = xg$, $(x, g)^{-1} = (xg, g^{-1})$ and

$(x,g)(xg,h) = (x,gh)$, and $\tilde{\lambda}$ consists of the measures $\epsilon_x \times \lambda$ (ϵ_x is the point mass at x).

EXAMPLE 2.3. *Discrete measured equivalence relations* (see [**17**]). Let X be an analytic Borel space and consider an equivalence relation on X which has countable equivalence classes and a Borel graph $R \subset X \times X$. The graph R inherits the trivial groupoid structure of $X \times X$ and this turns it into a Borel groupoid. The counting measures on the fibers R^x form a Borel Haar system λ. A σ-finite measure μ on X is quasi-invariant with respect to (R, λ) if and only if, for every Borel set $A \subset X$ such that $\mu(A) = 0$, we have $\mu([A]) = 0$, where $[A]$ is the saturation of A under the equivalence relation.

EXAMPLE 2.4. *Foliations* (see [**11, 12, 13**]). Let M be a manifold of dimension n and $\mathcal{F}$ be an integrable subbundle of the tangent space. A basic construction gives the holonomy groupoid $G(M, \mathcal{F})$ of the foliation $(M, \mathcal{F})$. It is a differentiable groupoid having M as its unit space. The Borel structure underlying it makes $G(M, \mathcal{F})$ into a Borel groupoid; a choice of a Borel density section for the vector bundle $\mathcal{F} \to M$ defines a Borel Haar system λ on $G(M, \mathcal{F})$. Any σ-finite measure μ in the Lebesgue class of M is quasi-invariant. One thus obtains a measured groupoid $(G(M, \mathcal{F}), \lambda, \mu)$.

2.2. Definition of measured amenability.

DEFINITION 2.5. We say that a measured groupoid (G, λ, μ) is *proper* if there exists a non-negative measurable function h on $G^{(0)}$ such that $(h \circ s)\lambda$ is almost identically one.

Thus, if (G, λ, μ) is proper, it has a measurable Haar system consisting of the probability measures $m^x = (h \circ s)\lambda^x$. One deduces that the quotient $G^{(0)}/G$ is an analytic measure space. Moreover, the family $\{m^x\}$ defines, according to the formula $m(f)(x) = m^x(f)$, an invariant mean

$$m : L^\infty(G, \mu \circ \lambda) \to L^\infty(G^{(0)}, \mu)$$

which is normal. Properness is a very stringent condition. In the case of a group G, it implies that the group is compact. In the case of an ergodic equivalence relation, it implies that there exists essentially a single orbit. Amenability is a weakening of properness. There are several ways to proceed, but they turn out to be equivalent.

DEFINITION 2.6. An *approximate invariant mean* for the measured groupoid (G, λ, μ) is a net (f_i) of non-negative measurable functions on G such that

(i) for all i and a.e. $x \in G^{(0)}$, we have $\int f_i d\lambda^x = 1$.
(ii) the functions $\gamma \mapsto \int |f_i(\gamma^{-1}\gamma') - f_i(\gamma')| d\lambda^x(\gamma')$ tend to the function zero in the $*$-weak topology of $L^\infty(G, \mu \circ \lambda)$.

There are alternate definitions of an approximate invariant mean. It could be a sequence rather than a general net. The condition (i) could have been replaced by the condition that the functions $x \mapsto \int f_i d\lambda^x$ belong to $L^\infty(G^{(0)}, \mu)$ and converge $*$-weakly to the function one. They all lead to same definition of amenability (see [**7**], 3.2.14, 3.2.15 and 6.1.3).

DEFINITION 2.7. An *invariant mean* for the measured groupoid (G, λ, μ) is a unital positive linear map

$$m : L^\infty(G, \mu \circ \lambda) \to L^\infty(G^{(0)}, \mu)$$

which is invariant in the sense that for all non-negative measurable f on G with $\lambda(f)$ bounded and all $\varphi \in L^\infty(G, \mu\circ\lambda)$, $m(f*\varphi) = f * m(\varphi)$, where $f*\varphi(\gamma) = \int f(\gamma_1)\varphi(\gamma_1^{-1}\gamma)d\lambda^{r(\gamma)}(\gamma_1)$ and $f*\psi(x) = \int f(\gamma)\psi\circ s(\gamma)d\lambda^x(\gamma)$.

DEFINITION 2.8. An *invariant measurable system of means* for the measured groupoid (G,λ,μ) is a family $\{m^x\}$ where m^x is a state on $L^\infty(G^x,\lambda^x)$ and x belongs to a co-null subset of $G^{(0)}$, such that

(i) for all $\varphi \in L^\infty(G,\mu\circ\lambda)$, the function $x \mapsto m^x(\varphi)$ is measurable.
(ii) for all $\varphi \in L^\infty(G,\mu\circ\lambda)$, the equality $\gamma m^{s(\gamma)}(\varphi) = m^{r(\gamma)}$ holds a.e.

THEOREM 2.9. ([**7**], Sections 3.1 and 3.2) *Let (G,λ,μ) be a measured groupoid. The following properties are equivalent:*

(i) *There exists an approximate invariant mean for (G,λ,μ).*
(ii) *There exists an invariant mean for (G,λ,μ).*
(iii) *There exists an invariant measurable system of means for (G,λ,μ).*

PROOF. We refer the reader to [**7**] for the proof. Let us just indicate that $(ii) \Rightarrow (i)$ by standard convexity arguments. To show $(i) \Rightarrow (iii)$, one can use the existence, due to Mokobodski [**31**], of a universally measurable limit LIM and define $m^x(\varphi) = LIM_n m_n^x(\varphi)$, where $m_n^x = f_n\lambda^x$. To show $(iii) \Rightarrow (ii)$, the obvious candidate for an invariant mean, namely $\tilde{m}(\varphi)(x) = m^x(\varphi)$, does not necessarily satisfy the required invariance condition; one exploits the fact that it sends $L^\infty(G,\mu\circ\lambda)^G$ onto $L^\infty(G^{(0)},\mu)^G$ (where the superscript G indicates the G-invariant elements) to obtain the existence of an invariant mean.

□

DEFINITION 2.10. We say that the measured groupoid (G,λ,μ) is *amenable* if it satisfies one of these equivalent conditions.

In the case of a group, our condition (ii), which is the existence of a topological left invariant mean on $L^\infty(G)$ ([**20**]), is one of the classical definitions of amenability. It is usual (but sometimes misleading because some authors ([**21, 10**]) give to it a different meaning) to say that the *action* of a group G on the measure space X is *amenable* when the semi-direct product groupoid $X \times G$ is amenable. From condition (i), it is immediate that all the actions of an amenable group are amenable. Non-amenable groups may admit interesting amenable actions. We shall meet some examples in the topological setting.

2.3. Measured equivalence relations. A groupoid G defines an equivalence relation on its unit space $X = G^{(0)}$ which has the image R of the map $(r,s): G \to X\times X$ as its graph. The isotropy at $x \in X$ is the group $G(x) = G_x^x$. When (G,λ,μ) is a measured groupoid, R is an analytic subgroupoid of $X\times X$ and possesses a Haar system α such that $\mu\circ\alpha$ is a pseudo-image of $\mu\circ\lambda$ (see [**23**]). The next result shows that the study of the amenability of (G,λ,μ) reduces to the amenability of the measured equivalence relation (R,α,μ) and of the isotropy subgroups. It was obtained previously by [**3, 19**] in the case of group actions.

THEOREM 2.11. ([**7**], 5.3.33) *Let (G,λ,μ) be a measured groupoid and (R,α,μ) its associated measured equivalence relation. Then (G,λ,μ) is amenable if and only if (R,α,μ) is amenable and almost every isotropy group $G(x)$ is amenable.*

For the proof, we refer again the reader to [**7**]. It is a particular case of a general result on groupoid extensions ([**7**], 5.3.32). Properly formulated, it is true that the amenability of the quotient and of the kernel implies the amenability of the groupoid. It is also true that a properly embedded subgroupoid of an amenable groupoid is amenable ([**7**], 5.3.22). The amenability of the quotient of a measured groupoid requires an additional condition.

The main result about the amenability of measured equivalence relations is the theorem of Connes-Feldman-Weiss [**14**] which states that a measured equivalence relation (R, λ, μ) with countable classes is amenable if and only if it is *hyperfinite*, i.e. it can be exhausted up to a null set by the union of an increasing sequence (R_n) of subequivalence relations with finite classes. The same proof applies to the general case and gives:

THEOREM 2.12. ([**14**]) *A measured equivalence relation (R, λ, μ) is amenable if and only if it is approximately proper, i.e. it can be exhausted up to a null set by the union of an increasing sequence (R_n) of subequivalence relations which are proper in the sense defined earlier.*

In the course of the proof, the following Følner property appears. In order to state it, we need a definition. Let (G, λ, μ) be a measured groupoid. We shall say that a Borel subset $K \subset G$ is *bounded* if there exists $M < \infty$ such that for *a.e.* $x \in G^{(0)}$, $\lambda^x(K) < M$ and $\lambda^x(K^{-1}) < M$.

DEFINITION 2.13. We say that the measured groupoid (G, λ, μ) satisfies the *Følner condition* if for each $\epsilon > 0$ and each bounded Borel subset $K \subset G$ of positive measure, there exists a bounded Borel subset $F \subset G$ such that $r(F)$ has positive measure, for each $x \in r(F)$, $\lambda^x(F_{r(K)})$is strictly positive and the inequality

$$\begin{aligned}&\textstyle\int 1_K(\gamma_1^{-1}\gamma)|1_F(\gamma) - 1_F(\gamma_1)|\delta(\gamma_1)^{-1}d\lambda^x(\gamma_1)d\lambda^x(\gamma)\\ &< \epsilon \textstyle\int 1_{F_{r(K)}}(\gamma_1)\delta(\gamma_1)^{-1}d\lambda^x(\gamma_1)\end{aligned}$$

holds for each $x \in r(F)$.

EXAMPLE 2.14. When G arises as the groupoid of germs of a finitely generated pseudo-group of partial homeomorphisms of a locally compact space X, the choice of a finite set of generators determines a bounded set $K \subset G$ which is called a bounded graph structure on G (see [**1, 28**]). For $P \subset G$, one defines its boundary with respect to K as $\partial_K P = \{\gamma \in P : \gamma K \not\subset P\}$. The above Følner condition can then be expressed, in the case of an invariant measure, as an isoperimetric inequality on the ratio of $\lambda^x(\partial_K P)/\lambda^x(P)$; in the general case, one has to introduce the module δ just as above. In the case of a foliation on a compact manifold endowed with an invariant transverse measure, the above Følner condition for the associated equivalence relation implies the usual Følner condition for almost every individual leaf (see also [**9**]).

We have the following:

PROPOSITION 2.15. ([**7**], 3.2.21 and 5.3.42) *Let (G, λ, μ) be a measured groupoid. If it is amenable, it satisfies the above Følner condition. If G is a group or an equivalence relation, the converse is true.*

It is likely that the converse is true in general.

2.4. Fixed point property. The original definition of an amenable measured groupoid given by R. Zimmer is expressed by a fixed point property. The definition we give below is essentially his, the difference is that we let the groupoid act on a field of Banach spaces rather than a single Banach space. More precisely, we shall assume that (E, L) is a separable measurable G-bundle of Banach spaces (cf. [**15, 16**]), i.e. E is a measurable field of Banach spaces $x \mapsto E_x$ over $(G^{(0)}, \mu)$ possessing a sequence $(g_n)_{n\in\mathbf{N}}$ of measurable sections such that for a.e. x, the set $\{g_n(x), n \in \mathbf{N}\}$ is total in E_x and L is a linear isometric representation of G in E. Thus, for $\gamma \in G$, $L(\gamma)$ is a linear isometry from $E_{s(\gamma)}$ onto $E_{r(\gamma)}$; this action is assumed to be measurable, in the sense that, if $x \mapsto f(x)$ is a measurable section of E over $G^{(0)}$, then $\gamma \mapsto L(\gamma)f \circ s(\gamma)$ is measurable section of r^*E over G.

DEFINITION 2.16. A measured groupoid (G, λ, μ) satisfies the *fixed point property* if for every separable measurable G-bundle (E, L) of Banach spaces over $(G^{(0)}, \mu)$ and every measurable G-field $x \mapsto C_x \subset E^*_{x,1}$ of non-empty $*$-weakly compact convex sets, there exists a G-invariant measurable section, i.e. an element $\varphi \in L^\infty(G^{(0)}, E^*)$ such that $\varphi(x) \in C_x$ for a.e. $x \in G^{(0)}$ and for a.e. $\gamma \in G$, ${}^tL(\gamma^{-1})\varphi \circ s(\gamma) = \varphi \circ r(\gamma)$.

The following theorem has been established in the case of group actions by R. Zimmer [**44, 45**] for discrete groups and by Adams-Elliott-Giordano [**3**] for locally compact groups. Our proof follows the same pattern.

THEOREM 2.17. ([**7**], 4.2.7) *Let (G, λ, μ) be a measured groupoid. Then the following statements are equivalent:*

(i) *(G, λ, μ) is amenable.*
(ii) *(G, λ, μ) satisfies the fixed point property.*

3. Amenable topological groupoids

3.1. Definition of topological amenability. We consider here a locally compact groupoid G . We assume that our locally compact spaces are Hausdorff. We do not assume that the topology of G is second countable; this assumption is only required when we use the general disintegration theorem ([**37**], 4.1) for the representations of a groupoid algebra; in particular this theorem and this assumption are not needed when G is the groupoid of the action of a locally compact group on a locally compact space. The definition of a continuous Haar system λ for G is as above, with the further requirements that λ^x is a Radon measure on G^x and that for every continuous compactly supported function f on G, the function $\lambda(f) : x \mapsto \lambda^x(f)$ is continuous. Just as in 2.1, continuous actions of locally compact groups and foliations provide basic examples of locally compact groupoids with continuous Haar system. The nicest kind of amenability is the following:

DEFINITION 3.1. We say that a locally compact groupoid G is *properly amenable* if it admits a continuous Haar system $m = \{m^x\}_{x\in G^{(0)}}$ consisting of probability measures m^x.

When G already admits a continuous Haar system, this condition is equivalent to the properness of G, in the sense that the map $(r, s) : G \to G^{(0)} \times G^{(0)}$ is proper. In the case of a group, this amounts to saying that the group is compact and, in the case of a group action, that the action is proper. Again, we look for a weakening of this condition.

DEFINITION 3.2. We say that a locally compact groupoid G is *topologically amenable* if it admits a *continuous approximate invariant mean*, i.e. a net $(m_i = \{m_i^x\}_{x\in G^{(0)}})$ of continuous systems of probability measures m_i^x on G^x which is approximately invariant in the sense that the function $\gamma \mapsto \|\gamma m_i^{s(\gamma)} - m_i^{r(\gamma)}\|_1$ tends to zero uniformly on the compact subsets of G.

In the case of the groupoid $G = X \times \Gamma$ of a continuous action of a group Γ on a space X, we identify G^x with Γ and view m_i^x as a probability measure on Γ; the approximate invariance condition becomes: $(x,\gamma) \mapsto \|\gamma m_i^{x\gamma} - m_i^x\|_1$ tends to zero uniformly on the compact subsets of $X \times \Gamma$.

When G admits a continuous Haar system λ, this condition can be expressed in terms of functions:

PROPOSITION 3.3. ([**7**], 2.2.6) *A locally compact groupoid with continuous Haar system (G,λ) is topologically amenable if and only if there exists a net (f_i) of non-negative continuous functions on G such that:*

(i) *For all i and for all $x \in G^{(0)}$, $\int f_i d\lambda^x = 1$.*
(ii) *The functions $\gamma \mapsto \int |f_i(\gamma^{-1}\gamma') - f_i(\gamma')| d\lambda^{r(\gamma)}(\gamma')$ tend to the function zero uniformly on the compact subsets of G.*

Again, this characterization admits variants. If G is second countable, a sequence could be used instead of a net. One could also replace (i) by the condition that the functions f_i have compact support and that the functions $x \mapsto \int f_i d\lambda^x$ tend to the function one uniformly on the compact subsets of $G^{(0)}$.

Another convenient and equivalent definition ([**7**], 2.2.13) of topological amenability for a locally compact groupoid with a continuous Haar system (G,λ) is the existence of a net (h_i) of positive type functions with compact support in G such that $h_{i|G^{(0)}} \leq 1$ for all i and $\lim_i h_i = 1$ uniformly on every compact subset of G.

An important property of topological amenability is its invariance under topological equivalence of groupoids. Recall ([**32**],2.1) that two locally compact groupoids G, H are called *equivalent* if there exists a locally compact space X equipped with a left action of G and a right action of H such that these actions commute, are free and proper, and the source [resp. range] map identifies the quotient space $G\backslash X$ [resp. X/H] to $H^{(0)}$ [resp. to $G^{(0)}$].

THEOREM 3.4. ([**7**], 2.2.17) *Let G, H be equivalent locally compact groupoids. Then G is topologically amenable if and only if H is topologically amenable.*

The following fact is used in proving the Baum-Connes conjecture for amenable locally compact groupoids with continuous Haar system; namely they admit *proper affine actions on continuous fields of Hilbert spaces.*

PROPOSITION 3.5. ([**8, 41**]) *Let (G,λ) be a topologically amenable locally compact groupoid with continuous Haar system. Then, there exist a continuous field $x \mapsto H_x$ of Hilbert spaces, a continuous action of G on it by linear isometries $L(\gamma) : H_{s(\gamma)} \to H_{r(\gamma)}$ and a continuous cocycle $b : \gamma \mapsto b(\gamma) \in H_{r(\gamma)}$ relative to this action such that the function $\gamma \in G \mapsto \|b(\gamma)\| \in \mathbf{R}^+$ is proper.*

3.2. Measurewise amenability.

DEFINITION 3.6. We say that a locally compact groupoid with continuous Haar system (G,λ) is *measurewise amenable* if, for every measure μ on $G^{(0)}$ quasi-invariant with respect to (G,λ), the measured groupoid (G,λ,μ) is amenable.

Clearly, topological amenability implies measurewise amenability. It turns out that the converse holds, at least under quite general assumptions.

THEOREM 3.7. ([**7**], 3.3.7) *Let (G, λ) be a locally compact groupoid with a continuous Haar system. Assume that G is equivalent to a second countable locally compact groupoid which has a continuous Haar system and countable orbits. Then, if it is measurewise amenable, it is also topologically amenable.*

PROOF. Again, we refer the reader to [**7**] for the proof, which is based on convexity arguments.

□

3.3. Some examples. We begin by an easy example of a topologically amenable action.

EXAMPLE 3.8. Let $\mathbf{F}_2$ be the free group on two generators a, b. Its boundary $X = \partial\mathbf{F}_2$ is the Cantor space of infinite reduced words $\dots x_1x_0$ in the alphabet $\{a, a^{-1}, b, b^{-1}\}$. The group acts by concatenation on the right. This action is topologically amenable. Indeed, the probability measures m_n^x on $\mathbf{F}_2$, defined for $n \in \mathbf{N}^*$ and $x \in X$ by

$$m_n^x(t) = \begin{cases} 1/n & \text{if } t = x_k \dots x_1x_0 \text{ with } k < n \\ 0 & \text{otherwise} \end{cases}$$

form a continuous approximate invariant mean for $X \times \mathbf{F}_2$.

By an explicit similar construction of a Borel approximate invariant mean, E. Germain shows, in the Appendix B of [**7**], that the action of a discrete word hyperbolic group Γ on its boundary $\partial\Gamma$ is measurewise (hence topologically) amenable. Measurewise amenability had been previously obtained by S. Adams in [**2**] by checking the fixed point property. Explicit construction of approximate invariant means for boundary actions can also be found in [**39**].

EXAMPLE 3.9. Let Γ be a discrete subgroup of $PSL(2, \mathbf{R})$. Its action on $P^1(\mathbf{R})$ is topologically amenable. Indeed, $P^1(\mathbf{R}) = PSL(2, \mathbf{R})/P$, where P is the subgroup of upper triangular matrices. The groupoid of the action of Γ on $P^1(\mathbf{R})$ is equivalent to the groupoid of the action of P on $\Gamma\backslash PSL(2, \mathbf{R})$, which is topologically amenable since P is amenable. More generally, the action of a discrete subgroup Γ of a connected semi-simple real Lie group G on G/P, where P is a minimal parabolic subgroup, is topologically amenable. Examples can be found in [**6**].

EXAMPLE 3.10. Let A be an $d \times d$-matrix with entries in $\{0, 1\}$, X_A the subspace of $\{1, \dots, d\}^{\mathbf{N}}$ consisting of all the words $\dots x_1x_0$ such that $A(x_{n+1}, x_n) = 1$ and $T_A : X_A \to X_A$ the associated subshift $T_A(x)_n = x_{n+1}$. Then

$$G = G(X_A, T_A) = \{(x, m - n, y) : x, y \in X_A, m, n \in \mathbf{N}, T_A^n(x) = T_A^m(y)\}$$

has a natural structure of locally compact groupoid admitting the counting measures as continuous Haar system. Its canonical cocycle is the map $(x, k, y) \in G \mapsto c(x, k, y) = k \in \mathbf{Z}$. Its topological amenability can be deduced from the amenability of $\mathbf{Z}$, the amenability of $c^{-1}(0)$, which is hyperfinite and a result on groupoid extensions. The same result holds for the groupoid $G(X, T)$ constructed from an arbitrary local homeomorphism T on a space X (see [**38**]).

4. Applications

We shall limit ourselves to the applications to the theory of C*-algebras.

4.1. Weak containment. Let (G, λ) be a locally compact groupoid with a continuous Haar system. We denote by $C_c(G, \lambda)$ the $*$-algebra of continuous compactly supported complex functions on G (see [**36, 33**]). The reduced norm of $f \in C_c(G, \lambda)$ is given by $\|f\|_{red} = \sup_{x \in G^{(0)}} \|L_x(f)\|$, where $L_x(f)$ is the operator on $L^2(G_x, \lambda_x)$ (we write $\lambda_x = (\lambda^x)^{-1}$) given by $L_x(f)\xi(\gamma) = \int f(\gamma\gamma_1)\xi(\gamma_1^{-1})d\lambda^x(\gamma_1)$. The full norm of $f \in C_c(G, \lambda)$ is given by $\|f\| = \sup \|L(f)\|$, where L runs over all non-degenerate representations of the $*$-algebra $C_c(G, \lambda)$ which are continuous for the inductive limit topology. The completion of $C_c(G, \lambda)$ with respect to these norms are called respectively the reduced C*-algebra $C^*_{red}(G, \lambda)$ and the full C*-algebra $C^*(G, \lambda)$. The inequality $\|f\|_{red} \leq \|f\|$ implies that $C^*_{red}(G, \lambda)$ is a quotient of $C^*(G, \lambda)$.

PROPOSITION 4.1. ([**7**], 6.1.8) *Let (G, λ) be a second countable locally compact groupoid with a continuous Haar system. If it is measurewise amenable, then $C^*_{red}(G, \lambda) = C^*(G, \lambda)$.*

PROOF. Again, we refer the reader to [**7**]. The proof uses the disintegration of representations. For a given quasi-invariant measure μ, the amenability of (G, λ, μ) is equivalent to the weak containment of the trivial representation in the regular representation (as representations of a C*-algebra $C^*(G, \lambda, \mu)$). In the case of group actions, we also refer the reader to [**4, 5**].

□

It is well known that in the case of a group, the converse is true. The abundance of quasi-invariant measures makes the general case more problematic.

REMARK 4.2. The assumption of second countability is used in the disintegration of representations. It can be by-passed, at least in some cases. For example, it can be shown directly that, when G is the groupoid of the action of a locally compact group on a locally compact space and without the second countability assumption, the topological amenability of G implies the equality of the reduced and the full C*-algebras.

4.2. Nuclearity. The main result about the nuclearity of a groupoid C*-algebra is:

THEOREM 4.3. ([**7**], 6.2.14) *Let (G, λ) be a second countable locally compact groupoid with a continuous Haar system.*

(i) *If it is measurewise amenable, then $C^*(G, \lambda)$ $(= C^*_{red}(G, \lambda))$ is a nuclear C*-algebra.*
(ii) *If for every $x \in G^{(0)}$, the isotropy subgroup $G(x)$ is discrete and if $C^*_{red}(G, \lambda)$ is nuclear, then (G, λ) is measurewise amenable.*

PROOF. The proof given in [**7**] uses the fact that a C*-algebra, all representations of which are hyperfinite, is nuclear. Just as in [**45, 46**], one shows that all the representations of an amenable measured groupoid are injective and that the injectivity of the regular representation and the existence of an invariant mean for the isotropy imply the amenability of the measured groupoid.

□

This theorem applies to the examples given at the end of the previous section. It gives the nuclearity of C*-algebras arising from boundary actions (Examples 3.1 and 3.2), Cuntz-Krieger algebras and graph algebras (Example 3.3).

REMARK 4.4. Again, the assumption of second countability can be by-passed, at least in some cases. It can be shown directly that, when G is the groupoid of the action of a group on a space and without the second countability assumption, the topological amenability of G implies the nuclearity of its C*-algebra.

4.3. Exactness. We shall be concerned here with the following generalization of amenability for a locally compact group.

DEFINITION 4.5. We say that a locally compact group G is *amenable at infinity* if it admits an amenable action on a compact space X, i.e. such that the semi-direct product groupoid $X \times G$ is topologically amenable.

EXAMPLE 4.6. Amenable groups, but also free groups, word hyperbolic groups or discrete subgroups of connected Lie groups are amenable at infinity, as shown in examples 3.1 and 3.2. M. Gromov has recently announced examples of finitely generated discrete groups which are not amenable at infinity.

There is a characterization of the amenability at infinity of a discrete group G which only invokes the group G. Indeed, G admits an amenable action on a compact space X if and only if its left action on the spectrum βG of the C*-algebra $l^\infty(G)$ is topologically amenable and this last condition can be expressed in terms of a net of functions on $G \times G$ (see [**7**], 5.2.3). N. Higson and J. Roe observed that, in the case of a finitely generated discrete group G, this is exactly the property A, introduced by G. Yu in [**43**], of the metric space G, (where G is endowed with a word-length metric associated to some finite set of generators).

PROPOSITION 4.7. (Higson-Roe [**26**]) *Let G be a finitely generated discrete group endowed with a word-length metric. Then, the following conditions are equivalent:*

(i) *The group G is amenable at infinity.*
(ii) *The metric space G has property A.*

We refer the reader to [**26, 27, 43**] for applications of this property to the Novikov conjecture and to [**40, 41, 42**] for the use of amenable groupoids in the Baum-Connes conjecture.

An important property enjoyed by the groups which are amenable at infinity is C*-exactness.

DEFINITION 4.8. ([**29**]) A locally compact group G is called *C*-exact* if the reduced crossed product construction transforms a short exact sequence of G-C*-algebras into a short exact sequence of C*-algebras.

If G is C*-exact, its reduced C*-algebra is exact. The converse holds for a discrete group. Amenability is a convenient way to prove C*-exactness:

PROPOSITION 4.9. ([**7**]) *Let G be a locally compact group. If it is amenable at infinity, then it is C*-exact.*

PROOF. Let us just prove the exactness of the reduced C*-algebra $C^*_{red}(G)$ and refer the reader to [**7**] for the complete statement. Let X be a compact amenable G-space. The map $f \in C_c(G) \mapsto \tilde{f} \in C_c(X \times G)$ such that $\tilde{f}(x,g) = f(g)$ extends to an injective homomorphism of $C^*_{red}(G)$ into $C^*_{red}(X \times G)$. This exhibits $C^*_{red}(G)$ as a sub-C*-algebra of a nuclear C*-algebra. Therefore, it is exact. □

References

1. S. Adams : Trees and amenable equivalence relations, *Ergodic Theory Dynam. Systems*, **10** (1990), 1–14.
2. S. Adams : Boundary amenability for word hyperbolic groups and an application to smooth dynamics of simple groups, *Topology*, **33** (1994),765–783.
3. S. Adams, G.A. Elliott and T. Giordano : Amenable actions of groups, *Trans. Amer. Math. Soc.*, **344** (1994), 803–822.
4. C. Anantharaman-Delaroche : Action moyennable d'un groupe localement compact sur une algèbre de von Neumann, *Math. Scand.*, **45** (1979), 289–304.
5. C. Anantharaman-Delaroche : Systèmes dynamiques non commutatifs et moyennabilité, *Math. Ann.*, **279** (1987), 297-315.
6. C. Anantharaman-Delaroche : Purely infinite C^*-algebras arising from dynamical systems, *Bull. Soc. Math. France*, **125** (1997), 199–225.
7. C. Anantharaman-Delaroche and J. Renault: *Amenable groupoids*, preprint 1998, to appear in Enseignement Mathématique.
8. M. Bekka, P.-A. Cherix, A. Valette : Proper affine isometric actions of amenable groups, *Novikov conjectures, index theorems and rigidity*, 1–4, Vol.2, London Mathematical Math. Soc. Lecture Notes Series 227, Cambridge Univ. Press 1993.
9. Y. Carrière and E. Ghys : Relations d'équivalence moyennables sur les groupes de Lie, *C. R. Acad. Sci. Paris*, **300** (1985), 677–680.
10. T. Ceccherini-Silberstein, R. Grigorchuk and P. de la Harpe : Amenability and paradoxes for pseudogroups and for discrete metric spaces, *Proc. Steklov Inst. Math.*, **224** (1999), 57–95.
11. A. Connes : Sur la théorie non commutative de l'intégration, *Algèbres d'opérateurs*, pp. 19–143, Lecture Notes in Math., **725**, Springer, Berlin, 1979.
12. A. Connes : A survey of foliations and operator algebras, *Operator algebras and applications*, Part I, 521–628, Proc. Sympos. Pure Math., **38**, Amer. Math. Soc., Providence, 1982.
13. A. Connes : *Noncommutative Geometry*, Academic Press, 1994.
14. A. Connes, J. Feldman, and B. Weiss : An amenable equivalence relation is generated by a single transformation, *J. Ergodic Theory and Dynamical Systems*, **1** (1981), 431–450.
15. J. Dixmier : *Les algèbres d'opérateurs dans l'espace hilbertien*, Gauthier-Villars, Paris, 1969.
16. R. Doran and J.M.G. Fell : *Representation of *-algebras, locally compact groups, and Banach *-algebraic bundles*, Vol. **1**, Academic Press, 1988.
17. J. Feldman and C.C. Moore : Ergodic equivalence relations, cohomology and von Neumann algebras I, *Trans. Amer. Math. Soc.*, **234** (1977), 289–324.
18. J. Feldman and C.C. Moore : Ergodic equivalence relations, cohomology and von Neumann algebras II, *Trans. Amer. Math. Soc.*, **234** (1977), 325–359.
19. V. Golodets and S. Sinel'shchikov : Amenable ergodic group actions and images of cocycles, *Dokl. Akad. Nauk SSSR*, **312** (1990), 1296-1299; transl. in *Soviet Math. Dokl.*, **41** (1990), 523–526.
20. F.P. Greenleaf : *Invariants means on topological groups*, Van Nostrand, New-York, 1969.
21. F.P. Greenleaf : Amenable actions of locally compact groups, *J. Funct. Anal.*, **4** (1969), 295–315.
22. M. Gromov : Hyperbolic groups in *Essays in group theory*, Gersten Ed., Math. Sci. Res. Inst. Publ., **8**, Springer-Verlag, New-York, 1987
23. P. Hahn : Haar measure for measure groupoids, *Trans. Amer. Math. Soc.*, **242** (1978), 1–33.
24. P. Hahn : The regular representation of measure groupoids, *Trans. Amer. Math. Soc.*, **242** (1978), 35–72.
25. N. Higson and G. Kasparov : Operator K-theory for groups which act properly and isometrically on Hilbert spaces, *Electronic Research Announcements*, A.M.S. **3** (1997), 131-141.
26. N. Higson and J. Roe : Amenable group actions and the Novikov conjecture, Preprint, Penn. State (1998).
27. N. Higson : Bivariant K-theory and the Novikov conjecture, Preprint, Penn. State (1999).
28. V. A. Kaimanovich : Amenability, hyperfinitness and isoperimetric inequalities, *C. R. Acad. Sci. Paris, Série* I, **325** (1997), 999–1004.
29. E, Kirchberg and S. Wassermann : Exact groups and continuous bundles of C*-algebras, Math. Ann., to appear.
30. A. Kumjian, D. Pask, I. Raeburn and J. Renault : Graphs, groupoids and Cuntz-Krieger algebras, J. Funct. Anal. **144** (1997), 505-541.

31. G. Mokobodski : Limites médiales, Exposé de P-A. Meyer, *Séminaire de Probabilités*, Strasbourg 1971-72, Lecture Notes in Math. **321**, 198–204, Springer, Berlin, 1973.
32. P. Muhly, J. Renault and D. Willams : Equivalence and isomorphism for groupoid C^*-algebras, *J. Operator Theory*, **17** (1987), 3–22.
33. P. Muhly : *Coordinates in operator algebras*, in preparation.
34. J. F. Plante : Foliations with measure preserving holonomy, *Ann. Math.*, **102** (1975), 327–361.
35. A. Ramsay : Virtual groups and group actions, *Adv. in Math.*, **6** (1971), 253–322.
36. J. Renault : *A groupoid approach to C^*-algebras*, Lecture Notes in Math., **793**, Springer, 1980.
37. J. Renault : Représentations des produits croisés d'algèbres de groupoïdes, *J. Operator Theory*, **18** (1987), 67–97.
38. J. Renault : Cuntz-like algebras, to appear in the Proceedings of the 17th Int. Conf. on Operator Theory, Timişoara, 1998.
39. G. Robertson and T. Steger : C^*-algebras arising from group actions on the boundary of a triangle building, *Proc. London Math. Soc.*, **72** (1996), 613–637.
40. G. Skandalis, J.-L. Tu and G. Yu : Coarse Baum-Connes conjecture and Groupoids, Preprint, Paris 7 (1999).
41. J-L. Tu : La conjecture de Baum-Connes pour les feuilletages moyennables, *K-theory*, **17**, $n^0 3$ (1999), 215–264.
42. J.-L. Tu : The Baum-Connes conjecture for amenable groupoids, Preprint, Paris 6 (1999).
43. G. Yu : The coarse Baum-Connes conjecture for spaces which admit a uniform embedding into Hilbert space, *Invent. Math.*, to appear.
44. R.J. Zimmer : Amenable ergodic group actions and an application to Poisson boundaries of random walks, *J. Funct. Anal.*, **27** (1978), 350–372.
45. R.J. Zimmer : Hyperfinite factors and amenable ergodic actions, *Invent. Math.*, **41** (1977), 23–31.
46. R.J. Zimmer : On the von Neumann algebra of an ergodic group action, *Proc. Amer. Math. Soc.*, **66** (1977), 289–293.

Département de Mathématiques, Université d'Orléans, B.P.6759, 45067 Orléans Cedex 2, France

E-mail address: claire@labomath.univ-orleans.fr

Département de Mathématiques, Université d'Orléans, B.P.6759, 45067 Orléans Cedex 2, France

E-mail address: renault@labomath.univ-orleans.fr

Contemporary Mathematics
Volume **282**, 2001

THE ROLE OF GROUPOIDS IN CLASSIFICATION THEORY: A NEW APPROACH

The UHF Algebra Case

Giulio Della Rocca and Masamichi Takesaki

ABSTRACT. We will examine the role of groupoids in the classification theory of UHF algebras, which has been once considered completed by the hand of J. Glimm [**GL**], but with the new approach based on groupoids we show that an important part of the classification has been neglected and that the missing part can be completed. As a byproduct, we show that the UHF-algebra A_∞ with dimension group $\mathbb{Q}$ has the property that $A_\infty \otimes \mathbb{K}$ admits an action of $\mathbb{Q}_+^*$ such that $(\tau \otimes \mathrm{Tr}) \circ \alpha_a = a(\tau \otimes \mathrm{Tr}), a \in \mathbb{Q}_+^*$, where τ is the canonical trace on A_∞.

§1. Introduction

The words groupoid and operator algebras are almost synonymous. So it is hard to decide what should be discussed in this article. But to emphasize the vital importance of the theory and the philosophy of groupoids in the context of analysis, we discuss the application of the idea of groupoids to the classification of UHF algebras as it would display clearly where the groupoid concept is needed and that the classification of UHF algebras due to Glimm, [**GL**], is incomplete in the new approach even though it has long been viewed as a complete theory. First, it should be emphasized that classification theories in analysis do not have a common back ground yet. Some people think the general classification is simply impossible due to the lack of Borel parametrizations as phrased in Mackey's Borel classification doctrine. But our experience indicates contrary. Let us begin by examining the so-called Mackey classification scheme.

In an old article [**M1**], G.W. Mackey proposed a strategy for classification theory which was needed for the classification of irreducible unitary representations of a separable locally compact group. In it, he asked to identify the set $\mathcal{X}$ of objects and the relevant equivalence relation " $\sim$ " among objects and form the quotient space $\mathcal{P} = \mathcal{X}/$" $\sim$ "; then to endow a standard Borel structure on $\mathcal{X}$ and the quotient Borel structure on $\mathcal{P}$. If it so happens that the parameter space $\mathcal{P}$ is standard, then

2000 *Mathematics Subject Classification.* Primary 46.
Research was supported in part by NSF grant DMS-9801324.

$\mathcal{P}$ can be given an explicit parametrization by means of a constructive method, which can be viewed as a satisfactory classification.

There have been a number of important cases of this successful classifications: the classification of irreducible unitary representations of a connected semi-simple Lie group, that of Lie algebras and that of UHF algebras. But there are important cases where $\mathcal{P}$ is not standard: the classification of irreducible representations of a countable discrete group with infinite conjugacy classes such as a finite permutation group S_∞ of $\mathbb{N}$, a free group F_n on n generators, $n \geq 2$, or $\mathrm{PSL}(n, \mathbb{Z}), n \geq 2$, (these groups are too important to ignore), and the isomorphism classification of factors.

Now, observe however in Mackey's scheme:

$$x \in \mathcal{X} \quad \underset{\text{Borel}}{\overset{f}{\longrightarrow}} \quad [x] \in \mathcal{P},$$

the class $[x] \in \mathcal{P}$ does not carry any information on $x \in \mathcal{X}$. But we know by experience that $\mathrm{Aut}(x)$ and $\mathrm{Iso}(x, y), x \sim y$, carry interesting informations on x. For example, the groups D_8, the dihedral group of order 8, and Q_8, the quotanionic group $\{1, \pm i, \pm j, \pm k\}$, have the identical list of the classes of irreducible unitary representations with identical tensor product decomposition rule, yet the classical Tanaka duality theorem asserts that they can be recovered by considering intertwiners. This is precisely the short coming of Mackey's approach. Therefore, the satisfactory classification theory must treat the equivalence relation " $\sim$ " itself as well. Now we realize that a standard Borel space with a Borel equivalence relation is a basic example of standard groupoids; even it has a noble name: a principal standard groupoid. If we take $\mathrm{Aut}(x)$ into consideration, then we should consider the groupoid:

$$\mathcal{G}_{\mathrm{obj}} = \bigcup_{\substack{x, y \in X \\ x \sim y}} \mathrm{Iso}(x, y)$$

with the obvious composition rule. It is then obvious that the invariant side ought to be a standard groupoid as well, not just a standard Borel set. So a classification means a Borel homomorphism f from the groupoid $\mathcal{G}_{\mathrm{obj}}$ into the groupoid $\mathcal{G}_{\mathrm{inv}}$ of invariants and intertwiners:

$$\mathcal{G}_{\mathrm{obj}} \quad \underset{\text{hom}}{\overset{f}{\longrightarrow}} \quad \mathcal{G}_{\mathrm{inv}}$$

such that $x \sim y$ in $\mathcal{G}_{\mathrm{obj}}, x, y \in X$, if and only if $f(x) \sim f(y)$ in $\mathcal{G}_{\mathrm{inv}}$. An attempt to show that every invariant $a \in \mathcal{G}_{\mathrm{inv}}^{(0)}$ appears as an invariant of some element of $\mathcal{G}_{\mathrm{obj}}^{(0)}$ leads naturally to the consideration whether the classification homomorphism f has a right inverse, viewed as a model construction, as a Borel groupoid homomorphism or not. It would be too naive to hope that this can be done or ought to be done always because an invariant is often given by an element of the quotient of a bigger object. For example, a cohomological invariant is an element of the quotient group $\mathrm{H} = \mathrm{Z}/\mathrm{B}$. In this case, the construction of a model might require to go back to Z as illustrated below:

It is rather doubtful if one can get a direct construction functor from a cohomological invariant:

$$h \in \mathrm{H} \twoheadrightarrow M(h) \in \mathcal{G}_{\mathrm{obj}}$$

Rather one gets

$$c \in \mathrm{Z} \longrightarrow M(c) \in \mathcal{G}_{\mathrm{obj}}$$
$$c \sim bc, \quad b \in \mathrm{B},$$

and the natural question is to find an association:

$$b \in \mathrm{B} \xrightarrow{?} \sigma_b \in \mathrm{Iso}(M(c), M(bc))$$
$$\text{with} \quad \sigma_{b_1 b_2} = \sigma_{b_1} \circ \sigma_{b_2}, \ b_1, b_2 \in \mathrm{B}.$$

But one should get instead

$$a \in \mathrm{C} \longrightarrow \sigma_a \in \mathrm{Iso}(M(c, (\partial a)c)$$
$$\text{with} \quad \sigma_{a_1 a_2} = \sigma_{a_1} \circ \sigma_{a_2}, \ a_1, a_2 \in \mathrm{C}.$$

Before going into the discussion of our objects, UHF algebras, let us examine the classification theory of AF algebras which has also been viewed as the theory is complete, [**Ef 3, El**], by means of the K_0-functor. But now with the new approach described above, we see that the present stage of the classification theory of AF-algebras is summerized in the following diagram:

CURRENT CLASSIFICATION THEORY OF AF-ALGEBRAS:

AF-Algebras	$\xrightarrow{K_0}$	Dimension Groups
Stable Isomorphisms	$\xleftarrow[?]{}$	Order Isomorphisms

But we should have instead:

COMPLETE CLASSIFICATION

Objects	Classification Functor $\longrightarrow$	Invariants
Morphisms	$\longleftarrow$ Model Construction Functor	Isomorphisms

Thus we realize that *a functorial model construction is missing* from the current classification theory!

§2. Dimension Group of a UHF Algebra

We consider only UHF algebras from now on which are *believed* to have been successfully classified *up to isomorphism* by J. Glimm in 1960. If one considers instead the stable isomorphism classification problem, then the situation is more complicated. Although the classification functor is given by the K_0-functor, the "inverse" model construction functor was *never questioned.*

To solve the problem, one must consider the question if the following exact sequence splits:

$$1 \longrightarrow \overline{\mathrm{Int}}(A \otimes \mathbb{K}) \longrightarrow \mathrm{Aut}(A \otimes \mathbb{K}) \xrightarrow{\mathrm{mod}} \mathrm{Aut}(D(A)) \longrightarrow 1$$

where A means UHF algebra, $\mathbb{K} = \mathcal{LC}(\mathfrak{H})$ is the ideal of compact operators on a separable infinite dimensional Hilbert space $\mathfrak{H}$ and $D(A)$ is the dimension group of A.

Let us recall the definition of a UHF algebra A, which is given by an inductive system of matrix algebras $\{M(k_n) = M_{k_n}(\mathbb{C}), \phi_n : n \in \mathbb{N}\}$ where $\phi_n : M_{k_n} \mapsto M_{k_{n+1}}, n \in \mathbb{N}$, is a unital homomorphism. Observe that the existence of a unital homomorphism $\phi : M_m \longrightarrow M_n$ is equivalent to a factorization $a = n/m \in \mathbb{N}$, so that ϕ can be viewed as a homomorphism $\phi : \mathbb{Z} \xrightarrow{\times a} \mathbb{Z}$. Thus to an inductive system $\{M(k_n), \phi_n : n \in \mathbb{N}\}$ there corresponds an inductive system of copies of the integer group $\{\mathbb{Z}, \phi_n : n \in \mathbb{N}\}$.

Now suppose a sequence $\{a_n\}$ of positive integers is given and consider an inductive sequence of the integer group:

$$\mathbb{Z} \xrightarrow{\times a_1} \mathbb{Z} \xrightarrow{\times a_2} \mathbb{Z} \xrightarrow{\times a_3} \mathbb{Z} \xrightarrow{\times a_4} \mathbb{Z} \xrightarrow{\times a_5} \cdots \longrightarrow \mathbb{Z} \xrightarrow{\times a_n} \mathbb{Z} \xrightarrow{\times a_{n+1}} \cdots .$$

Let $D = \varinjlim\{\mathbb{Z}, \{a_n\}\}$ be the inductive limit group. Since each map preserves the positivity in $\mathbb{Z}$, the group D is a totally ordered group.

The next lemma shows how the group looks like.

LEMMA 2.1. *The following diagram is an isomorphism of the inductive systems in the diagram:*

$$\begin{array}{ccccccccccc}
\mathbb{Z} & \xrightarrow{\times a_1} & \mathbb{Z} & \xrightarrow{\times a_2} & \mathbb{Z} & \xrightarrow{\times a_3} & \mathbb{Z} & \xrightarrow{\times a_4} & \mathbb{Z} & \longrightarrow \\
\downarrow{\times 1} & & \downarrow{\times \frac{1}{a_1}} & & \downarrow{\times \frac{1}{a_1 a_2}} & & \downarrow{\times \frac{1}{a_1 a_2 a_3}} & & \downarrow{\times \frac{1}{a_1 a_2 a_3 a_4}} & \\
\mathbb{Z} & \xrightarrow{\times 1} & \frac{1}{a_1}\mathbb{Z} & \xrightarrow{\times 1} & \frac{1}{a_1 a_2}\mathbb{Z} & \xrightarrow{\times 1} & \frac{1}{a_1 a_2 a_3}\mathbb{Z} & \xrightarrow{\times 1} & \frac{1}{a_1 a_2 a_3 a_4}\mathbb{Z} & \longrightarrow
\end{array}$$

which gives the oredered isomorphism:

$$D \cong \bigcup_{k=0}^{\infty} \left(\frac{1}{a_0 a_1 \cdots a_k} \right) \mathbb{Z} \subset \mathbb{Q}$$

where $a_0 = 1$.

Let $\mathfrak{D}$ be the space of ordered subgroups of $\mathbb{Q}$. Fix $D \in \mathfrak{D}$. By Lemma 2.1, a sequence $\{r_k\}$ of positive generators in D can be arranged in such a way that

$$\begin{aligned}
&D = \langle r_1, r_2, \cdots, r_k, \cdots \rangle; \\
&\mathbb{Z} = D_0 \subset D_{k-1} = \langle r_{k-1} \rangle \subsetneq D_k = \langle r_k \rangle \quad \text{for} \quad k = 1, 2, \cdots, n,
\end{aligned}$$

where n can be infinite. So we have $r_k \mid r_{k-1}$, i.e. $r_{k-1} = a_k r_k$ with $a_k \in \mathbb{N}$, so that $r_k = 1/(a_1 \cdots a_k)$. Thus we obtain an inductive system:

$$\mathbb{Z} \xrightarrow{\times a_1} \mathbb{Z} \xrightarrow{\times a_2} \mathbb{Z} \xrightarrow{\times a_3} \mathbb{Z} \xrightarrow{\times a_4} \cdots,$$

which identifies the group D with the inductive limit:

$$D = \varinjlim\{\mathbb{Z}, \{a_k\}\}.$$

DEFINITION 2.2. We call the sequence $\{r_n\}$ of rational numbers of $(*)$ a *generating sequence* of D and $\{a_n\}$ the *accompanied ratio sequence.*

Neither a generating sequence $\{r_n\}$ nor its accompanied sequence $\{a_n\}$ is uniquely determined by D. For example, any permutation of the sequence $\{a_n\}$ does not affect on the limit group D. So we need the next lemma to distinguish two members of $\mathfrak{D}$.

LEMMA 2.3. *If $\{r_k\}$ and $\{s_k\}$ represents two members C and D of $\mathfrak{D}$ in such a way that*

$$\begin{aligned}
C &= \langle r_1, r_2, \cdots \rangle, \quad \langle r_1, \cdots, r_n \rangle = \langle r_n \rangle; \\
D &= \langle s_1, s_2, \cdots \rangle, \quad \langle s_1, \cdots, s_n \rangle = \langle s_n \rangle,
\end{aligned}$$

then the necessary and sufficient condition for $C = D$ is that

i) *for every* $n \in \mathbb{N}$ *there exists* $m \in \mathbb{N}$ *such that* $r_m \mid s_n$, *i.e.* $\langle s_n \rangle \subset \langle r_m \rangle$;
ii) *for every* $k \in \mathbb{N}$ *there exists* $\ell \in \mathbb{N}$ *such that* $s_\ell \mid r_k$, *i.e.* $\langle r_k \rangle \subset \langle s_\ell \rangle$.

We omit the proof which is straghtforward. An immediate consequence is the following:

LEMMA 2.4. i) (Telescoping) *If* $\{r_n\}$ *is a generating sequence of* $D \in \mathfrak{D}$, *then any subsequence* $\{r_{n_k} : k \in \mathbb{N}\}$ *is also a generating sequence of* D.

ii) (Refinement) *If* $\{r_n\}$ *is a generating sequence of* $D \in \mathfrak{D}$, *then a sequence* $\{s_n\}$ *obtained by inserting new terms in between two consecutive terms of* $\{r_n\}$, *which must satisfy the divisibility condition on each consecutive terms, is also a generating sequence of* D.

Combining Lemmas 2.1 through 2.4, we obtain the following results:

THEOREM 2.5. (J. Glimm) *Each group* D *of* $\mathfrak{D}$ *corresponds uniquely to the list of primes with multiplicity possibly infinite, which appear as prime factors of terms in the list of the accompanied sequence of a generating sequence.*

REMARK 2.6. We will consider here the UHF algebras whose generating sequence has only prime with infinite multiplicity. In fact the automorphisms of those that have a "finite part" leave such a part invariant. The same happens to the corresponding dimension group; the details will be given in §4. This makes them not interesting for our porpuses. Let us denote by $\mathbb{UHF}$ the subset of UHF algebras specified above, i.e., the set of UHF algebras whose accompanied sequence consists of primes with infinite multiplicity.

Given any prime number p, define infinite C*-tensor product

$$M(p^\infty) \cong M(p) \otimes M(p) \otimes \cdots M(p) \otimes \cdots .$$

The dimension group associated to $M(p^\infty)$ is $\mathbb{Z}[1/p]$, i.e., the abelian group of the ring generated by $\mathbb{Z}$ and $1/p$. See [Ef 3] appendix of Chapter 2. For any given finite number of increasing primes : $p_1, p_2, \cdots, p_r$, to the UHF algebra

$$M(p_1^\infty, p_2^\infty, \cdots, p_r^\infty) \cong M(p_1^\infty) \otimes M(p_2^\infty) \otimes \cdots \otimes M(p_r^\infty)$$

there corresponds the dimension group $\mathbb{Z}[1/p_1, 1/p_2, \cdots, 1/p_r]$. Which is, in turn, isomorphic to $\mathbb{Z}[1/m]$, where $m = \operatorname{lcm}\{p_1, p_2, ..., p_r\} = p_1 p_2 \cdots p_r$.

Similarly, each increasing infinite sequence $\{p_i\}$ of primes determines uniquely the UHF algebra:

$$M(p_1^\infty, p_2^\infty, \cdots) = \prod_{j=1}^{\infty}{}^{\otimes} M(p_j^\infty)$$

given by the infinite. See [Gl 1]. The dimension group is the subgroup of $\mathbb{Q}$ given by the abelian group of the ring generated by $\mathbb{Z}$ and $\{1/p_j : j \in \mathbb{N}\}$.

If $\{p_i\}$ is the list of all primes in the increasing order, the corresponding UHF algebra is called *the universal* UHF *algebra*, denoted by A_∞. The dimension group is of course $\mathbb{Q}$.

The discussion following Glimm's theorem is the classical classification, presented in this form by Effros and others, see [Ef 3]. This is the point where our analysis diverts from the classical approach. On the universal UHF algebra A_∞ as well as on any UHF algebra, there is a unique trace τ with $\tau(1) = 1$. This trace τ fixes the associated scale, see [Ef 3], of the dimension group $\mathbb{Q}$, i.e.,

$\Gamma = \tau(\mathrm{Proj}(A_\infty))$. In the case of the stabilized universal UHF algebra $A_\infty \otimes \mathbb{K}$, a semi-finite trace is no longer uniquely determine by the algebra itself. This means that when we view $\mathbb{Q}$ as the stabilized dimension group it has no special "place" for 1. In fact the scale is $\mathbb{Q}^+$. Let us make a closer look at the dimension group $\mathbb{Q}$.

§3. Dimension group $\mathbb{Q}$

LEMMA 3.1. *For the additive group $\mathbb{Q}$ of rational numbers, we have the following statements*:

i) *If a and b are non-zero elements of $\mathbb{Q}$, then $a\mathbb{Z} \cap b\mathbb{Z} \neq \{0\}$;*

ii) *If π is a homomorphism of a subgroup D of $\mathbb{Q}$ into $\mathbb{Q}$, then there exists $\alpha_\pi \in \mathbb{Q}$ such that $\pi(x) = \alpha_\pi x,\ x \in D$.*

iii) *If $a = s/p \in \mathbb{Q}, p > 0$, and $b = t/q \in \mathbb{Q}, q > 0$, are both in the reduce form, then the subgroup $\langle a, b\rangle = a\mathbb{Z}+b\mathbb{Z}$ generated by a and b is of the form $c\mathbb{Z}$ with $c \in \mathbb{Q}$ of the form: $c = u/r$ where r is the least common multiple $\mathrm{lcm}(p, q)$ of p and q and u is the greatest common divisor $\mathrm{gcd}(tp_1, sq_1)$ of $tp_1 = rt/q$ and $sq_1 = rs/p$.*

iv) *For $a \in \mathbb{Q}$ and $b \in \mathbb{Q}$, $\langle a\rangle = a\mathbb{Z} = \langle b\rangle = b\mathbb{Z}$ occurs if and only if $a = \pm b$.*

PROOF. i) Let $p/q = a/b$ be the reduced fraction of the ratio a/b with $p, q \in \mathbb{Z}$ and $q > 0$. Then we have $0 \neq qa = pb \in a\mathbb{Z} \cap b\mathbb{Z}$.

ii) We may and do assume that $\pi(D) \neq \{0\}$, so that we can choose an element $a \in D$ with $\pi(a) \neq 0$. As $\mathbb{Q}$ is torsion free, we have $\ker(\pi) \cap a\mathbb{Z} = \{0\}$. Thus the statement (i) implies that $\ker(\pi) = \{0\}$, i.e. the homomorphism π is injective. For any other element $b \in D$ we set $\pi(a)/\pi(b) = k/\ell$ with $k, \ell \in \mathbb{Z}$. Then

$$\pi(\ell a) = \ell\pi(a) = k\pi(b) = \pi(kb).$$

The injectivity of π yields that $\ell a = kb$, i.e. $a/b = k/\ell = \pi(a)/\pi(b)$. Thus $\pi(a)/a = \pi(b)/b$, which means that the ratio $\pi(a)/a$ is independent of the choice of $a \in D$. Writing it α_π, we see that $\pi(x) = \alpha_\pi x, x \in D$.

iii) Let d be the greatest common divisor of p and q. Then with $p_1 = p/d$ and $q_1 = q/d$ we have $r = p_1q_1d$ and for an arbitrary pair $m, n \in \mathbb{Z}$

$$\begin{aligned} ma + nb &= \frac{ms}{p_1 d} + \frac{nt}{q_1 d} = \frac{msq_1 + ntp_1}{p_1q_1d} = \frac{mku + n\ell u}{r} \\ &= \frac{(mk + n\ell)u}{r} \in c\mathbb{Z}, \end{aligned}$$

where $k = sq_1/u$ and $\ell = tp_1/u$. Since k and ℓ are relatively prime, we have $k\mathbb{Z} + \ell\mathbb{Z} = \mathbb{Z}$.

iv) Suppose $\langle a\rangle = \langle b\rangle$. Then we have $a = kb$ and $b = \ell a$ for some $k, \ell \in \mathbb{Z}$. Thus $k\ell = 1$ which is possible only when $k = \pm 1$. This completes the proof.

Q.E.D.

REMARK 3.2. In (iii), as s and q are independent as well as t and p, u and r can fail to be relatively prime.

DEFINITION 3.3. The number α_π will be called the *modulus* of π and denoted by $\mathrm{mod}(\pi)$. We will call the c in (iii) the greatest common divisor of a and b and write $\mathrm{gcd}(a, b)$ since it is determined uniquely up to the sign by (iv). We choose positive c for positive a and b.

Let $\mathfrak{D}^*$ be the set of all subgroups of $\mathbb{Q}$ which contains the integer group $\mathbb{Z}$ and also Prim be the set of primes. Enumerate Prim in the increasing order $\{p_n : n \in \mathbb{N}\}$. Observe that

$$\mathbb{Q} = \bigcup_{\substack{p, \cdots, r \in \text{Prim}, \\ k, \cdots, \ell \in \mathbb{Z}}} \left(\frac{1}{p}\right)^k \cdots \left(\frac{1}{r}\right)^\ell \mathbb{Z}.$$

DEFINITION 3.4. We say that a rational number $a \in \mathbb{Q}$ *divides* another rational number $b \in \mathbb{Q}$ if $b/a \in \mathbb{Z}$ and write $a \mid b$.

Of course $a \mid b$ if and only if $\langle a \rangle \supset \langle b \rangle$.

LEMMA 3.5. *If $a = p/q \in \mathbb{Q}$ and $b = r/s \in \mathbb{Q}$ are in the reduced from, then the necessary and sufficient condition for $a \mid b$ is that $p \mid r$ and $s \mid q$.*

PROOF. It is trivial that the condition is sufficient for the divisibility of b by a. Suppose $b/a = n \in \mathbb{Z}$. Then we have

$$n = \frac{b}{a} = \frac{qr}{ps} \quad \Rightarrow \quad nps = qr.$$

As $(p, q) = 1$ and $(r, s) = 1$, there exist integers $x, y, u, v \in \mathbb{Z}$ such that

$$px + qy = 1 \quad \text{and} \quad ru + sv = 1.$$

Now we get

$$\begin{aligned} r &= r(px + qy) = rpx + rqy = rpx + npsy = (rx + nsy)p \in p\mathbb{Z}; \\ q &= q(ru + sv) = qru + qsv = npsu + qsv = (npu + qv)s \in s\mathbb{Z}. \end{aligned}$$

This completes the proof. Q.E.D.

THEOREM 3.6. *Let $\{p_1, \cdots, p_n, \cdots\}$ be the list of primes corresponding to a member $D \in \mathfrak{D}^*$. If the multiplicity of every prime in the list is infinite, then*

$$D \cong \mathbb{Z}\left[\frac{1}{p_1}, \cdots, \frac{1}{p_n}, \cdots\right].$$

We say that a group $D \in \mathfrak{D}$ has *purely infinite mulitplicity* if the group satisfies the assumption of the last theorem. If the list of primes is finite, we say that the group D is of *finite type*.

THEOREM 3.7. *For the groups in $\mathfrak{D}^*$ of finite type with purely infinite multiplicity, we have*

$$\mathbb{Z}\left[\frac{1}{p_1}, \cdots, \frac{1}{p_n}\right] = \mathbb{Z}\left[\frac{1}{p_1 \cdots p_n}\right].$$

We are now able to present the groupoid of the invariants.

LEMMA 3.8. *Let $\mathfrak{D}$ be the collection of non-trivial subgroups of the additive group $\mathbb{Q}$. Let $\widetilde{\mathcal{G}}$ be the set of all pairs $(D_1, D_2) \in \mathfrak{D} \times \mathfrak{D}$ such that $D_1 \cong D_2$. Consider*

$$\mathcal{G} = \bigcup_{(D_2, D_1) \in \widetilde{\mathcal{G}}} (D_2, D_1) \times \mathrm{Iso}(D_1, D_2)$$

and denote element of $\mathcal{G}$ *by* (D_2, α, D_1) *if* $\alpha \in \mathrm{Iso}(D_1, D_2)$. *We conclude that* $\mathcal{G}$ *is a standard Borel groupoid. To each element* $(D_2, \alpha, D_1) \in \mathcal{G}$, *there corresponds uniquely* $\mathrm{mod}(\alpha) \in \mathbb{Q}^*$ *such that* $\alpha(x) = \mathrm{mod}(\alpha)x$, $x \in D_1$.

PROOF. The only question is the Borel structure of $\mathcal{G}$ as the groupoid property of $\mathcal{G}$ is obvious. But this is again easy because $\mathcal{G}$ is readily identified with the standard Borel space $\mathbb{Q}^* \times \mathfrak{D}$. Q.E.D.

The details on the Borel structure of $\mathfrak{D}$ and the other spaces involved will be given in §5. According to the new approach to explicit construction of a *model construction functor*, the above groupoid has to be outlined in greater details. So far we investigated the "objects", now we concentrate on the "arrows".

§4 Isomorphism between Dimension Groups

LEMMA 4.1.

$$\mathrm{Aut}\left(\mathbb{Z}\left[\frac{1}{p}\right]\right) = \{p^s : s \in \mathbb{Z}\}.$$

PROOF. Let $\alpha_0 \in \mathrm{Aut}(\mathbb{Z}[1/p])$ such that $\mathrm{mod}(\alpha_0) = 1/p$. We claim that

$$\mathrm{Aut}\left(\mathbb{Z}\left[\frac{1}{p}\right]\right) = \{\alpha_0^n : n \in \mathbb{Z}\}.$$

To this end, suppose that we have an element $\alpha \in \mathrm{Aut}(\mathbb{Z}[1/p])$ which is not an integer power of α_0. Let $a = \mathrm{mod}(\alpha) \in \mathbb{Q}^*$. As $a\mathbb{Z}[1/p] = \mathbb{Z}[1/p]$, a must be an element of $\mathbb{Z}[1/p]$, i.e. $a = q/p^s$, $s \in \mathbb{Z}$. Let $\beta = \alpha_0^s \circ \alpha$. Then $\beta \in \mathrm{Aut}(\mathbb{Z}[1/p])$ and $\mathrm{mod}(\beta) = q$. We may and do assume that p and q are relatively prime. As β is surjective, we have an element m/p^r such that $qm/p^r = 1$ and $(p, m) = 1$. But this is impossible. Q.E.D.

LEMMA 4.2. *For any* $r \in \mathbb{N}$,

$$\mathrm{Aut}\left(\mathbb{Z}\left[\frac{1}{p_1}, \frac{1}{p_2}, \cdots, \frac{1}{p_r}\right]\right) \cong \{p_1^{r_1} \cdots p_k^{r_k} : r_1, \cdots, r_k \in \mathbb{Z}\}.$$

PROOF. We need only to show that

$$a\mathbb{Z}\left[\frac{1}{p_1}, \cdots, \frac{1}{p_n}\right] = \mathbb{Z}\left[\frac{1}{p_1}, \cdots, \frac{1}{p_n}\right], \quad a \in \mathbb{Q}^*, \tag{4.1}$$

if and only if $a = p_1^{s_1} p_2^{s_2} \cdots p_n^{s_n}$ with $(s_1, \cdots, s_n) \in \underbrace{\mathbb{Z} \oplus \cdots \oplus \mathbb{Z}}_{n \text{ times}}$.

The "if" part is trivial. We only prove the "only if" part. So assume that $a \in \mathbb{Q}^*$ satisfies (4.1). Then certainly a is in the group, so that a is of the form: $a = m/(p_1^{s_1} \cdots p_n^{s_n})$. As in the last lemma, we must have an element $b = n/(p_1^{r_1} \cdots p_n^{s_n}) \in \mathbb{Z}[1/p_1, \cdots, 1/p_n]$ so that

$$ab = 1, \quad \text{i.e.} \quad \frac{m}{p_1^{s_1} p_2^{s_2} \cdots p_n^{s_n}} \cdot \frac{n}{p_1^{r_1} \cdots p_n^{r_n}} = 1.$$

This is possible only when $m = 1 = n$ and $s_1 = -r_1, \cdots, s_n = -r_n$.

Q.E.D.

Let $G \in \mathfrak{D}$ and $H \in \mathfrak{D}$ be both of finite type. Then they have the form:

$$G = \left\langle a, \mathbb{Z}\left[\frac{1}{p_1}, \cdots, \frac{1}{p_k}\right]\right\rangle;$$
$$H = \left\langle b, \mathbb{Z}\left[\frac{1}{q_1}, \cdots, \frac{1}{q_\ell}\right]\right\rangle$$

with $p_i, q_j \in \text{Prim}$. Assume that

$$a \notin \mathbb{Z}\left[\frac{1}{p_1}, \cdots, \frac{1}{p_k}\right] \text{ and } b \notin \mathbb{Z}\left[\frac{1}{q_1}, \cdots, \frac{1}{q_\ell}\right],$$

and write

$$a = \frac{a'}{p(p_1^{s_1} \cdots p_k^{s_k})} \quad \text{and} \quad b = \frac{b'}{q(q_1^{r_1} \cdots q_\ell^{r_k})}$$

in the reduced form, i.e. $p_i \nmid p, 1 \leq i \leq k$, $q_j \nmid q, 1 \leq j \leq \ell$, and the fractions are reduced. Since $(p, p_1^{s_1} \cdots p_k^{s_k}) = 1$ and $(q, q_1^{r_1} \cdots q_\ell^{r_\ell}) = 1$, we can find integers u, v, x, y so that

$$u(p_1^{s_1} \cdots p_k^{s_k}) + vp = 1 \quad \text{and} \quad x(q_1^{r_1} \cdots q_\ell^{r_\ell}) + yq = 1.$$

With these integers we rewrite a and b in the following form:

$$a = a'\left(\frac{u}{p} + \frac{v}{p_1^{s_1} \cdots p_k^{s_k}}\right);$$
$$b = b'\left(\frac{x}{q} + \frac{y}{q_1^{r_1} \cdots q_\ell^{r_\ell}}\right).$$

Therefore we summerize

$$G = \left\langle \frac{a}{p}, \mathbb{Z}\left[\frac{1}{p_1}, \cdots, \frac{1}{p_k}\right]\right\rangle \quad \text{and} \quad H = \left\langle \frac{b}{q}, \mathbb{Z}\left[\frac{1}{q_1}, \cdots, \frac{1}{q_\ell}\right]\right\rangle$$

with the additional condition that

$$p_i \nmid p, \quad 1 \leq i \leq k; \qquad q_j \nmid q, \quad 1 \leq j \leq \ell.$$

LEMMA 4.3. *The two groups G and H agree, i.e.*

$$\left\langle \frac{a}{p}, \mathbb{Z}\left[\frac{1}{p_1}, \cdots, \frac{1}{p_k}\right]\right\rangle = \left\langle \frac{b}{q}, \mathbb{Z}\left[\frac{1}{q_1}, \cdots, \frac{1}{q_\ell}\right]\right\rangle$$

if and only if $\{p_1, \cdots, p_k\} = \{q_1, \cdots, q_\ell\}, p = q$ and there exists $m \in \mathbb{Z}$ such that

$$\frac{a}{p} \equiv \frac{mb}{p} \mod \mathbb{Z}\left[\frac{1}{p_1}, \cdots, \frac{1}{p_k}\right], \quad (m, p) = 1$$

provided that no p_i divides p and no q_j divides q and $p > 0, q > 0$.

PROOF. The "if" part is trivial. So we prove only the "only if" part. Suppose that the both sides of the above expressions are equal. Suppose that one of p's, say p_1, is not in the second list $\{q_1, \cdots, q_\ell\}$. Then for large enough $m \in \mathbb{N}$, p^{-m} cannot be in the right hand side. Hence the p-list is contained in the q-list. By symmetry, we conclude that the p-list and q-list are the same.

Since a/p is in H, we have

$$\frac{a}{p} = \frac{nb}{q} + \frac{u}{p_1^{r_1} \cdots p_k^{r_k}};$$
$$aq(p_1^{r_1} \cdots p_k^{r_k}) = (nb(p_1^{r_1} \cdots p_k^{r_k}) + uq)p.$$

Thus as $(a, p) = 1$, we have $p \mid q(p_1^{r_1} \cdots p_k^{r_k})$. As p has no p_i-factor, $p \mid q$. By symmetry, we have $q \mid p$. Thus we conclude $p = q$ as we assume $p, q > 0$ for the denominators and

$$\frac{a}{p} = \frac{nb}{p} + \frac{u}{(p_1^{r_1} \cdots p_k^{r_k})}.$$

By symmetry we also have

$$\frac{b}{p} = \frac{ma}{p} + \frac{v}{(p_1^{s_1} \cdots p_k^{s_k})} = \frac{mnb}{p} + \frac{mu}{(p_1^{r_1} \cdots p_k^{r_k})} + \frac{v}{(p_1^{s_1} \cdots p_k^{s_k})};$$
$$\frac{(1 - mn)b}{p} = \frac{mu}{(p_1^{r_1} \cdots p_k^{r_k})} + \frac{v}{(p_1^{s_1} \cdots p_k^{s_k})},$$

which implies that $p \mid (1 - mn)$, i.e. $mn \equiv 1 \bmod p$. Q.E.D.

REMARK 4.4 As p and $p_1^{r_1} \cdots p_k^{r_k}$, $r_1, \cdots, r_k \geq 0$, are relatively prime, we can write an element $g \in G$ uniquely in the reduced form:

$$g = \frac{m}{p^\varepsilon p_1^{r_1} \cdots p_k^{r_k}}, \quad \varepsilon = 0, 1, \ r_1, \cdots, r_k \in \mathbb{Z}.$$

THEOREM 4.5.

$$\operatorname{Aut}\left(\left\langle \frac{a}{p}, \mathbb{Z}\left[\frac{1}{p_1}, \cdots, \frac{1}{p_k}\right]\right\rangle\right) = \{p_1^{r_1} \cdots p_k^{r_k} : r_1, \cdots, r_k \in \mathbb{Z}\}.$$

PROOF. We identify an automorphism with its modulus. So we have $\operatorname{Aut}(G) = \{\alpha \in \mathbb{Q}^* : \alpha G = G\}$. If $\alpha = p_1^{r_1} \cdots p_k^{r_k}$, then

$$\alpha\left(\frac{m}{p_1^{s_1} \cdots p_k^{s_k}}\right) = \frac{m}{p_1^{s_1 - r_1} \cdots p_k^{r_k - s_k}} \in G.$$

As $\alpha^{-1} = p_1^{-r_1} \cdots p_k^{-r_k}$ is of the same form as α, we have $\alpha^{-1}G \subset G$.

Conversely, suppose $\alpha \in \operatorname{Aut}(G)$. Then we have $\alpha = \alpha(1) \in G$. So it has the reduced form: $\alpha = m/(p^\varepsilon p_1^{r_1} \cdots p_k^{r_k}), \varepsilon = 0, 1$. If $\varepsilon = 1$, then $\alpha^2(1) = m^2/(p^2 p_1^{2r_1} \cdots p_k^{2r_k}) \notin G$. Thus $\varepsilon = 0$. Now we conclude that $m = 1$ because

$$\alpha^{-1} = \frac{p_1^{r_1} \cdots p_k^{r_k}}{m} \in G.$$

Q.E.D.

We now move onto subgroups of $\mathbb{Q}$ of the general type, i.e. not necessarily of finite type. Each $D \in \mathfrak{D}$ gives rise to a generalized integers: $p_1^{n_1} p_2^{n_2} \cdots$. We split the list of primes in the generalized integers as the set p_j of primes with infinite multiplicity and the set $\{q_k\}$ of primes with finite multiplicity. The group D is of the following form:

$$D = \left\langle \frac{1}{q_1^{n_1}}, \cdots, \frac{1}{q_j^{n_j}}, \cdots; \mathbb{Z}\left[\frac{1}{p_1}, \cdots, \frac{1}{p_k}, \cdots\right]\right\rangle.$$

LEMMA 4.6.

$$\mathrm{Aut}_+(D) = \{p_1^{r_1} \cdots p_k^{r_k} : k \in \mathbb{N}, r_i \in \mathbb{Z}\},$$

where $\mathrm{Aut}_+(D)$ *means the group of positivity preserving automorphisms of* D.

PROOF. As before, each $\alpha = p_1^{r_1} \cdots p_k^{r_k} \in \mathbb{Q}_+^*$ gives rise to an automorphism of D by multiplication. Suppose $\alpha \in \mathbb{Q}_+^*$ is given so that $\alpha D = D$. As $\alpha = \alpha(1) \in D$, it is of the form: $\alpha = m/(qp_1^{r_1} \cdots p_k^{r_k})$ with q a finite product of q_j's. Thus q and $p_1^{r_1} \cdots p_k^{r_k}$ are relatively prime and also the fraction is reduced. Now we have

$$\alpha^2(1) = \frac{m^2}{q^2 p_1^{2r_1} \cdots p_k^{2r_k}} \in D,$$

which yields that q must be one. Looking at α^{-1}, we conclude that $m = 1$. Q.E.D.

§5 Borel structures

This section concerns of the Borel structures on the three spaces mentioned in the introduction : $\mathfrak{D}$, $\mathfrak{Z}$ and $\mathfrak{A}$.

THEOREM 5.1. *The space* $\mathfrak{D}$ *of ordered subgroups of* $\mathbb{Q}$ *is a compact metric space, in particular a Polish space.*

PROOF. First, the space $\mathfrak{X} = \{0,1\}^{\mathbb{Q}}$ of all subsets of the countable set $\mathbb{Q}$ is a compact metric space, i.e. the Cantor set. A member $G \in \mathfrak{X}$ is a subgroup of $\mathbb{Q}$ if and only if

$$\chi_G(s+t) \geq \chi_G(s)\chi_G(t), \quad s, t \in \mathbb{Q};$$
$$\chi_G(-s) = \chi_G(s), \quad s \in \mathbb{Q}.$$

The function $G \in \mathfrak{X} \mapsto \chi_G(s) \in \{0,1\}, s \in \mathbb{Q}$, is a continuous function of G, so that $\mathfrak{D}$ is a closed subset of the compact metric space $\mathfrak{X}$. Thus, $\mathfrak{D}$ must be a Polish space. This proves the theorem. Q.E.D.

Now we move on to the collection of inductive systems of matrix algebras or that of copies of the integer group $\mathbb{Z}$. The dimension group of a matrix algebra is equipped with the corresponding scale, i.e. the range of the usual trace Tr.

Let us recall that the two systems have the same ϕ_n which is the multiplication times a_n. As already mentioned $a_0 = 1$. We assume that $M(k_0) = \mathbb{C}$. Therefore $k_1 = a_1$, $k_2 = a_2 a_1$ and $k_n = a_n a_{n-1} \cdots a_2 a_1$.

Thus in both cases,the two systems can be construct from $\{a_n, n \in \mathbb{N}\}$ Therefore we get to the following statement:

THEOREM 5.2. i) *The space of inductive system of matrix algebras is naturally endowed with a standard Borel structure.*

ii) *The space of inductive system of copies of* $\mathbb{Z}$ *is naturally endowed with a standard Borel structure.*

PROOF. As remarked above each point of $\mathfrak{Z}$ is identified with a sequence $\{a_n : n \in \mathbb{N}\}$. Therefore

$$\mathfrak{Z} = \mathbb{N}^{\mathbb{N}}$$

It is well known that $\mathbb{N}^{\mathbb{N}}$ is a Polish Space. This completes the proof. Q.E.D.

To handle the last space, $\mathfrak{A}$, we need few lemmas.

Any UHF algebra can be viewed as a C*-subalgebra of A_∞ with extra conditions. We are going to prove that the set $S(A_\infty)$ of C*-subalgebras of A_∞ is a Borel subset of the collection $F(A_\infty)$, equipped with the Effros Borel structure, see [Ke 1] and [Ef 1], of closed subsets of the Polish space A_∞, which is known to be standard.

The following is a basic fact about the Effros-Borel space.

THEOREM 5.3. (The Selection theorem for $F(X)$) (Kuratowski-Ryll-Nardzewski) *Let X be a Polish space. There is a sequence of Borel functions $f_n : F(X) \to X$, such that for non-empty $F \in F(X)$, $\{f_n(F)\}_{n\in\mathbb{N}}$ is dense in F.*

See [Ke 1; Theorem 12.13 page 76]. From the above theorem follows:

LEMMA 5.4. *For any separable* C*-*algebra A, the set $S(A)$ of* C*-*sub-algebras of A is a Borel set in the standard Borel space $F(A)$ of all non-empty closed subsets of A, and therefore, it is a standard Borel space.*

PROOF. See also [Ke 1; 12.19 page 79].

Let $C \in F(A)$. Then $C \in S(A)$ if and only if with the choice functions $\{f_n\}$ of Theorem 4.3,

1. $f_n(C) + f_m(C) \in C, \quad m, n \in \mathbb{N}$;
2. $f_n(C) \cdot f_m(C) \in C, \quad m, n \in \mathbb{N}$;
3. $f_n(C)^* \in C, \quad n \in \mathbb{N}$.

   These are equivalent to the following set of conditions:
4. $\sup_{n,m} \inf_k ||(f_n(C) + f_m(C)) - f_k(C)|| = 0$;
5. $\sup_{n,m} \inf_k ||f_n(C) \cdot f_m(C) - f_k(C)|| = 0$;
6. $\sup_n \inf_k ||f_n(C)^* - f_k(C)|| = 0$.

It is easily seen that the above three conditions are given by Borel functions. This completes the proof. Q.E.D.

LEMMA 5.5. *Let* $\mathrm{Conv}(\mathfrak{H})$ *be the set of all non-empty bounded closed convex subset of a Hilbert space $\mathfrak{H}$. The map: $C \in \mathrm{Conv}(\mathfrak{H}) \mapsto x_0(C) \in C$ is a continuous map from the complete metric space $\mathrm{Conv}(\mathfrak{H})$ into $\mathfrak{H}$ where $x_0(C)$ is the unique point in C such that $\|x_0(C)\| = \inf_{x\in C} \|x\|$ and the distance $d(C, D)$ between two sets $C, D \in \mathrm{Conv}(\mathfrak{H})$ is defined by the following:*

$$\begin{aligned} d(C, D) &= \max\{\sup_{y\in D} d(y, C), \sup_{x\in C}(x, D)\} \\ &= \max\left\{\sup_{y\in D} \inf_{u\in C} \|y - u\|, \sup_{x\in C} \inf_{v\in D} \|x - v\|\right\}. \end{aligned}$$

PROOF. Let a sequence $\{C_n\}$ in $\mathrm{Conv}(\mathfrak{H})$ converges to $C \in \mathrm{Conv}(\mathfrak{H})$. We claim that $\lim_{n\to\infty} \|x_0(C_n)\| = \|x_0(C)\|$. Take $\varepsilon > 0$. If $d(C_n, C) < \varepsilon$, then there exist points $x_n \in C_n$ and $y_n \in C$ such that $\|x_0(C) - x_n\| < \varepsilon$ and $\|x_0(C_n) -$

$y_n\| < \varepsilon$. As $\|x_n\| \geq \|x_0(C_n)\|$, $\|x_0(C)\| + \varepsilon \geq \|x_0(C_n)\|$. Also the inequality $\|y_n\| \geq \|x_0(C)\|$ implies $\|x_n(C)\| + \varepsilon \geq \|x_0(C)\|$. Thus we have the convergence: $\lim_{n\to\infty} \|x_0(C_n)\| = \|x_0(C)\|$.

Now look at the middle point $z_n = (x_0(C) + y_n)/2 \in C$ and compute:

$$\begin{aligned}\|z_n\|^2 &= \frac{\|x_0(C)\|^2 + \|y_n\|^2}{2} - \left\|\frac{x_0(C) - y_n}{2}\right\|^2 \\ &\leq \frac{\|x_0(C)\|^2 + (\|x_0(C_n)\| + \varepsilon)^2}{2} - \frac{(\|x_0(C) - x_0(C_n)\| - \varepsilon)^2}{4}.\end{aligned}$$

Thus unless $\lim_{n\to\infty} \|x_0(C) - x_0(C_n)\| = 0$, the right hand side of the above inequality becomes smaller than $\|x_0(C)\|$, which contradicts to the fact that $z_n \in C$. Q.E.D.

On the UHFalgebra A_∞, we have a uniqe trace, say τ. We consider the Hilbert space $\mathfrak{H} = L^2(A_\infty, \tau)$. For each $x \in A_\infty$ and $B \in S(A_\infty)$ consider the weakly closed convex closure $C_B(x)$ of the orbit $\{\mathrm{Ad}(u(x)) : u \in U(B)\}$ of each $x \in A_\infty$ within $L^2(A_\infty, \tau)$. Denote the n-th unitary choice function by $u_n(B) = f_n(U(B)), n \in \mathbb{N}$.

LEMMA 5.6. *The correspondence*: $(x, B) \in A_\infty \times S(A_\infty) \mapsto C_B(x) \in \mathrm{Conv}(\mathfrak{H})$ *is a Borel function.*

PROOF. Embed $S(A_\infty)$ into the Cartesian product $U(A_\infty)^{\mathbb{N}}$ by: $B \in S(A_\infty) \mapsto \{u_n(B) : n \in \mathbb{N}\} \in U(A_\infty)^{\mathbb{N}}$ as a Borel subset. For each sequence $U = \{u_n\} \in U(A_\infty)^{\mathbb{N}}$ and a point $x \in A_\infty$, let $C_U(x)$ be the weakly closed convex closure of $\{\mathrm{Ad}(u_n)(x) : n \in \mathbb{N}\}$ in $\mathfrak{H}$. We claim that the map: $(x, U) \in A_\infty \times U(A_\infty)^{\mathbb{N}} \mapsto C_U(x) \in \mathrm{Conv}(\mathfrak{H})$ is continuous. Let G be a strongly open set in $\mathcal{L}(\mathfrak{H})$. If $C_U(x) \cap G \neq \emptyset$, then there exist convex coefficients: $c_i \geq 0$ and $\sum_{i=1}^n c_i = 1$ such that $\sum_{i=1}^n c_i u_i x u_i^* \in G$. As G is open, there exists an open neighborhood $\mathcal{U}$ of (x, U) in $A_\infty \times U(A_\infty)^{\mathbb{N}}$ such that $\sum_{i=1}^n c_i v_i y v_i^* \in G$ for every $(y, V) = (y, \{v_n\}) \in \mathcal{U}$. Therefore $\{(x, U) \in A_\infty \times U(A_\infty)^{\mathbb{N}} : C_U(x) \cap G \neq \emptyset\}$ is open. Thus our assertion follows and hence so does the lemma as well. Q.E.D.

COROLLARY 5.7. *In the standard representation of* A_∞ *on* $L^2(A_\infty, \tau)$, *the map*: $(x, B) \in A_\infty \times S(A_\infty) \mapsto \mathcal{E}_{B' \cap \mathcal{R}_0}(x) \in \mathcal{R}_0$ *is a Borel function relative to the Borel structure of the strong operator topology in* $\mathcal{R}_0$, *where* $\mathcal{R}_0$ *is the von Neumann algebra generated by* A_∞ *on* $L^2(A_\infty, \tau)$ *which is an approximately finite dimensional factor of type* II_1 *and* $\mathcal{E}_{B' \cap \mathcal{R}_0}$ *is the conditional expectation from* $\mathcal{R}_0$ *onto* $B' \cap \mathcal{R}_0$ *relative to the trace* τ.

LEMMA 5.8. i) *The subset* $\mathfrak{X} = \{B \in S(A_\infty) : \mathcal{E}_{B' \cap \mathcal{R}_0}(A_\infty) \subset A_\infty\}$ *is a Borel subset of* $S(A_\infty)$ *and hence a standard Borel space.*

ii) *The map*: $B \in S(A_\infty) \mapsto B^c = B' \cap A_\infty \in S(A_\infty)$ *is a Borel map.*

PROOF. i) Fix a norm dense sequence $\{x_n : n \in \mathbb{N}\}$ in A_∞. Then we have

$$\mathfrak{X} = \{B \in S(A_\infty) : \sup_m \inf_n \|\mathcal{E}_{B' \cap \mathcal{R}_0}(x_m) - x_n\| = 0\}.$$

The norm function: $x \in \mathcal{R}_0 \mapsto \|x\| \in \mathbb{R}$ is lower semi-continuous relative to the strong operator topology. Therefore the defining functions of the above $\mathfrak{X}$ are all Borel functions of $B \in S(A_\infty)$.

ii) For each $x \in A_\infty$, the condition that $x \in B^c$ is equivalent to the statement: $[x, u_n(B)] = 0, n \in \mathbb{N}$. Set $\varphi(x, B) = \sup_n \|[x, u_n(B)]\| \leq 2\|x\|$ for each $x \in A_\infty$.

Then φ is a positive real valued Borel function on $A \times S(A_\infty)$, which is lower semi-continuous on x for each fixed $B \in S(A_\infty)$. Hence the subset $R = \{(x, B) \in A_\infty \times S(A_\infty) : x \in B^c\} = \varphi^{-1}(0)$ is Borel, so that if U is a non-empty open subset of A_∞, then the subset $\mathfrak{X}_U = \{B \in S(A_\infty) : B^c \cap U \neq \emptyset\}$ is analytic. Now the complement of $\mathfrak{Y}_U = \complement\mathfrak{X}_U$ is given by

$$\mathfrak{Y}_U = \{B \in S(A_\infty) : B^c \cap U = \emptyset\}.$$

But we have the following chain of equivalences:

$$B \in \mathfrak{Y}_U \Leftrightarrow \bigcup_{k=1}^{\infty} \varphi^{-1}\left(\left[\frac{1}{k}, \infty\right)\right) \supset U \times \{B\}$$

Q.E.D.

LEMMA 5.9. *If M is a finite dimensional* C*-*algebra, then the set* Iso(M, A) *of unital isomorphisms of M into any fixed unital separable* C*-*algebra A is a Polish space under the metric:*

$$d(\rho, \varphi) = \|\rho - \varphi\| = \sup\{\|\rho(x) - \varphi(x)\| : x \in M, \|x\| \leq 1\}.$$

PROOF. The space $L(M, A)$ of all bounded linear transformations from M into A is a separable Banach space as $\dim M < \infty$, which can be identified with the injective Banach space tensor product: $M^* \otimes_\lambda A$. It is easy to see that Iso(M, A) is a closed subset of $L(M, A)$. Q.E.D.

We fix a countable dense subset, say Iso$(M, A)_0$, of Iso(M, A) for each M and A.

THEOREM 5.10. *The subset* $\mathbb{UHF}$ *of* UHF-*subalgebras of A_∞ is a Borel subset of the standard Borel space $S(A_\infty)$.*

PROOF.
Consider the following function $\mathfrak{A}(\cdot)$:

$$\mathfrak{A}(A) = \sup_n \inf\{\sup_{1 \leq i \leq n} \operatorname{dist}(f_i(A), \varphi(M(k))) : \varphi \in \operatorname{Iso}(M, A_\infty)_0, k \in \mathbb{N}\},$$

with the choice functions f_i of Theorem 5.3. Then $\mathbb{UHF} = \mathfrak{A}^{-1}(0)$. As only a countably many Borel functions involved in the definition of $\mathfrak{A}(\cdot)$, $\mathfrak{A}(\cdot)$ is a Borel function over $S(A_\infty)$. This completes the proof. Q.E.D.

To relate a UHF algebra to A_∞ in a straightforward way, we need to restrict the set $\mathbb{UHF}$ of UHF-subalgebras further. We simply remind the reader that the classification of UHF-subalgebras is a big open area of operator algebras as a counter part in C*-algebras of the subfactor theory. So let $\mathbb{UHF}_0$ be the set of all unital UHF-subalgebras A of A_∞ such that with $A^c = A' \cap A_\infty$,

$$\text{(A)} \quad \begin{cases} A_\infty = A \otimes A^c; \\ A^c \cong A_\infty. \end{cases}$$

REMARK 5.11. There exists a UHF-subalgebra A of A_∞ such that $A_\infty = A \otimes A^c$ but A^c is not a UHF-algebra. Cf. [EK]

REMARK 5.12. Notice that restricting ourselves to $\mathbb{UHF}_0$ we will have canonical embeddings for each pair $A, B \in \mathbb{UHF}_0$:

$$A \hookrightarrow A_\infty :$$
$$\mathrm{Iso}(A,\ B) \hookrightarrow \mathrm{Aut}(A_\infty);$$
$$\mathrm{Iso}(A \otimes \mathbb{K},\ B \otimes \mathbb{K}) \hookrightarrow \mathrm{Aut}(A_\infty \otimes \mathbb{K}).$$

The above will be discussed in details in the next section.

THEOREM 5.13. *The set $\mathbb{UHF}_0$ is a standard Borel space.*

PROOF. First, we fix a norm dense sequence $\{x_n\}$ in A_∞. For each $r = (r_1, \cdots, r_\ell) \in \mathbb{Q}^{<\mathbb{N}}$ set

$$r(B) = \sum_{k=1}^{\ell} r_k \mathcal{E}_{B' \cap \mathcal{R}_0}(x_k) f_k(B).$$

Now we can identify $\mathbb{UHF}_0$ in the following way:

$$\begin{aligned} \mathfrak{X} &= \Big\{B \in \mathbb{UHF} : \mathcal{E}_{B' \cap \mathcal{R}_0}(A_\infty) \subset A_\infty\};\\ \mathfrak{Y} &= \{B \in \mathbb{UHF} : \inf_{r \in \mathbb{Q}^{<\mathbb{N}}} \|x_m - r(B)\| = 0, m \in \mathbb{N}\};\\ \mathfrak{Z} &= \{B \in \mathbb{UHF} : B^c \cong A_\infty\};\\ \mathbb{UHF}_0 &= \mathfrak{X} \cap \mathfrak{Y} \cap \mathfrak{Z}. \end{aligned}$$

Thus $\mathbb{UHF}_0$ is a Borel subset of $\mathbb{UHF}$. Q.E.D.

§6 The Model Construction (Borel) Functor

We are now arriving at the point to construct the "inverse" functor. We want to stress that it is not merely the inverse of the classifying functor but it provides a model construction in the functorial way, which serves as the inverse of the classifying functor.

REMARK 6.1. Note here the difference of our setting and the classical one. Recall $\mathfrak{D}^* = \{D \in \mathfrak{D} : D \supset \mathbb{Z}\}$. In the category of UHF, unstablized, every isomorphism ϕ means a unital isomorphism which means that $\mathrm{mod}(\phi) = 1$, i.e.

$$\begin{aligned} &\mathrm{mod}(\phi) = 1, \phi \in \mathrm{Aut}(A_\infty);\\ D \cong C \text{ in } \mathfrak{D}^* \quad &\Leftrightarrow \quad D = C, \quad D, C \in \mathfrak{D}^*;\\ &\mathrm{Iso}_{\mathfrak{D}^*}(D, C) = \{1\}. \end{aligned}$$

On the contrary, look at the automorphism $\alpha = p \in \mathrm{Aut}(\mathbb{Z}[1/p])$, the multiplication by p. Which one, if any is the corresponding automorphism of $M(p^\infty)$? We will give th answer in the following discussion.

Given $A \in \mathbb{UHF}_0$, we want to identify the stabilized algebra $A \otimes \mathbb{K}$ inside $A_\infty \otimes \mathbb{K}$. Namely, by assumption A_∞ has the tensor product decomposition:

$$A_\infty = (A' \cap A_\infty) \otimes A,$$

so that

$$A_\infty \otimes \mathbb{K} = (A' \cap A_\infty) \otimes A \otimes \mathbb{K}$$

naturally. In the sequel, we denote $A' \cap A_\infty$ by A^c. Trivially, we can identify $\mathrm{Aut}(A \otimes \mathbb{K})$ with the subgroup

$$\{\mathrm{id}\} \otimes \mathrm{Aut}(A \otimes \mathbb{K}) \subset \mathrm{Aut}(A_\infty \otimes \mathbb{K})$$

relative to the decomposition above. Now observe that there is no canonical inclusion: $A_\infty \hookrightarrow A_\infty \otimes \mathbb{K}$ nor canonical trace $\widetilde{\tau}$ on $A_\infty \otimes \mathbb{K}$. For example,

$$A_\infty \otimes \mathbb{K} \cong A_\infty \otimes M_n(\mathbb{C}) \otimes \mathbb{K}$$

and

$$A_\infty \cong A_\infty \otimes M_n(\mathbb{C})$$

as well as

$$\mathbb{K} \cong M_n(\mathbb{C}) \otimes \mathbb{K}.$$

To fix an inclusion: $A_\infty \hookrightarrow A_\infty \otimes \mathbb{K}$, we choose a minimal projection, say e, and identify A_∞ with $A_\infty \otimes \mathbb{C}e$. This identification also fixes a trace $\widetilde{\tau}$ on $A_\infty \otimes \mathbb{K}$ by $\tau(a) = \widetilde{\tau}(a \otimes e), a \in A_\infty$. This is also fixed by specifying the value $\widetilde{\tau}(1 \otimes e) = u \in \mathbb{Q}_+^*$. We will call this *normalization.*

With this identification in mind, we write $\widetilde{A}_\infty = A_\infty \otimes \mathbb{K}$. In this stabilized situation as we have observed, there is no canonical trace, i.e. there is no canonical identification of a dimension group with a subgroup of $\mathbb{Q}$. This forces us to expand the collection $\mathfrak{D}^*$, i.e. subgroups of $\mathbb{Q}$ containing $\mathbb{Z}$, to the collection $\mathfrak{D}$ of all subgroups of $\mathbb{Q}$.

However every member $D \in \mathfrak{D}$ is isomorphic to a member of $\mathfrak{D}^*$ under the multiplication by the inverse of any fixed non-zero element u of D_+ which is precisely the normalization. Thus the results concerning the automorphism groups remain valid without adjustment.

Before going into the general case, we want to study the correspondence between $\mathrm{Aut}(M(p^\infty) \otimes \mathbb{K})$ and $\mathrm{Aut}(\mathbb{Z}[1/p]) = \{p^s : s \in \mathbb{Z}\}$ for p prime. Once again look at the automorphism $\alpha = p \in \mathrm{Aut}(\mathbb{Z}[1/p])$ the multiplication by p and, this time, successfully find the corresponding automorphism of $M(p^\infty) \otimes \mathbb{K}$. This is the same as finding an automorphism of $M(p^\infty) \otimes \mathbb{K}$ which multiplies the trace $\widetilde{\tau}$ by the factor p.

To this end, we need to make a little detour into the crossed product construction of UHF algebras. Let $X = \{0, 1, \cdots, p-1\}^{\mathbb{N}}$ and $\widetilde{X} = \mathbb{Z} \times X$. Then X is a totally disconnected compact abelian group. With $\mu(\{k\}) = 1/p, 0 \le k \le p-1$, we define the product measure $\mu = \prod_{n \in \mathbb{N}}^{\otimes} \mu_n$ on X and $\widetilde{\mu} =$ (The counting measure) $\times\, \mu$. Now let the countable group $G = \mathbb{Z} \times \sum_{n \in \mathbb{N}}^{\oplus} (\mathbb{Z}_p)_n$ act on $\widetilde{X}$ by translation. Then we have $M(p^\infty) \otimes \mathbb{K} \cong C_0(\widetilde{X}) \rtimes (\mathbb{Z} \times G)$. This allows us to identify $M(p^\infty) \otimes \mathbb{K}$ with $C_0(\widetilde{X}) \rtimes G$. We now define a homeomorphism T of $\widetilde{X}$ as follows:

$$T(n, \{x(k) : k \in \mathbb{N}\}) = (pn + x(1), \{x(k+1) : k \in \mathbb{N}\})$$

for each $(n, \{x(k) : k \in \mathbb{N}\} \in \widetilde{X}$. This T commutes with the action of G. Thus it generates an automorphism α of $C_0(\widetilde{X}) \rtimes (\mathbb{Z} \times G)$. It is not hard to see that $\widetilde{\tau} \circ \alpha = p\widetilde{\tau}$ because the transformation T multiplies the measure $\widetilde{\mu}$ by the factor p.

For $p_1, \cdots, p_k \in \mathrm{Prim}$, we construct automorphisms

$$\alpha_1 \in \mathrm{Aut}(M(p_1^\infty) \otimes \mathbb{K}), \cdots, \alpha_k \in \mathrm{Aut}(M(p_k^\infty) \otimes \mathbb{K})$$

following the above procedure. Now we construct the action of $\mathrm{Aut}(\mathbb{Z}[1/p_1, \cdots, 1/p_k])$ on $M(p_1^\infty, \cdots, p_k^\infty) \otimes \mathbb{K}$ by the following:

$$\alpha_{p_1^{r_1} \cdots p_k^{r_k}} = \alpha_1^{r_1} \otimes \cdots \otimes \alpha_k^{r_k} \in \mathrm{Aut}((M(p_1^\infty) \otimes \mathbb{K}) \otimes \cdots \otimes (M(p_k^\infty) \otimes \mathbb{K})).$$

Summarizing the above discussion, we reach the following lifting theorem:

THEOREM 6.2. *If $p_1, p_2, \cdots, p_k$ are distinct prime numbers, then the automorphism group of the dimension group $D(M(p_1^\infty, \cdots, p_k^\infty))$ acts on the stabilized* UHF *algebra $M(p_1^\infty, \cdots, p_k^\infty) \otimes \mathbb{K}$.*

Generalizing the above arguments, we can prove the next result:

THEOREM 6.3. *The positive multiplicative group $\mathbb{Q}_+^*$ acts on the stabilized universal* UHF *algebra $A_\infty \otimes \mathbb{K}$.*

PROOF. Let X be the space of double sequences:

$$X = \left\{ \{x(k, \ell)\} : x(k, \ell) = 0, 1, \cdots, p_k - 1, \ell \in \mathbb{N} \right\}$$

and $\widetilde{X} = \mathbb{Z}^{<\mathbb{N}} \times X$, where the group structure of $\mathbb{Z}^{<\mathbb{N}}$ is given as a subgroup of $\mathbb{Z}^{\mathbb{N}}$, i.e. each element z of $\mathbb{Z}^{<\mathbb{N}}$ is identified with th corresponding sequence which places 0's on the tale of z. The group $G = \mathbb{Z}^{<\mathbb{N}} \times \sum_{\ell \in \mathbb{N}, k \in \mathbb{N}}^{\oplus} (\mathbb{Z}_{p_k})_\ell$ acts on $\widetilde{X}$ as a topological transformation group. The crossed product algebra $C_0(\widetilde{X}) \rtimes G$ is identified with
$A_\infty \otimes \mathbb{K}$. For each j let T_j be the transformation given by:

$$T_j(\{y(n)\}, \{x(k, \ell)\}) = \begin{cases} (p_j y(j) + x(j, 1)), \{x(j, \ell + 1)\}; \\ (y(n), \{x(k, \ell)\}) \quad \text{for } (n, k) \neq (j, j). \end{cases}$$

This transformation T_j commutes with the action of G, hence it gives rise to an automorphism α_j of $A_\infty \otimes \mathbb{K}$ which multiplies the trace $\widetilde{\tau}$ by p_j. For distinct j and k, the automorphisms α_j and α_k commutes. Thus the correspondence

$$p_1^{r_1} \cdots p_k^{r_k} \in \mathrm{Aut}_+(\mathbb{Q}) = \mathbb{Q}_+^* \mapsto \alpha_1^{r_1} \circ \cdots \circ \alpha_k^{r_k} \in \mathrm{Aut}(\widetilde{\mathfrak{A}}_\infty)$$

is an injective homomorphism such that

$$\widetilde{\tau} \circ \alpha_1^{r_1} \circ \cdots \circ \alpha_k^{r_k} = (p_1^{r_1} \cdots p_k^{r_k}) \widetilde{\tau}.$$

Q.E.D.

We denote this action of $\mathbb{Q}_+^*$ on $\widetilde{\mathfrak{A}}_\infty$ by γ.

COROLLARY 6.4. *The exact sequence:*

$$1 \longrightarrow \overline{\mathrm{Int}}(A_\infty \otimes \mathbb{K}) \longrightarrow \mathrm{Aut}(A_\infty \otimes \mathbb{K}) \overset{\mathrm{mod}}{\longrightarrow} \mathbb{Q}_+^* \longrightarrow 1$$

splits.

COROLLARY 6.5. *The crossed product* C^**-algebra* $A = (A_\infty \otimes \mathbb{K}) \rtimes \mathbb{Q}_+^*$ *is a nuclear* C^**-algebra such that the cyclic representation* π_φ *induced by the weight* $\varphi = \widetilde{\tau} \circ \mathcal{E}$ *generates an approximately finite dimensional factor of type* III_1 *where* $\mathcal{E}$ *is the conditional expectation of* A *onto* $(A_\infty \otimes \mathbb{K})$ *associated with the crossed product.*

For each generalized integer $p_1^{r_1} \cdots p_k^{r_k} \cdots$, the UHF algebra $M(p_1^{r_1} \cdots, p_k^{r_k} \cdots)$ is naturally identified with the corresponding tensor product component of $A_\infty = \prod_{p \in \mathrm{Prim}}^{\otimes} M(p^\infty)$.
However, as we are going to consider an action of the groupoid $\mathcal{G}$ on $\mathbb{UHF}$ and the action of $\mathbb{Q}_+^*$ was constructed via crossed product, we need to relate the above correspondence to the crossed product setting. Recall that X can be naturally identified with $\prod_{k\in\mathbb{N}} \mathbb{Z}_{p_k}^{\mathbb{N}}$ and the group G is also identified with $\sum_{k\in\mathbb{N}}^{\oplus} \mathbb{Z}_{p_k}^{<\mathbb{N}}$. Each component $\{\mathbb{Z}_{p_k}, \mathbb{Z}_{p_k}\}$ gives rise to an algebra isomorphic to $M(p_k)$ as $C(\mathbb{Z}_{p_k}) \rtimes \mathbb{Z}_{p_k} \cong M(p_k)$.

This observation leads us to look at the collection $\mathfrak{G}_{\mathrm{UHF}_0}$ of subgroupoids $\mathcal{H}$ of $\mathcal{G}$ such that

1. $C^*(\mathcal{H})$ is a UHF algebra;
2. $\mathcal{H}$ decomposes $\mathcal{G}$ into the product groupoid $\mathcal{H} * \mathcal{K}$ such that $C^*(\mathcal{K})$ is UHF.

It is now clear that each $\mathcal{H} \in \mathfrak{G}_{\mathrm{UHF}_0}$ gives $C^*(\mathcal{H}) \in \mathbb{UHF}_0$ canonically. Clearly this correspondence is a functor. Now given a member

$$D = D(p_1^{r_1} \cdots p_k^{r_k} \cdots) \in \mathfrak{D}^*,$$

we associate the subgroupoid $\mathcal{H}(D)$ by

$$\mathcal{H}(D) = (\mathbb{Z}^{<\mathbb{N}} \times \sum_{k\in\mathbb{N}}{}^{\oplus} (\mathbb{Z}_{p_k})^{r_k}) \ltimes (\mathbb{Z}^{<\mathbb{N}} \times \prod_{k\in\mathbb{N}} (\mathbb{Z}_{p_k})^{r_k}) \subset \mathcal{G}$$

where if $r_j = \infty$ then the first $(\mathbb{Z}_{p_j})^{r_j}$ must be replaced by $\mathbb{Z}_{p_j}^{<\mathbb{N}}$. Thus we associate in turn the corresponding UHF algebra $A(D) = C^*(\mathcal{H})$. For a general $(D, u) \in \mathfrak{D}$, we take

$$\mathcal{H}(D, u) = T_u(\mathcal{H}(u^{-1}D)); \quad A(D, u) = \gamma_u(A(D)) \in \mathbb{UHF}_0.$$

We now state our main result:

THEOREM 6.6. *There exists a natural functor from the groupoid* $\widetilde{\mathcal{G}} = \mathbb{Q}_+^* \ltimes \mathfrak{D}$ *into the groupoid* $\mathbb{UHF}_0$ *of* UHF *subalgebras of* A_∞ *satisfying the requirement* (A) *:*

$$(D, u) \in \tilde{\mathfrak{D}} \mapsto A(D, u) \in \mathbb{UHF}_0;$$
$$g \in \mathbb{Q}_+^* \mapsto \gamma_g \in \mathrm{Aut}(A_\infty \otimes \mathbb{K}).$$

References

[B] Bourbaki,N., *General Topology*, Herman, Paris, 1966.

[B 1] Bratelli,O., *Inductive limits of finite dimensional* C*-*algebras*, Trans. Amer. Math. Soc. **171** (1972), 195-234.

[Brd] Broder,A, *Introduction to Function Algebras Benjamin, New York,* (1969).

[C 1] Connes, A., *une classification des factors of type* III, Ann. Sci. Ecole Norm. Sup., **6** (1973), 133-252.

[C 2] Connes, A., *On the classification of von Neumann algebras and their automorphisms,*, Symposia Math., **20** (1976), 435-478..

[C 3] Connes, A., *Sur la theorie non commutative de l'integration, Algebres d'Operateurs Lecture Notes in Math., Springer-Verlag,* **725** (1979), 19-143.

[Ef 1] Effros, E.G., *The Borel space of von Neumann algebras on a separable Hilbert space*, Pacific Journal of Math, **15, n.4,** (1965), 1153-1164.

[Ef 2] Effros, E.G., *Transformation group and* C* *algebras*, Ann.Math. **81,** (1965), 38-55.

[Ef 3] Effros,E.G., *Dimensions and* C*-*algebras*, Regional conference series in mathematics; **no.46,** (1981), American Mathematical Society.

[EHS] Effros, E.G., Handelman,D. and Shen,C.L., *Dimension groups and their affine representations*, Amer. J. Math. **102** (1980), 385-407.

[El] Elliott, G., *On the classification of inductive limits of sequences of semi-simple finite dimensional algebras*, J. Algebra, **38 1976**, 29-44.

[EK] Evan, D. and Kishimoto, A., *Compact group actions on* UHF *algebras obtained by folding the interval*, J. Funct. Analysis, **98** (1991), 346-360.

[GL 1] Glimm, J., *On a certain class of operator algebras*, Trans. Amer. Math. Soc., **95** (1960), 318-340.

[GL 2] Glimm, J., *Type* I C*-*algebras*, Ann. Math. **73** (1961), 572-612.

[H] Hjorth, G., *Classification and orbit equivalence relations*, Providence, R.I., American Mathematical Society, 2000.

[H.k.] Hjorth, G. and Kechris, A.S., *Recent developments in the theory of Borel reducibility*, To appear, 2000.

[KST] Katayama,Y., Sutherland, C.E. and Takesaki, M., *the structure of the automorphism group of a factor and cocycle conjugacy of discrete group action*, Invent. Math., (1998), 331-380.

[Ke 1] Kechris, A.S., *Classical descriptive set theory*, vol. 156, Springer-Verlag, 1995.

[Kn 1] Knapp, A.W., *Group representation and harmonic analysis from Euler to Langlands*, Notices of the Amer. Math. Soc., **43, no. 5** (1996), 545.

[L] Lang, S., *Algebra Second edition (1984), Addison-Wesley..*

[M 1] Mackey, G.W., *Borel structure in groups and their duals,*, Trans. Amer. Math. Soc. **85** ((1957)), 134-165.

[M 2] Mackey, G.W., *Ergodic theory and virtual groups*, Math. Ann. **166** (1966), 187-207.

[S] Sutherland, C.E., *A Borel parametrization of Polish groups*, Publ. RIMS Kyoto Univ. **21, No. 6** (1985), 1067-1086.

[T 1] Takesaki, M., *A duality in the representation theory of* C*-*algebras*, Ann. Math., **82** (1967), 370-382.

[T 2] Takesaki, M., *A classification theory based on groupoids*, Proc. of the US-Japan Seminar, Kyoto, 1983, Pittman Research Notes in Math. Ser., **123** (1986), 400-410.

Department of Mathematics, University of California, Los Angeles, California 90095-1555

E-mail address: `gdrocca@math.ucla.edu`

Department of Mathematics, University of California, Los Angeles, California 90095-1555

E-mail address: `mt@math.ucla.edu`

Contemporary Mathematics
Volume **282**, 2001

Bundles Over Groupoids

Paul S. Muhly

ABSTRACT. We offer a perspective that helps to view the theory of Fell bundles over groupoids and, more generally, product systems more easily. As an application, we prove a generalization of Kumjian's stabilization theorem.

1. Introduction

Our objective in this note is to promote a *perspective* that we have found very useful when thinking about C^*-algebras associated with bundles over groupoids. Our view is that the notion of Fell bundle and more generally product systems over a groupoid or a subcategory of a groupoid should be viewed in terms of subcategories of the category of C^*-algebras where the morphisms are C^*-correspondences. We have found this perspective to be very helpful in analyzing C^*-algebras that come from Fell bundles and in thinking about more general types of product systems. We shall show how this perspective helps to clarify some work of S. Yamagami [**36**] and we shall use it to generalize to arbitrary groupoids Kumjian's stabilization theorem [**20**], which he proved in the context of the reduced C^*-algebra of an r-discrete groupoid. We shall also indicate how the notion of a product system leads to a number of non-self-adjoint algebras that have been of interest in recent years.

2. C^*-Correspondences

DEFINITION 1. *Let A and B be two C^*-algebras. A C^*-*correspondence *from A to B is a (right) Hilbert B-module E endowed with a C^*-representation $\pi : A \to \mathcal{L}(E)$.*

We follow Lance [**22**] for the notation and general theory of Hilbert C^*-modules. So here, and throughout, $\mathcal{L}(E)$ denotes the C^*-algebra of bounded adjointable operators on E. Thus, a correspondence from A to B is nothing but an A, B-bimodule $E = {}_AE_B$ endowed with a B-valued inner product so that with respect to the associated norm E is complete and so that the A-action is given by continuous adjointable operators.

Let ${}_AE_B$ and ${}_BF_C$ be two C^*-correspondences. Their *composition* is simply their (internal) tensor product ${}_AE_B \otimes_B {}_BF_C$. This is an A, C-correspondence with

2000 *Mathematics Subject Classification.* Primary 46L08, 47L45, 46M18, 47L55, 47L65; Secondary 22D25.

C-valued inner product given by the formula

$$\langle \xi_1 \otimes \eta_1, \xi_2 \otimes \eta_2 \rangle_C = \langle \eta_1, \langle \xi_1, \xi_2 \rangle_B \eta_2 \rangle_C.$$

The left action of A is the obvious one.

Composition of correspondences is not quite associative: $E \otimes (F \otimes G)$ is only isomorphic to $(E \otimes F) \otimes G$. By an *isomorphism* from ${}_AX_B$ to ${}_AY_B$ we mean, of course, a bijective linear map $U : {}_AX_B \to {}_AY_B$ that preserves all the structure. Because we wish to think categorically and because isomorphisms in a category have a well-defined meaning, we shall refer to 'isomorphism' in the sense just presented as '*unitary equivalence*'. The collection of C^*-algebras with unitary equivalence classes of correspondences between them constitutes a category, which we shall denote by $\mathbf{C}$. However, as is customary, we shall blur the distinction between a correspondence and its unitary equivalence class.

As a first example, we note that $\mathbf{C}$ generalizes the usual category $\mathbf{A}^o$, where $\mathbf{A}$ has the collection of C^*-algebras as objects and usual C^*-homomorphisms as morphisms, and where $\mathbf{A}^o$ denotes the opposite category of $\mathbf{A}$. Indeed, if $\varphi : A \to B$ is a C^*-homomorphism from A to B, then B becomes an A, B-correspondence, which we denote by ${}_\varphi B$, as follows: The Hilbert C^*-module structure on B is the obvious one, $\langle b_1, b_2 \rangle := b_1^* b_2$, and the left action of A on B is given by φ, $a \cdot b := \varphi(a)b$. The reason that $\mathbf{C}$ generalizes $\mathbf{A}^o$ is that ${}_\varphi B \otimes {}_\psi C$ is naturally isomorphic to ${}_{\psi\circ\varphi}C$, not ${}_{\varphi\circ\psi}C$. Thus the transformation F that is the identity on the collection of C^*-algebras and carries a C^*-homomorphism $\varphi : A \to B$ to ${}_\varphi B$ is a *contravariant* functor from $\mathbf{A}$ to $\mathbf{C}$. Also, it should be noted that if we emphasize the distinction between a correspondence and its unitary equivalence class, then at the level of $\mathbf{A}$ we are distinguishing between homomorphisms and conjugacy classes of homomorphisms. (Recall that two homomorphisms $\varphi_1, \varphi_2 : A \to B$ are *conjugate* in case there is a unitary u in the multiplier algebra of B such that $\varphi_2(a) = u\varphi_1(a)u^*$ for all $a \in A$.) To say the same thing differently, the correspondences ${}_{\varphi_1}B$ and ${}_{\varphi_2}B$ from A to B are unitarily equivalent if and only if the homomorphisms φ_1 and φ_2 are conjugate.

REMARK 2. *It is possible to get around the "waffling" we are doing about the distinction between correspondences and their unitary equivalence classes. More accurately, it is possible to account for it systematically and to incorporate the whole structure within the context of what MacLane calls a bicategory* [**23**, cf. Section XII.7]. *We shall not pursue this here, since at this point it does not seem to serve any of our immediate purposes.*

We identify a C^*-algebra A with the so-called *regular* correspondence it determines. This is just the correspondence ${}_\iota A$, where $\iota : A \to A$ is the identity map. Of course a correspondence ${}_\varphi A$ is unitarily equivalent to ${}_\iota A$ precisely when φ is a (generalized) inner automorphism.

DEFINITION 3. *An A, B-correspondence ${}_AE_B$ is* invertible *in case there is a B, A-correspondence ${}_BF_A$ such that ${}_AE_B \otimes_B {}_BF_A$ is unitarily equivalent to ${}_\iota A$ and ${}_BF_A \otimes_A {}_AE_B$ is unitarily equivalent to ${}_\iota B$.*

We learned the following proposition in conversations with David Blecher and from the preprint [**34**, Proposition 4.1]. It seems, however, to be found in a number of places in the literature - at least implicitly. See, in particular, [**9**].

PROPOSITION 4. *An A, B-correspondence ${}_AE_B$ is invertible if and only if ${}_AE_B$ is an equivalence bimodule in the sense of Rieffel* [**32, 33**]. *In this event, Rieffel's dual module* [**32**], $({}_AE_B)^*$, *serves as an inverse.*

Thus an A, B-correspondence $E = {}_AE_B$ is invertible if and only if E also has an A-valued inner product ${}_A\langle\cdot,\cdot\rangle$ such that

(1) the closed linear span of the inner products, ${}_A\langle\xi,\eta\rangle$, $\xi,\eta \in A$, is A;
(2) the closed linear span of the inner products, $\langle\xi,\eta\rangle_B$, $\xi,\eta \in B$, is B; and
(3) $\xi\langle\eta,\zeta\rangle_B = {}_A\langle\xi,\eta\rangle\zeta$, $\xi,\eta,\zeta \in E$.

Of course, the prototypical example of an invertible correspondence is the one between B and $\mathcal{K}(E_B)$, where E_B is any (full) Hilbert C^*-module over B, and where $\mathcal{K}(E_B)$ is the algebra of generalized compact operators on E_B. We have ${}_{\mathcal{K}(E_B)}\langle\xi,\eta\rangle = \xi\otimes\eta^*$, where $\xi\otimes\eta^*(\zeta) = \xi\langle\eta,\zeta\rangle$ by definition. So 3., above, is satisfied by definition. Condition 1. is satisfied also by definition, and 2. is satisfied by hypothesis.

EXAMPLES AND AVATARS 5. (1) *Let A and B be C^*-algebras. Then a homomorphism, φ say, from A into $M(B)$ may also be regarded as a correspondence from A to B. The formulas used to define ${}_\varphi B$ work as well here and we shall continue to use the notation ${}_\varphi B$. The point of calling attention to this generalization is to note that composition of correspondences is better behaved than composition of homomorphisms into multiplier algebras. Indeed, if $\varphi : A \to M(B)$ and $\psi : B \to M(C)$, then $\psi\circ\varphi$ may not make sense, since the range of φ may intersect B only at 0. However, while ${}_{\psi\circ\varphi}C$ won't make sense in this case, ${}_\varphi B\otimes_B {}_\psi C$ (which ought to be ${}_{\psi\circ\varphi}C$) always makes sense.*

(2) *C^*-correspondences play a key role in KK-theory. Indeed, if A and B are C^*-algebras, then $KK(A,B)$ is defined to be certain equivalence classes of pairs $({}_AE_B, F)$ where $F \in \mathcal{L}(E_B)$ has certain properties that need not concern us here. In KK-theory one focuses on the equivalence classes rather than on the modules ${}_AE_B$ themselves.*

(3) *Suppose A is a C^*-algebra and that B is a C^*-subalgebra. Suppose, also, that $\Phi : A \to B$ is a conditional expectation. Then ${}_AE_B := A$ with the obvious bimodule structure becomes a C^*-correspondence from A to B when it is endowed with the B-valued inner product defined by the formula $\langle\xi,\eta\rangle_B = \Phi(\xi^*\eta)$. Correspondences of this form lie at the heart of Jones index theory.*

(4) *Suppose M and N are two von Neumann algebras. Connes defines a correspondence from M to N to be a Hilbert space H endowed with commuting representations of M and N^o, where N^o is the opposite von Neumann algebra of N* [**12**]. *Thanks to* [**6**], *this really is the same as the notion we are discussing, provided the notion of a Hilbert C^*-module is replaced by that of a* self-dual *C^*-module with normal actions.*

(5) *Suppose that A is a C^*-algebra. The collection of unitary equivalence classes of invertible correspondences from A to A is a group, called the* Picard group *of A and denoted $Pic(A)$. This is a much studied object in a variety of contexts. For example, Brown, Green and Rieffel* [**8**] *showed that if A has a countable approximate unit, in particular if A is separable, then $Pic(A)$ is isomorphic to $Out(A\otimes\mathcal{K}(H)) := Aut(A\otimes\mathcal{K}(H))/Ginn(A\otimes$*

$\mathcal{K}(H)$), where H is a separable, infinite dimensional Hilbert space and $Ginn(A \otimes \mathcal{K}(H))$ denotes the automorphisms of $A \otimes \mathcal{K}(H)$ implemented by unitaries in $A \otimes \mathcal{K}(H)$. Phillips and Raeburn [**29**, Remarks 2.23] *show that if X is a locally compact Hausdorff space, then $Pic(C_0(X))$ is naturally isomorphic to the semidirect product, $Homeo(X) \ltimes H^1(X, \mathcal{S})$, where $Homeo(X)$ is the homeomorphism group of X and $H^1(X, \mathcal{S})$ is the first sheaf cohomology group associated with the sheaf $\mathcal{S}$ of continuous, circle-valued functions on X. In still another direction, let M be a II_1-factor. Let τ_∞ be the canonical trace on $M \otimes B(H)$, where H is separable, infinite dimensional Hilbert space. If θ is an automorphism of either M or $M \otimes B(H)$, then we shall write $[\theta]$ for the corresponding element in the associated outer automorphism group. Then the analysis of the Fundamental group of M, $\mathcal{F}(M)$, found in* [**18**, p. 990], *shows that the map from $Out(M)$ to $Out(M \otimes B(H)) = Pic(M)$ defined by $[\theta] \to [\theta \otimes \iota]$ maps onto the subgroup $\{[\psi] \mid \tau_\infty \circ \psi = \psi\}$ and the quotient is isomorphic to $\mathcal{F}(M)$. Thus we have a natural short exact sequence:*

$$1 \to Out(M) \to Pic(M) \to \mathcal{F}(M) \to 1.$$

Of course in this analysis, when we identify $Pic(M)$ with $Out(M \otimes B(H))$, we must be speaking of self-dual correspondences in the definition of $Pic(M)$.

3. Fell Bundles

For the remainder of this note, G we denote a second countable, locally compact, Hausdorff groupoid with Haar system λ. We will be studying bundles over G and related spaces. We denote our bundles by fraktur letters and the fibres of a bundle by the corresponding Roman letters. Thus, if $\mathfrak{B}$ is a bundle over G and if $\gamma \in G$, then the fibre of $\mathfrak{B}$ over γ will be denoted by $B(\gamma)$. We shall use [**16**] for our basic reference about Banach bundles. The following definition defines the notion of a *Fell bundle* over G. The term is due to Alex Kumjian [**20**] because it generalizes to groupoids concepts originating in the work of Fell [**15**]. In the setting of groupoids, the idea goes back to the preprint of Yamagami [**36**]. The definition may seem long and technical. However, what it is basically saying is this: A Fell bundle over G is a Banach bundle $\mathfrak{B}$ over G that is endowed with a structure of a subgroupoid of the category **C**, whose objects are C^*-algebras and whose morphisms are C^*-correspondences, in such a way that the bundle map $p : \mathfrak{B} \to G$ is a functor. Of course, this can't be quite correct because **C** consists of unitary equivalence classes of correspondences and whatever they are, the fibres of $\mathfrak{B}$ must be Banach spaces. Therefore in the actual definition, we have to adjust for this. However, thinking in these terms makes the details of the following definition quite clear.

DEFINITION 6. *A* Fell bundle *over G is a second countable Banach bundle $\mathfrak{B} \xrightarrow{p} G$ such that*

(1) *For each $u \in G^{(0)}$, $B(u)$ is a C^*-algebra and for each $\gamma \in G$, $B(\gamma)$ is an invertible $B(r(\gamma))$, $B(s(\gamma))$-correspondence.*
(2) *For each $(\alpha, \beta) \in G^{(2)}$, there is a unitary equivalence U^α_β between $B(\alpha) \otimes_{B(s(\alpha))} B(r(\beta))$ and $B(\alpha\beta)$ that defines an associative bilinear*

product m *on* $\mathfrak{B}^{(2)} := \{(b_1, b_2) \in \mathfrak{B} \times \mathfrak{B} \mid (p(b_1), p(b_2)) \in G^{(2)}\}$ *by the formula* $m(b_1, b_2) = U^{p(b_1)}_{p(b_2)}(b_1 \otimes b_2)$.

(3) *There is an involution on* $\mathfrak{B}$, $b \to b^*$, *such that*
   (a) $p(b^*) = p(b)^{-1}$,
   (b) $b \to b^*$ *is conjugate linear,*
   (c) $(m(b_1, b_2))^* = m(b_2^*, b_1^*)$, *and*
   (d) *for* $b_1, b_2 \in B(\gamma)$, $m(b_1^*, b_2) = \langle b_1, b_2 \rangle_{B(s(\gamma))}$ *while* $m(b_1, b_2^*) = {}_{B(r(\gamma))}\langle b_1, b_2 \rangle$.

(4) *The maps* m *and the involution are continuous.*

It should be noted that the space $\mathfrak{B}^{(2)}$ is the pull back to $G^{(2)}$ of $\mathfrak{B}$ under the multiplication map from $G^{(2)}$ to G and has a natural structure of a Banach bundle on $G^{(2)}$. Thus requiring the continuity of m makes perfectly good sense. From now on, we shall drop the notation m and simply write $b_1 b_2$ for $m(b_1 b_2)$. Also, a number of easy observations fall out of the definition. For example, we see that $B(\gamma)^*$ is isomorphic to the inverse of the $B(r(\gamma))$, $B(s(\gamma))$-correspondence $B(\gamma)$. Also, we see that formulas like $\|b^*b\|_{B(s(\gamma))} = \|b\|^2_{B(\gamma)}$, $b^*b \geq 0$ in $B(s(\gamma))$, etc., hold in $\mathfrak{B}$.

We note that what we call a Fell bundle is sometimes called a *saturated* Fell bundle, where the saturation assumption is the assumption that $B(\gamma)$ is full as a Hilbert $B(r(\gamma))$-module and as a Hilbert $B(s(\gamma))$-module. It is sometimes necessary to relax the fullness conditions. The simplest example of a non-saturated Fell bundle that arises in our work occurs with the Fell bundles associated Cuntz-Pimsner algebras. We discuss this briefly in Section 6 just before Definition 16.

EXAMPLES 7.

(1) Let $\mathfrak{A} \xrightarrow{\pi} G^{(0)}$ be an ordinary C^*-bundle over $G^{(0)}$. Assume that G acts on $\mathfrak{A}$ by a homomorphism $\alpha : G \to Iso(\mathfrak{A})$. Assume, too, that α is continuous in the sense that $\tilde{\alpha} : s_*(\mathfrak{A}) \to r_*(\mathfrak{A})$ is continuous, where $s_*(\mathfrak{A})$ and $r_*(\mathfrak{A})$ are the pull back bundles, $s_*(\mathfrak{A}) := \{(a, \gamma) \mid a \in A(s(\gamma))\}$ and $r_*(\mathfrak{A}) := \{(a, \gamma) \mid a \in A(r(\gamma))\}$, and where $\tilde{\alpha}(a, \gamma) = (\alpha_\gamma(a), \gamma)$. If we set $\mathfrak{B} := s_*(\mathfrak{A})$, with $p(a, \gamma) = \gamma$, then we obtain a Fell bundle over G where the product and involution are given by the formulas
$$(a_1, \gamma_1) \cdot (a_2, \gamma_2) = (\alpha_{\gamma_2^{-1}}(a_1)a_2, \gamma_1\gamma_2)$$
and
$$(a, \gamma)^* := (\alpha_\gamma(a^*), \gamma^{-1}).$$
This Fell bundle $\mathfrak{B}$ is called the *crossed product of* $\mathfrak{A}$ *by* G and is denoted $\mathfrak{B}(\mathfrak{A}, \alpha)$.

(2) Suppose E is a twist over G in the sense of Kumjian [**19**]. Then E is a principal circle bundle over G. The associated line bundle $\mathfrak{B}$ over G is a Fell bundle. (See [**20**].)

(3) In [**31**], Renault considers a situation that generalizes simultaneously the two preceding examples. Consider a groupoid extension
$$G^{(0)} \to S \to \Sigma \to G \to G^{(0)}$$
where S is a (locally compact) bundle of abelian groups over $G^{(0)}$ (with a Haar system). We suppose that the other conditions required of S and Σ in [**31**, p.70] are satisfied. Suppose, also, that $\mathfrak{A} \to G^{(0)} = \Sigma^{(0)}$ is a C^*-bundle on which Σ acts by $\{\alpha_\sigma\}_{\sigma\in\Sigma}$ in such a way that the restriction

of the action to S is given by a homomorphism χ of S into the bundle of multipliers of $\mathfrak{A}$ such that

$$\alpha_s(a) = \chi(s)a\chi(s)^{-1},\ (s,a) \in S * \mathfrak{A}$$

and

$$\chi(\sigma s\sigma^{-1}) = \alpha_\sigma(\chi(s)),\ (\sigma, s) \in \Sigma * S.$$

Then S acts on the bundle $\Sigma * \mathfrak{A}$ over Σ by the formula $(\sigma, a) \cdot s = (\sigma s, \alpha_{s^{-1}}(a))$. (Really, $\mathfrak{A}$ in $\Sigma * \mathfrak{A}$ is the pull back of $\mathfrak{A}$ to Σ by the source map, but we don't indicate this in the notation because it would involve confusing multiple uses of 's'.) The quotient $\Sigma * \mathfrak{A}/S$ has the structure of a Fell bundle over $\Sigma/S \simeq G$.

Given a Fell bundle $\mathfrak{B}$ over G, we write $C_c(G, \mathfrak{B})$ for the space of continuous compactly supported sections of $\mathfrak{B}$ and give it the inductive limit topology. Then $C_c(G, \mathfrak{B})$ becomes a topological $*$-algebra when endowed with the product

$$f * g(\gamma) := \int f(\alpha)g(\alpha^{-1}\gamma)\, d\lambda^{r(\gamma)}(\alpha)$$

and involution

$$f^*(\gamma) := f(\gamma^{-1})^*.$$

A moment's reflection will reveal that these operations are well defined and that, indeed, $C_c(G, \mathfrak{B})$ is a topological $*$-algebra, as claimed.

DEFINITION 8. *A* representation *of the pair $(G, \mathfrak{B})$ is a triple (π, H, μ), where μ is a quasi-invariant measure on $G^{(0)}$; $H = \{H_u\}_{u \in G^{(0)}}$ is a Borel Hilbert bundle on $G^{(0)}$; and $\pi : \mathfrak{B} \to End(H)$ is a Borel $*-$ functor, i.e., for all $b \in B(\gamma)$, $\pi(b) \in B(H_{s(\gamma)}, H_{r(\gamma)})$, $\pi(b_1)\pi(b_2) = \pi(b_1 b_2)$ a.e. $\nu^{(2)} := \int_{G^{(0)}} \lambda^u \times \lambda_u\, d\mu(u)$, and $\pi(b^*) = \pi(b)^*$.*

Given a representation (π, H, μ) of $(G, \mathfrak{B})$, one obtains a $*$-representation $\sigma : C_c(G, \mathfrak{B}) \to B\left(\int^\oplus H_u\, d\mu\right)$ by the formula

$$(\sigma(f)\xi, \eta) = \int (\pi(f(\gamma)\xi(s(\gamma)), \eta(r(\gamma))\, d\nu_0(\gamma),$$

where as usual, $\nu := \int \lambda^u\, d\mu(u)$, $D := d\nu/d\nu^{-1}$ (which makes sense by the assumption that μ is quasi-invariant), and where $\nu_0 := D^{-1/2}\nu$. Then σ is called the *integrated form* of (π, H, μ). It is not hard to see that σ acts by bounded operators on $\int^\oplus H_u\, d\mu$ and that σ is continuous with respect to the inductive limit topology on $C_c(G, \mathfrak{B})$ and the weak operator topology on $B(\int^\oplus H_u\, d\mu)$. The following "converse" assertion is Yamagami's generalization [**36**] of Renault's disintegration theorem [**31**].

THEOREM 9. *Given a Hilbert space K and a dense linear manifold K_0 endowed with an* algebraic *representation $\sigma : C_c(G, \mathfrak{B}) \to Lin(K_0)$ such that*

(1) *$f \to (\xi, \sigma(f)\eta)$ is continuous,*
(2) *$(\xi, \sigma(f)\eta) = (\sigma(f^*)\xi, \eta)$, and*
(3) *σ is nondegenerate in the sense that the linear span of the vectors of the form $\sigma(f)\xi$, $f \in C_c(G, \mathfrak{B})$ and $\xi \in K_0$ is dense in K,*

then there is a representation (π, H, μ) and a Hilbert space isomorphism U from $\int H_u \, d\mu$ to K such that $U\tilde{\sigma}(f)U^{-1} = \sigma(f)$, where $\tilde{\sigma}$ is the integrated form of (π, H, μ).

It follows that for each $f \in C_c(G, \mathfrak{B})$ the supremum of the norms $\|\sigma(f)\|$ taken over all algebraic representations σ of $C_c(G, \mathfrak{B})$ satisfying the hypotheses of Theorem 9 is finite and defines a C^*-norm on $C_c(G, \mathfrak{B})$. The completion of $C_c(G, \mathfrak{B})$ in this norm is denoted $C^*(G, \mathfrak{B})$ and is called the *C^*-algebra of the bundle $\mathfrak{B}$.*

In the examples above, $C^*(G, \mathfrak{B}(\mathfrak{A}, \alpha))$ is denoted $\mathfrak{A} \rtimes_\alpha G$, when the Fell bundle $\mathfrak{B}(\mathfrak{A}, \alpha)$ is determined by an action α of G on a C^*-bundle $\mathfrak{A}$ over $G^{(0)}$. When the Fell bundle comes from a twist E over G, the C^*-algebra is denoted $C^*(G; E)$. The C^*-algebra associated with the third example is denoted by $C^*(G, \Sigma, \mathfrak{A}, \lambda)$ in [**31**].

4. Morita Equivalence of Fell Bundles

The notion of Morita equivalence of Fell bundles over groupoids makes perfectly good sense and was first proposed by Yamagami in [**36**]. Our objective in this section is to discuss this notion from the perspective of correspondences.

Suppose that for $i = 1, 2$, $\mathfrak{B}_i \xrightarrow{p_i} G_i$ is a Fell bundle over G_i and think of $\mathfrak{B}_i$ as a groupoid and p_i as a functor. Then, essentially, an *equivalence* between $(G_1, \mathfrak{B}_1)$ and $(G_2, \mathfrak{B}_2)$ is a pair $(X, \mathfrak{E})$ where X is a G_1, G_2-equivalence in the sense of [**24**] and where $\mathfrak{E} \xrightarrow{p} X$ is a Banach bundle such that $\mathfrak{E}$ serves as an equivalence between the groupoids $\mathfrak{B}_1$ and $\mathfrak{B}_2$. Of course, all the various structures and mappings involved must be compatible.

The details of the definition are somewhat involved. We offer them as a series of three points, with commentary.

DEFINITION 10. *For $i = 1, 2$ suppose $\mathfrak{B}_i$ is a Fell bundle over G_i. A $(G_1, \mathfrak{B}_1)$, $(G_2, \mathfrak{B}_2)$-*equivalence *is a pair $(X, \mathfrak{E})$ consisting of a G_1, G_2-equivalence X and a Banach bundle $\mathfrak{E} \xrightarrow{p} X$ such that the following points I – III are satisfied.*

I: Each fibre $E(x) := p^{-1}(x)$ is a $B_1(r(x))$, $B_2(s(x))$-equivalence.

Here, we use the standard notation when discussing equivalence of groupoids: $r : X \to G_1^{(0)}$ is the continuous open map that is used to define the left G_1-action on X, and $s : X \to G_2^{(0)}$ is the continuous open map that is used to define the right G_2-action.

The justification for point I is that if we view the bundles $\mathfrak{B}_i$ as groupoids, then the unit space $\mathfrak{B}_i^{(0)}$ of $\mathfrak{B}_i$ is nothing but $\mathfrak{B}_i | G_i^{(0)}$. Therefore, if they are to act on $\mathfrak{E}$ we must have maps r and s from $\mathfrak{E}$ onto $\mathfrak{B}_1^{(0)}$ and $\mathfrak{B}_2^{(0)}$, respectively, so that the following diagram commutes.

$$\begin{array}{ccccc} \mathfrak{B}_1^{(0)} & \xleftarrow{r} & \mathfrak{E} & \xrightarrow{s} & \mathfrak{B}_2^{(0)} \\ p_1 \downarrow & & \downarrow p & & \downarrow p_2 \\ G_1^{(0)} & \xleftarrow[r]{} & X & \xrightarrow[s]{} & G_2^{(0)} \end{array}$$

This means, simply, that each fibre of $\mathfrak{E}$, $E(x) := p^{-1}(x)$, must be a $B_1(r(x))$, $B_2(s(x))$-bimodule. Other conditions, to be given in a moment, require that, in fact, each $E(x)$ is a $B_1(r(x))$, $B_2(s(x))$-equivalence.

II: For each $(\gamma, x) \in G_1 * X$ $(= \{(\gamma, x) \in G \times X \mid s(\gamma) = r(x)\})$ there is a contractive, $B_1(r(\gamma))$, $B_2(s(x))$-correspondence map V_x^γ from $B_1(\gamma) \otimes_{B(s(x))} E(x)$ to $E(\gamma x)$ that defines a continuous action of $\mathfrak{B}_1$ on $\mathfrak{E}$. A similar statement must hold for $\mathfrak{B}_2$ acting to the right on $\mathfrak{E}$. The right and left actions must commute.

This means that if π_1 and π_2 denote the left and right projections of $G_1 * X$ on G_1 and X, respectively, and if $m_1 : G_1 * X \to X$ is the map describing the action of G_1 on X, while $m_2 : \pi_1^*(\mathfrak{B}_1) \otimes \pi_2^*(\mathfrak{E}) \to \mathfrak{E}$ is defined by $m_2(b_1 \otimes e) = V_{p(e)}^{p_1(b_1)}(b_1 \otimes e)$, then (m_1, m_2) is a continuous bundle map (over m_1) from $\pi_1^*(\mathfrak{B}_1) \otimes \pi_2^*(\mathfrak{E})$ to $\mathfrak{E}$ that satisfies the obvious algebraic relations that are necessary to have an action. For example, it is necessary to satisfy the equation

$$V^{p_1(b_2)}_{p(V^{p_1(b_1)}_{p(e)}(b_1 \otimes e))}(b_2 \otimes (V^{p_1(b_1)}_{p(e)}(b_1 \otimes e)) = V^{p_1(b_2 b_1)}_{p(e)}((U^{p_1(b_2)}_{p_1(b_1)}(b_2 \otimes b_1)) \otimes e),$$

where $U^{p_1(b_2)}_{p_1(b_1)}$ denotes the isomorphism from $B_1(p_1(b_2)) \otimes_{B_1(s(b_2))} B_1(p_1(b_1))$ to $B_1(p_1(b_2)p_1(b_1))$ that defines the multiplication on $\mathfrak{B}_1$. Thus, if we simply write be for $m_2(b \otimes e)$, this equation expresses the associativity of the action: $b_2(b_1 e) = (b_2 b_1)e$.

In practice, the continuity is checked by showing that if b is a basic continuous section of $\mathfrak{B}_1$ and ξ is a basic continuous section of $\mathfrak{E}$, and if the $\tilde{b}$ and $\tilde{\xi}$ are the associated sections in the pull backs, $\pi_1^*(\mathfrak{B}_1)$ and $\pi_2^*(\mathfrak{E})$, then $\tilde{b}\tilde{\xi}$ is a continuous section of $\mathfrak{E}$. Also, it should be noted that while V_x^γ is only required to be contractive, the assumption that it defines an action of $\mathfrak{B}_1$ on $\mathfrak{E}$ forces it to be a correspondence isomorphism.

To present the third condition, we need some additional observations. First, recall that G_1 is naturally isomorphic to $X *_s X^o / G_2$, while G_2 is naturally isomorphic to $G_1 \backslash X^o *_r X$, where X^o denotes the opposite, G_2, G_1-equivalence. We shall therefore write elements of G_1 as ordered pairs $[x, y]_1$ such that $[x\gamma, y]_1 = [x, \gamma y]_1$, $\gamma \in G_2$ and we shall write elements of G_2 as ordered pairs $[x, y]_2$ such that $[x\gamma, y]_2 = [x, \gamma y]_2$, $\gamma \in G_1$. We shall write π_1 and π_2 for the left and right projection maps from $X *_s X^o$ onto X and from $X^o * X$ onto X, and we shall write $q_1 : X *_s X^o \to G_1$ and $q_2 : X^o * X \to G_2$ for the quotient maps. Also, we shall use the same notation for a bilinear map and the corresponding linear map on the tensor product it induces. That is, if E_1, E_2, and F are vector spaces, and if $\langle \cdot, \cdot \rangle$ is a bilinear map from $E_1 \times E_2$ to F, then we identify $\langle \cdot, \cdot \rangle$ with the linear map $\phi : E_1 \otimes E_2 \to F$ defined by the formula $\phi(e_1 \otimes e_2) = \langle e_1, e_2 \rangle$.

III: There are continuous bundle maps $[[\cdot,\cdot]]_1 : \pi_1^*(\mathfrak{E}) \otimes \pi_2^*(\mathfrak{E}^*) \to \mathfrak{B}_1$ and $[[\cdot,\cdot]]_2 : \pi_1^*(\mathfrak{E}^*) \otimes \pi_2^*(\mathfrak{E}) \to \mathfrak{B}_2$ over q_1 and q_2, respectively, such that if we keep the same notation, $[[\cdot,\cdot]]_1$ and $[[\cdot,\cdot]]_2$, for the maps on fibres, then $[[\cdot,\cdot]]_1$ induces a unitary equivalence between the $B_1(r(x))$, $B_1(s^o(y))$-equivalences $B_1([x,y])$ and $E(x) \otimes_{B_2(s(x))} E(y)^*$, while $[[\cdot,\cdot]]_2$ induces a unitary equivalence between the $B_2(r^o(x))$, $B_1(s(y))$-equivalences $E(x)^* \otimes_{B_1(r(y))} E(y)$ and $B_2([x,y])$. Further, the pairings must be linked by the equation

$$[[\xi, \eta]]_1 \zeta = \xi [[\eta, \zeta]]_2,$$

for the appropriate choices of ξ, η and ζ, where, for $i = 1, 2$, we identify $[[\xi, \eta]]_i$ with an element of $\mathfrak{B}_i$ according to the isomorphisms just postulated.

An alternative way of saying that $(X, \mathfrak{E})$ is a $(G_1, \mathfrak{B}_1)$, $(G_2, \mathfrak{B}_2)$-equivalence, and the approach taken by Kumjian in [**20**], is to say that if $L = \begin{bmatrix} G_1 & X \\ X^o & G_2 \end{bmatrix}$ is the linking groupoid for G_1, G_2, and X represented as 2×2 matrices, as is customary, and if $\mathfrak{L}$ is the bundle over L such that $\mathfrak{L}|G_1 = \mathfrak{B}_1$, $\mathfrak{L}|X = \mathfrak{E}$, $\mathfrak{L}|G_2 = \mathfrak{B}_2$, and $\mathfrak{L}|X^o = \mathfrak{E}^*$ (the bundle whose fibres are $E(x)^*$, $x \in X^o$), then $\mathfrak{L}$ carries natural operations making it a Fell bundle over L.

We write $C_c(X, \mathfrak{E})$ for the space of compactly supported, continuous sections of $\mathfrak{E}$ on X. Then, with the operations we are about to define, $C_c(X, \mathfrak{E})$ has the structure of a $C_c(G_1, \mathfrak{B}_1)$, $C_c(G_2, \mathfrak{B}_2)$-bimodule with $C_c(G_1, \mathfrak{B}_1)$- and $C_c(G_2, \mathfrak{B}_2)$-valued inner products. For these formulas, ξ and η are in $C_c(X, \mathfrak{E})$, $\varphi \in C_c(G_1, \mathfrak{B}_1)$, while $\psi \in C_c(G_2, \mathfrak{B}_2)$.

$$\varphi\xi(x) := \int \varphi(\gamma)\xi(\gamma^{-1}x)\, d\lambda_1^{r(x)}(\gamma)$$

$$\xi\psi(x) := \int \xi(x\gamma)\psi(\gamma^{-1})\, d\lambda_2^{s(x)}(\gamma)$$

$${}_{C_c(G_1,\mathfrak{B}_1)}\langle \xi, \eta\rangle(\gamma) := \int [[\xi(\gamma x\alpha), \eta(\alpha^{-1}x)]]_1\, d\lambda_2^{s(x)}(\alpha)$$

$$\langle \xi, \eta\rangle_{C_c(G_2,\mathfrak{B}_2)}(\gamma) := \int [[\xi(x\beta^{-1}), \eta(\beta^{-1}x\gamma)]]_2\, d\lambda_1^{r(x)}(\beta)$$

Of course, the point of all this is the following theorem of Yamagami [**36**] that generalizes the main result of [**24**].

THEOREM 11. *The operations just defined endow $C_c(X, \mathfrak{E})$ with the structure of a pre-equivalence $C_c(G_1, \mathfrak{B}_1)$, $C_c(G_2, \mathfrak{B}_2)$-bimodule whose completion is a $C^*(G_1, \mathfrak{B}_1)$, $C^*(G_2, \mathfrak{B}_2)$-equivalence bimodule.*

5. Examples

As our first illustration of Theorem 11, we show how it relates to an old result from [**11**] and [**13**].

EXAMPLE 12. *If G is a group acting strongly continuously on C^*-algebras B_1 and B_2 and if E is a B_1, B_2-equivalence on which G acts strongly continuously by $\mathbb{C}$-linear isometries $\{V_g\}_{g\in G}$ such that*

$${}_{B_1}\langle V_g\xi, V_g\eta\rangle = \alpha_g({}_{B_1}\langle \xi, \eta\rangle)$$

and

$$\langle V_g\xi, V_g\eta\rangle_{B_2} = \beta_g(\langle \xi, \eta\rangle_{B_2}),$$

where $\{\alpha_g\}_{g\in G}$ (resp. $\{\beta_g\}_{g\in G}$) denote the actions of G on B_1 and B_2, respectively, then the crossed product C^*-algebras $B_1 \rtimes_\alpha G$ and $B_2 \rtimes_\beta G$ are strongly Morita equivalent.

PROOF. Form the Fell bundles $\mathfrak{B}(B_1, \alpha)$ and $\mathfrak{B}(B_2, \beta)$ as in the first of Examples 7 and let $\mathfrak{E} = E \times G$ be the trivial bundle associated with E. Then by Theorem 11 it suffices to show that $(G, \mathfrak{E})$ may be endowed with the structure of a $(G, \mathfrak{B}(B_1, \alpha))$, $(G, \mathfrak{B}(B_2, \beta))$-equivalence. For this, it is perhaps easiest simply to modify the proof given of Theorem 1 in [**13**]. Form the linking algebra L of E suggestively denoted matricially by the equation

$$L = \begin{bmatrix} B_1 & E \\ E^* & B_2 \end{bmatrix}$$

and observe that $\gamma : G \to Aut(L)$ defined by the formula

$$\gamma_g\left(\begin{bmatrix} b_1 & \xi \\ \eta^* & b_2 \end{bmatrix}\right) = \begin{bmatrix} \alpha_g(b_1) & V_g(\xi) \\ (V_g\eta)^* & \beta_g(b_2) \end{bmatrix}$$

gives a strongly continuous action of G on L. The crossed product bundle $\mathfrak{B}(L,\gamma)$ is then a Fell bundle over G, which may be written suggestively as

$$\mathfrak{B}(L,\gamma) = \begin{bmatrix} \mathfrak{B}(B_1,\alpha) & \mathfrak{E} \\ \mathfrak{E}^* & \mathfrak{B}(B_2,\beta) \end{bmatrix}.$$

The definition of the actions of $\mathfrak{B}(B_1,\alpha)$ and $\mathfrak{B}(B_2,\beta)$ and the inner products $[[\cdot,\cdot]]_i$, $i = 1,2$, are immediate from this, as are the verification of the conditions I – III in Definition 10. □

REMARK 13. *This result generalizes easily to groupoids acting on C^*-bundles over their unit spaces. It generalizes, in fact, to the setting of the third example in Examples 7. When this is done, Theorem 11 gives a slightly different proof of* [**31**, Corollaire 5.4].

As a second illustration, we outline how a result of Phil Green's fits into the setting described here. We learned it from John Quigg.

EXAMPLE 14. *Suppose that G is a locally compact group acting on a C^*-algebra A, say by an action α, and that H is a closed subgroup of G. Then it is well known that the transformation group groupoid $G \times G/H$ is Morita equivalent to H and that G serves as a $(G\times G/H)$, H - equivalence. If we let G act on $A\times G/H$ via the diagonal action, $t\cdot(a,sH) = (\alpha_t(a), tsH)$, then $G\times(A\times G/H)$ becomes a Fell bundle $\mathfrak{B}_1$ over $G \times G/H$ in the obvious fashion. The C^*-algebra $C^*(G\times G/H,\mathfrak{B}_1)$ was first studied by Takesaki in* [**35**]. *Using the action α, the trivial bundle $\mathfrak{E} = G \times A$ over G can be equipped with the structure of an equivalence between $\mathfrak{B}_1$ and the crossed product bundle $\mathfrak{B}_2 = H \times A$ over H, where the action of H on A is obtained by restricting α to H. Utilizing Theorem 11, one concludes that $C_c(G,\mathfrak{E})$ may be completed to yield an equivalence between $C^*(G\times G/H,\mathfrak{B}_1)$ and the crossed product $A \rtimes_\alpha H$. This is done in section 2 of* [**17**].

Our next goal is to indicate how to generalize to general groupoids, Kumjian's stabilization theorem [**20**]. For this purpose, we fix a Fell bundle $\mathfrak{B}$ over a groupoid G and we let $V(u) := \int_{G_u}^{\oplus} B(\gamma)\,d\lambda_u(\gamma)$. Note that the direct integral of Hilbert C^*-modules is a perfectly straightforward matter. One can take the space $C_c(G,\mathfrak{B})$ and for each $u \in G^{(0)}$ give it the $B(u)$-valued inner product $\langle\xi,\eta\rangle_{B(u)} := \int \xi(\gamma)^*\eta(\gamma)\,d\lambda_u(\gamma)$. Observe that $\gamma \to \xi(\gamma)^*\eta(\gamma)$, $\gamma \in G_u$, is in $C_c(G_u, B(u))$, and so all of this makes sense. The completion of $C_c(G,\mathfrak{B})$ is a Hilbert $B(u)$-module that is denoted $\int_{G_u}^{\oplus} B(\gamma)\,d\lambda_u(\gamma)$. The $V(u)$, $u \in G^{(0)}$, are the fibres of a Banach bundle $\mathfrak{V}$ over $G^{(0)}$. Each section $\xi \in C_c(G,\mathfrak{B})$ determines a section $\tilde{\xi}$ to $\mathfrak{V}$. Namely, $\tilde{\xi}(u) := \xi|G_u$. The topology on $\mathfrak{V}$ is that determined by $\{\tilde{\xi} \mid \xi \in C_c(G,\mathfrak{B})\}$.

The bundle $\mathfrak{V}^*$ has fibres $V(u)^*$, which, in turn, can be viewed as $\int_{G^u}^{\oplus} B(\gamma)\,d\lambda^u$. Indeed,

$$\begin{aligned} V(u)^* &= \left(\int_{G_u}^{\oplus} B(\gamma)\, d\lambda_u(\gamma) \right)^* \\ &= \int_{G_u}^{\oplus} B(\gamma)^*\, d\lambda_u(\gamma) \\ &= \int_{G_u}^{\oplus} B(\gamma^{-1})\, d\lambda_u(\gamma) \\ &= \int_{G^u}^{\oplus} B(\gamma)\, d\lambda^u. \end{aligned}$$

The map that sends $\xi \otimes b \in V(r(\gamma)) \otimes_{B(r(\gamma))} B(\gamma)$ to ξb, where $\xi b(\alpha) = \xi(\alpha) b$, $s(\alpha) = r(\gamma)$, extends to a Hilbert $B(s(\gamma))$-module isomorphism from $V(r(\gamma)) \otimes_{B(r(\gamma))} B(\gamma)$ onto $V(s(\gamma))$, that we shall denote by U_γ. In fact, the U_γ assemble to give a bundle isomorphism $\tilde{U}$ from $r^*(\mathfrak{V}) \otimes_{\mathfrak{B}(0)} \mathfrak{B}$ onto $s^*(\mathfrak{V})$, since $\tilde{U}$ maps continuous sections to continuous sections. Likewise, there is a bundle isomorphism $\tilde{V}$ from $\mathfrak{B} \otimes_{\mathfrak{B}(0)} s^*(\mathfrak{V}^*)$ onto $r^*(\mathfrak{V}^*)$ which on fibres carries $b \otimes \eta^*$ to $b\eta^*$. So when we compose these, we obtain a bundle isomorphism

$$\begin{aligned} s^*(\mathcal{K}(\mathfrak{V})) &\simeq s^*(\mathfrak{V}) \otimes_{\mathfrak{B}(0)} s^*(\mathfrak{V}^*) \\ &\overset{\tilde{U}^{-1}\otimes I}{\simeq} (r^*(\mathfrak{V}) \otimes_{\mathfrak{B}(0)} \mathfrak{B}) \otimes_{\mathfrak{B}(0)} s^*(\mathfrak{V}^*) \\ &\simeq r^*(\mathfrak{V}) \otimes_{\mathfrak{B}(0)} (\mathfrak{B} \otimes_{\mathfrak{B}(0)} s^*(\mathfrak{V}^*)) \\ &\overset{I\otimes\tilde{V}}{\simeq} r^*(\mathfrak{V}) \otimes_{\mathfrak{B}(0)} s^*(\mathfrak{V}^*) \\ &\simeq r^*(\mathcal{K}(\mathfrak{V})). \end{aligned}$$

This isomorphism will be denoted by $\tilde{\alpha}$. It gives the desired action $\{\alpha_\gamma\}_{\gamma \in G}$ of G on $\mathcal{K}(\mathfrak{V})$ via the formula $\alpha_\gamma = U_\gamma^{-1} \otimes V_\gamma$.

Now, let $X = G$, with the right and left actions of G and let $\mathfrak{E} = r^*(\mathfrak{V}) \otimes_{\mathfrak{B}(0)} \mathfrak{B}$, i.e., let $E(\gamma) = V(r(\gamma)) \otimes_{B(r(\gamma))} B(\gamma)$, and let the topology of $\mathfrak{E}$ be determined by sections of the form $\tilde{\xi} \otimes \eta$, where $\xi, \eta \in C_c(G, \mathfrak{B})$. Then the obvious identification of $(V(r(\gamma)) \otimes B(\gamma))^*$ with $B(\gamma^{-1}) \otimes V(r(\gamma))^*$ allows us to define a $\mathfrak{B}(\mathcal{K}(\mathfrak{V}), \alpha)$-valued pairing $[[\cdot,\cdot]]_1$ on $\mathfrak{E}$ because $\mathcal{K}(V(r(\gamma)))$ is naturally isomorphic to $(V(r(\beta)) \otimes B(\beta)) \otimes (B(\alpha^{-1}) \otimes V(r(\alpha))^*)$ via the multiplication on $\mathfrak{B}$, when γ is written as $\gamma = \beta\alpha^{-1}$. Similarly, when γ is written as $\gamma = \alpha^{-1}\beta$, so that $r(\alpha) = r(\beta)$, we find that $E(\alpha)^* \otimes E(\beta)$ is naturally isomorphic to $(B(\alpha^{-1}) \otimes V(r(\alpha))^*) \otimes (V(r(\beta)) \otimes B(\beta))$, which in turn is isomorphic to $B(\alpha^{-1}\beta) = B(\gamma)$. This is gives the $\mathfrak{B}$-valued pairing $[[\cdot,\cdot]]_2$ on $\mathfrak{E}$. It is now very easy to check that $(G, \mathfrak{E})$ is a $(G, \mathfrak{B}(\mathcal{K}(\mathfrak{V}), \alpha))$, $(G, \mathfrak{B})$-equivalence, and so the following theorem is an immediate result of Theorem 11.

THEOREM 15. *(Kumjian's Stabilization Theorem* [**20**, Corollary 4.5]*) If $\mathfrak{B}$ is a Fell bundle over a locally compact groupoid G and if $\mathfrak{V}$ and $\mathfrak{E}$ are the bundles constructed above, then $(G, \mathfrak{E})$ has the structure of a $(G, \mathfrak{B}(\mathcal{K}(\mathfrak{V}), \alpha), G)$, $(G, \mathfrak{B})$-equivalence and so $C_c(G, \mathfrak{E})$ has the structure of a pre-equivalence bimodule whose completion is an $C^*(G, \mathcal{K}(\mathfrak{V}) \rtimes_\alpha G)$, $C^*(G, \mathfrak{B})$-equivalence.*

We like to think of Theorem 15 as the groupoid equivalent of the stabilization theorem of Packer and Raeburn [**28**, Theorem 3.4] on twisted crossed product C^*-algebras. To explain why, let G be a locally compact group. A *twisted action* of G on a C^*-algebra A consists of a pair of Borel maps $\alpha : G \to Aut(A)$ and $u : G \times G \to UM(A)$, where $UM(A)$ is the space of unitary multipliers of A and where the Borel structure on $Aut(A)$ comes from the topology of pointwise norm convergence and the Borel structure on $UM(A)$ comes from the strict topology. These maps are related by certain equations that need not concern us here. Associated with each twisted action of G on A one obtains a C^*-algebra that we shall denote by $A \rtimes_{\alpha,u} G$. The point to be made, and was made by Busby [**10**], is that when G is second countable, there is a bijective correspondence between twisted actions of G on separable C^*-algebras *satisfying certain smoothness conditions* and separable Fell bundles over G. The correspondence implements an isomorphism between the twisted crossed product and the associated Fell bundle C^*-algebra. On the other hand, Packer and Raeburn show in Theorem 3.4 of [**28**] that there is an ordinary action β of G on the stabilized algebra $A \otimes \mathcal{K}(L^2(G))$ such that $A \rtimes_{\alpha,u} G$ and $A \otimes \mathcal{K}(L^2(G)) \rtimes_\beta G$ are strongly Morita equivalent. One could get their theorem directly from Theorem 15 and the analysis of Busby [**10**], if one knew that every twisted action was exterior equivalent to smooth action in Busby's sense. However, Exel and Laca [**14**] showed that *given* the Packer-Raeburn stabilization theorem [**28**, Theorem 3.4], one *can* realize $A \rtimes_{\alpha,u} G$ as the C^*-algebra of a Fell bundle over G and therefore the smoothness hypothesis in [**10**] is not necessary.

There is an enormous and rapidly growing literature on twisted crossed products, Morita equivalence, and Fell bundles over groups. We hope this note will aid in the development of an extension of that theory to groupoids.

6. Product Systems

We have been discussing Fell bundles over groupoids, so that the fibres are *invertible* correspondences. On the other hand, in a few very special, but highly ramified, cases, one encounters bundles of correspondences over *semigroups*. We conclude with a brief discussion of this and how the correspondence *qua* morphism picture helps to understand things.

We fix a C^*-algebra A and a correspondence E from A to A. Remember that E should be viewed as a generalized endomorphism of A. Just as an ordinary endomorphism of a C^*-algebra can be extended to an invertible endomorphism of a bigger C^*-algebra (at least if it is injective), the same can be done with E. This was done by Pimsner in [**30**] as follows. Assume the left action of A on E is faithful. Then the map from $\mathcal{L}(E^{\otimes n})$ to $\mathcal{L}(E^{\otimes n+1})$ defined by the formula $T \mapsto T \otimes I$ is injective. Let $\mathcal{A}$ be the closure of $\cup_{n=0}^{\infty}\mathcal{K}(E^{\otimes n})$ in the inductive limit of the $\mathcal{L}(E^{\otimes n})$ and define $\mathcal{E}$ to be $E \otimes_\iota \mathcal{A}$, where $\iota : A \to \mathcal{A}$ is the natural imbedding. Then $\mathcal{E}$ is an invertible correspondence from $\mathcal{A}$ to $\mathcal{A}$, except that the left $\mathcal{A}$-valued inner product on $\mathcal{E}$ may not be full. In fact, the discussion on pages 194 and 195 shows that the $\mathcal{A}$-valued inner product is full precisely when the left action of A on E is by compact operators, i.e., when $\pi(A) \subseteq \mathcal{K}(E)$.

If $B(n)$ is defined to be $\mathcal{E}^{\otimes n}$ when $n > 0$, A, when $n = 0$, and $(\mathcal{E}^*)^{\otimes(-n)}$, when $n < 0$, then we obtain a Fell bundle $\mathfrak{B}$ over $\mathbb{Z}$. (The topology is determined by all the finitely non-zero sections). The bundle $\mathfrak{B}$ is saturated precisely when the $\mathcal{A}$-valued inner product on $\mathcal{E}$ is full, which, as we have just mentioned, is the same as when the left action of A on E is by compact operators. The C^*-algebra $C^*(\mathbb{Z}, \mathfrak{B})$ is the Cuntz-Pimsner algebra, $\mathcal{O}(E)$. This is not the way Pimsner defined $\mathcal{O}(E)$, but he showed that this was an equivalent formulation in Section 2 of [**30**].

DEFINITION 16. *For $i = 1, 2$, let E_i be a correspondence from A_i to A_i, and let X be an A_1, A_2-equivalence. We say that E_1 is (strongly) Morita equivalent to E_2 (via X) and write $E_1 \underset{X}{\sim} E_2$ in case the correspondence from A_1 to A_2, $E_1 \otimes_{A_1} X$ is unitarily equivalent to $X \otimes_{A_2} E_2$.*

If E_1 is strongly Morita equivalent to E_2 via X, then X implements, in an obvious way, a strong Morita equivalence between $\mathcal{A}_1$ and $\mathcal{A}_2$, and for each $n \in \mathbb{Z}$, a strong Morita equivalence between $\mathcal{E}_1^{\otimes n}$ and $\mathcal{E}_2^{\otimes n}$. These can be combined in the obvious fashion to yield a Morita equivalence between the corresponding bundles $\mathfrak{B}_1$ and $\mathfrak{B}_2$ in the sense of Definition 10. The following result, then, is a consequence of Theorem 11. It was proved in the special case when each E_i is invertible in [**1**]. The general case is covered in [**27**].

THEOREM 17. *If X implements a strong Morita equivalence between two correspondences E_1 and E_2, then X can be promoted to a $\mathfrak{B}_1$,$\mathfrak{B}_2$-bundle equivalence and, consequently, the C^*-algebras $\mathcal{O}(E_1)$ and $\mathcal{O}(E_2)$ are strongly Morita equivalent.*

There is a conditioned converse to this theorem that is proved in [**27**], which asserts that if the correspondences E_1 and E_2 are "aperiodic" in a sense that generalizes the notion of aperiodicity for self-homeomorphisms of compact spaces, and if certain non-self-adjoint subalgebras of $\mathcal{O}(E_1)$ and $\mathcal{O}(E_2)$ are strongly Morita equivalent in the sense of [**7**], then E_1 and E_2 are Morita equivalent in the sense of Definition 16. This theorem generalizes a venerable theorem of Arveson [**2**] on algebraic conjugacy invariants for ergodic measure preserving transformations.

The non-self-adjoint subalgebras stand in the same relation to $\mathcal{O}(E)$ as the disc algebra stands in relation to the continuous functions on the unit circle. They have been our primary objects of study in recent years and have brought us to study groupoid theoretic generalizations. We conclude with a description of these and a brief discussion of some challenges associated with them. Some of the theory is necessarily very technical. We therefore omit some technical hypotheses in order not to get bogged down.

We fix once and for all a subcategory P of our basic groupoid G. We assume P is wide in the sense that $G^{(0)} \subseteq P$, and we assume that P is the closure of its interior. Actually, most of our work has been done in the case when G is r-discrete and principal and P satisfies $P \cup P^{-1} = G$. This means that G is effectively an equivalence relation on $G^{(0)}$ and our assumption on P implies that P totally orders each equivalence class. In general, we think of P as a partial order, hence the notation.

DEFINITION 18. *A* product system *over P is a Banach bundle $\mathfrak{B} \xrightarrow{p} P$ such that $\mathfrak{B}$ is a subcategory of $\mathbf{C}$ and p is a functor.*

The notion of product system is due to Arveson [**4**], who studied product systems over the additive semigroup $\mathbb{R}_+$. The fibres of his bundles are Hilbert spaces, i.e., correspondences from $\mathbb{C}$ to $\mathbb{C}$. We note, too, that his bundles are merely endowed with a Borel structure and not a topology. However, this distinction need not concern us.

Of course the definition requires the same kind of interpretation that the notion of Fell bundle requires, i.e., we have to distinguish between the fibres as Banach spaces and as unitary equivalence classes of C^*-correspondences. In particular, the definition really requires that $B(u) := p^{-1}(u)$, $u \in G^{(0)}$, be a C^*-algebra, so that $\mathfrak{B}^{(0)} := \mathfrak{B}|G^{(0)}$ is a bundle of C^*-algebras. Also, then, each fibre $B(\gamma) := p^{-1}(\gamma)$, $\gamma \in P$, must be a $B(r(\gamma))$, $B(s(\gamma))$-correspondence such that when $(\alpha, \beta) \in P^{(2)} \subseteq G^{(2)}$, $B(\alpha\beta) \sim_u B(\alpha) \otimes_{B(s(\alpha))} B(\beta)$.

Observe that $C_c(P, \mathfrak{B})$ is an algebra under pointwise addition and scalar multiplication and the product $*$ defined by the formula

$$f * g(\gamma) := \int_{\alpha \leq \gamma} f(\alpha) g(\alpha^{-1}\gamma)\, d\lambda^{r(\gamma)}(\alpha),$$

where we write $\alpha \leq \gamma$ precisely when $\gamma = \alpha\beta$, for some $\beta \in P$. That this is an algebra is easy to check. Also, the operations are continuous in the inductive limit topology. This algebra is not selfadjoint, however.

Next we form the *Fock space* $L^2(P, \mathfrak{B})$ associated with $\mathfrak{B}$. This is the Hilbert C^*-module over $C_0(G^{(0)}, \mathfrak{B}^{(0)})$ obtained by completing $C_c(P, \mathfrak{B})$ in the inner product given by the formula

$$\langle \xi, \eta \rangle(u) := \int_{P \cap G_u} \langle \xi(\gamma), \eta(\gamma) \rangle_{B(s(\gamma))}\, d\lambda_u(\gamma).$$

Observe that we may represent $C_c(P, \mathfrak{B})$ on $L^2(P, \mathfrak{B})$ by the formula $\lambda(f)\xi = f * \xi$, obtaining a homomorphism of $C_c(G, \mathfrak{B})$ in $\mathcal{L}(L^2(P, \mathfrak{B}))$. The norm closure of $\lambda\,(C_c(P, \mathfrak{B}))$ is a non-self-adjoint subalgebra of $\mathcal{L}(L^2(P, \mathfrak{B}))$ that we denote by $\mathcal{T}_+(L^2(P, \mathfrak{B}))$ and call the *tensor algebra* of the bundle $(P, \mathfrak{B})$. The reason for this terminology is that in the special case when $G = \mathbb{Z}$, $P = \mathbb{Z}_+$, and the fibres of $\mathfrak{B}$ are given by the tensor powers of a single correspondence E from $B(0)$ to $B(0)$, then the algebra $\mathcal{T}_+(L^2(P, \mathfrak{B}))$ is the tensor algebra $\mathcal{T}_+(E)$ defined and studied in [**26**]. The C^*-subalgebra of $\mathcal{L}(L^2(P, \mathfrak{B}))$ generated by $\mathcal{T}_+(L^2(P, \mathfrak{B}))$ is called the *Toeplitz algebra* of $(P, \mathfrak{B})$ and is denoted by $\mathcal{T}(L^2(P, \mathfrak{B}))$. The terminology derives from the work in [**26**] and from elsewhere. In his setting [**5**], Arveson calls $\mathcal{T}(L^2(P, \mathfrak{B}))$ the spectral C^*-algebra of the product system $\mathfrak{B}$.

Given a product system $\mathfrak{B}$ over P, each set $G^u \cap P$ has the structure of a directed set if $P \cup P^{-1} = G$ (and more generally). In this situation, we may, for each $u \in G^{(0)}$, form

$$\mathcal{L}(u) := \varinjlim_{\gamma \in G^u} \mathcal{L}(B(\gamma))$$

and in each $\mathcal{L}(u)$, let $\mathcal{A}(u)$ be the C^*-algebra generated by $\cup_{\gamma \in G^u} \mathcal{K}(B(\gamma))$. When this is done, we may form $\mathcal{B}(\gamma) := B(\gamma) \otimes_{B(s(\gamma))} \mathcal{A}(s(\gamma))$, obtaining an invertible correspondence $\mathcal{B}(\gamma)$ from $\mathcal{A}(r(\gamma))$ to $\mathcal{A}(s(\gamma))$. (As we mentioned earlier, $\mathcal{B}(\gamma)$ need not be full, but we not shall dwell upon that here.) For $\gamma \in P^{-1}$, we define $\mathcal{B}(\gamma) := \mathcal{B}(\gamma^{-1})^*$. Then, when $P \cup P^{-1} = G$, we obtain a Fell bundle $\tilde{\mathfrak{B}}$ over G and we define $\mathcal{O}(P, \mathfrak{B}) := C^*(G, \tilde{\mathfrak{B}})$. Our long-term objective is:

GOAL 19. *Determine conditions under which the C^*-envelope of $\mathcal{T}_+(L^2(P, \mathfrak{B}))$ is $\mathcal{O}(P, \mathfrak{B})$.*

The notion of C^*-envelope of a subalgebra of a C^*-algebra was introduced by Arveson in [**3**], where it is shown how the representation theory and structure theory, generally, of the subalgebra is intimately tied to its C^*-envelope. In [**26**], we showed that the C^*-envelope of $\mathcal{T}_+(E)$ is $\mathcal{O}(E)$ under very mild hypotheses on E. This contains the prototypical example that generated the theory developed by Arveson [**3**], namely the disc algebra. Indeed, in this case, E is simply $\mathbb{C}$, as is A. The tensor algebra, then, is completely isometrically isomorphic to the disc algebra and the C^*-envelope is $C(\mathbb{T})$, which is $\mathcal{O}(E)$ in this case.

The special case when G is r-discrete and amenable, P is (essentially) arbitrary, and $\mathfrak{B}$ is the line bundle associated with a twist, as in the 2nd of the Examples 7, was worked out in [**25**]. There, it was shown that the C^*-envelope of $\mathcal{T}_+(L^2(P,\mathfrak{B}))$ is $\mathcal{O}(P,\mathfrak{B})$. We have been able to conjoin the analyses from [**25**] and [**26**] to calculate the situation in some more general situations, but our goal, in any substantially general form, is still far off.

One particular problem of general groupoid theoretic interest that has originated in our pursuit of it is this:

PROBLEM 20. *If G is an amenable r-discrete principal groupoid, does there exist an abelian group A and an A-valued cocycle $c : G \to A$ such that $c^{-1}(0) = G^{(0)}$?*

Note that amenability is necessary for such a cocycle to exist. We allow the cocycle c to be a coboundary, but we really are interested in the situation when c is not a coboundary. Thus, we would be happy to assume something like the minimality of G. Also, we would like to know if we can specialize the group A to $\mathbb{R}$.

The role in our theory of such a cocycle is the analogue of the gauge automorphism group in the theory of $\mathcal{O}(E)$ and in Arveson's theory of product systems [**4, 5**].

References

[1] B. Abadie, S. Eilers, and R. Exel, *Morita equivalence for crossed products by Hilbert C^*-bimodules*, Trans. Amer. Math. Soc. **350** (1998), 3043–3054.

[2] Wm. Arveson, *Operator algebras and measure preserving transformations,* Acta Math. **118** (1967), 95–109.

[3] Wm. Arveson, *Subalgebras of C^*-algebras*, Acta Math. **123** (1969), 141–224.

[4] Wm. Arveson, *Continuous analogues of Fock space,* Mem. A.M.S. **80 no. 3** (1989).

[5] Wm. Arveson, *Continuous analogues of Fock space II: the spectral C^*-algebra,* J. Funct. Anal. **90** (1990), 138–205.

[6] M. Baillet, Y. Denizeau, and J-F. Havet, *Indice d'une esperance conditionelle,* Compositio Math. **66** (1988), 199–236.

[7] D. Blecher, P. Muhly, and V. Paulsen, *Categories of operator modules - Morita equivalence and projective modules*, Mem. A.M.S. **143, no. 681** (1989).

[8] L. Brown, P. Green, and M. Rieffel, *Stable isomorphisms and strong Morita equivalence of C^*-algebras,* Pac. J. Math. **71** (1977), 349–363.

[9] L. Brown, J. Mingo, and N-T. Shen, *Quasi-multipliers and embeddings of Hilbert C^*-bimodules,* Can. J. Math. **46** (1994), 1150–1174.

[10] R. Busby, *On the equivalence of twisted group algebras and Banach $*$-algebraic bundles,* Proc. Amer. Math. Soc. **37** (1973), 142-148.

[11] F. Combes, *Crossed products and Morita equivalence,* Proc. London Math. Soc. (3) **49** (1984), 289–306.

[12] A. Connes, *Noncommutative Geometry,* Academic Press, Inc., San Diego, 1994.

[13] R. Curto, P. Muhly, and D. Williams, *Cross products of strongly Morita equivalent C^*-algebras,* Proc. Amer. Math. Soc. **90** (1984), 528–530.

[14] R. Exel and M. Laca, *Continuous Fell bundles associated to measurable twisted actions,* Proc. Amer. Math. Soc. **125** (1997), 795–799.

[15] J. Fell, *An extension of Mackey's method to Banach $*$-algebraic bundles,* No. **90**, Memoirs of the Amer. Math. Soc., 1969.

[16] J. Fell and R. Doran, *Representation of $*$-Algebras, Locally Compact Groups, and Banach $*$-Algebraic Bundles, Vol. 1*, vol. **125**, of *Pure and Applied Mathematics,* Academic Press, Inc., San Diego, 1988.

[17] P. Green, *The local structure of twisted covariance algebras,* Acta. Math. **140** (1978), 191–250.

[18] R. Kadison and J. Ringrose, *Fundamentals of the theory of operator algebras (Vol. II: Advanced Theory),* Graduate Studies in Mathematics, v. **16**, Amer. Math. Soc., 1997.

[19] A. Kumjian, *On C^*-diagonals,* Canadian J. Math. **38** (1986), 969–1008.

[20] A. Kumjian, *Fell bundles over groupoids,* Proc. Amer. Math. Soc. **126** (1998), 1115–1125.

[21] A. Kumjian, P. Muhly, J. Renault, and D. Williams, *The Brauer group of a locally compact groupoid,* Amer. J. Math. **120** (1998), 901–954.

[22] E. C. Lance, *Hilbert C^*-modules: A toolkit for operator algebraists,* vol. **124** of *London Mathematical Society Lecture Note Series,* Cambridge Univ. Press, Cambridge, 1994.

[23] S. MacLane, *Categories for the Working Mathematician, 2nd ed.,* Springer Verlag, New York, 1998.

[24] P. Muhly, J. Renault, and D. Williams, *Equivalence and isomorphism of groupoid C^*-algebras,* J. Operator Theory **17** (1987), 3–22.

[25] P. Muhly and B. Solel, *Hilbert modules over operator algebras,* Mem. A.M.S. **117 no. 559** (1995).

[26] P. Muhly and B. Solel, *Tensor algebras over C^*-correspondences (Representations, dilations, and C^*-envelopes),* J. Functional Anal. **158** (1998), 389–457.

[27] P. Muhly and B. Solel, *Morita equivalence and tensor algebras,* to appear in Proc. London Math. Soc.

[28] J. Packer and I. Raeburn, *Twisted crossed products of C^*-algebras,* Math. Proc. Camb. Phil. Soc. **106** (1989), 293–311.

[29] J. Phillips and I. Raeburn, *Automorphisms of C^*-algebras and second Čech cohomology,* Indiana U. Math. J. **29** (1980), 799–822.

[30] M. Pimsner, *A class of C^*-algebras generalizing both Cuntz-Krieger algebras and crossed products by $\mathbb{Z}$,* in *Free Probability Theory,* D. Voiculescu, Ed., Fields Institute Communications **12**, 189-212, Amer. Math. Soc., Providence, 1997.

[31] J. Renault, *Représentation des produits croisés d'algèbres de groupoïdes,* J. Operator Theory **18** (1987), 67–97.

[32] M. Rieffel, *Induced representations of C^*-algebras,* Adv. Math. **13** (1974), 176–257.

[33] M. Rieffel, *Applications of strong Morita equivalence to transformation group C^*-algebras,* in vol. **38** of *Proc. Sympos. Pure Math.,* pages 299–310. Amer. Math. Soc., 1982.

[34] J. Schweizer, *Characterizing the simplicity of Cuntz-Pimsner Algebras,* preprint 1998.

[35] M. Takesaki, *Covariant representations of C^*-algebras and their locally compact automorphism groups,* Acta Math. **119** (1967), 273–302.

[36] S. Yamagami, *On the ideal structure of C^*-algebras over locally compact groupoids,* preprint 1987.

Department of Mathematics, University of Iowa, Iowa City, IA 52242-0001
E-mail address: muhly@math.uiowa.edu

Contemporary Mathematics
Volume **282**, 2001

Groupoids and Foliations

André Haefliger

ABSTRACT. To each foliation of codimension q is associated a dynamic object of dimension q, its (transverse étale) holonomy groupoid well defined only up to equivalence. Its importance was recognized at the earliest stage of foliation theory. In this paper we review some basic invariants of equivalence classes of étale groupoids and relate them to natural problems in foliation theory.

As the (transverse) holonomy groupoid of a foliation is an étale groupoid, we shall mainly be concerned in this talk by étale groupoids. We first recall the basic definitions and the notations we shall use.

1. Definitions and terminology

1.1 ETALE GROUPOIDS. A groupoid $(\mathcal{G}, T)$ (or simply $\mathcal{G}$) is a small category $\mathcal{G}$ with set of objects T such that all the elements of $\mathcal{G}$ are invertible. We have two projections: the source projection $s : \mathcal{G} \to T$ and the target projection $t : \mathcal{G} \to T$. The composition gg' of two elements g and g' of $\mathcal{G}$ is defined if $s(g) = t(g')$. For $x \in T$, the group of elements of $\mathcal{G}$ with source and target x is denoted $\mathcal{G}_x$ and its unit element is denoted 1_x. The orbit of $x \in T$ is the subset of points $y \in T$ such that there is $g \in \mathcal{G}$ with $s(g) = x$ and $t(g) = y$.

A topological groupoid $(\mathcal{G}, T)$ is a groupoid such that $\mathcal{G}$ and T are topological spaces, the composition and passage to inverses are continuous, the inclusion $x \mapsto 1_x$ is a homeomorphism $T \to \mathcal{G}$ onto its image. An *étale groupoid* is a topological groupoid $(\mathcal{G}, T)$ such that the projections s and t are étale maps, i.e. are locally homeomorphisms. The inclusion $T \to \mathcal{G}$ identifies T with the open subspace of units of $\mathcal{G}$. The space of orbits is the quotient space $\mathcal{G}\backslash T$ of T by the equivalence relation whose classes are the orbits. We like to view the map $\mathcal{G} \times_T T \to T$ associating to a pair (g, x) with $s(g) = x$ the point $t(g) \in T$ as a continuous action (see the next example).

1.2 EXAMPLES.

(a) Let Γ be a discrete group acting on a topological space T through homeomorphisms, the action of $\gamma \in \Gamma$ on $x \in T$ being denoted $\gamma.x$. The associated

1991 *Mathematics Subject Classification.* Primary 57R30, 58H05; Secondary 55N91, 58H10.
Key words and phrases. Foliations, groupoids, cohomology, classifying spaces.

étale groupoid is $(\Gamma \ltimes T, T)$, where $\Gamma \ltimes T$ is the product of Γ (endowed with the discrete topology) with T, the source projection sending (γ, x) to x and the target projection sending (γ, x) to $\gamma.x$. The composition $(\gamma, x)(\gamma', x')$ is defined if $x = \gamma.x'$ and is equal to $(\gamma\gamma', x')$.

(b) Let $\mathcal{P}$ be a pseudogroup of partial homeomorphims of a topological space T (the elements of $\mathcal{P}$ are homeomorphisms from open sets of T onto open sets of T). Let $(\mathcal{H}, T)$ be the associated groupoid of germs, where $\mathcal{H}$ is the space of the germs of the elements of $\mathcal{P}$ at the points of T (with the usual topology of germs), the source and target projection associating to a germ its source and target. Such an étale groupoid is called a groupoid of germs. One can recover $\mathcal{P}$ from $\mathcal{H}$, namely the elements of $\mathcal{P}$ are partial homeomorphims of T obtained by composing continuous sections of the source map above open sets of T with the target projection. In other words, the notions of pseudogroups and of groupoids of germs are equivalent notions.

To every étale groupoid $(\mathcal{G}, T)$ we can associate a groupoid of germs $(\mathcal{H}, T)$ of partial homeomorphisms of T together with a surjective continuous homomorphism

$$\mathcal{G} \to \mathcal{H}$$

projecting to the identity of T as follows. To an element $g \in \mathcal{G}$ with source x, we associate the germ at x of the composition with the target projection of a continuous local section $\tilde{g} : U \to \mathcal{G}$ of the source projection such that $\tilde{g}(x) = g$.

(c) Given an étale groupoid $(\mathcal{G}, T)$ and an open cover $\mathcal{U} = \{U_i\}_{i \in I}$ of T, we can construct an étale groupoid $(\mathcal{G}_{\mathcal{U}}, T_{\mathcal{U}})$ called the *localization of* $\mathcal{G}$ *over* $\mathcal{U}$. Here $T_{\mathcal{U}}$ is the disjoint union of the open sets U_i, namely the pairs (i, x) with $i \in I$ and $x \in U_i$. The elements of $\mathcal{G}_{\mathcal{U}}$ are the triples (i, g, j) with $s(g) \in U_j$ and $t(g) \in U_i$, the source and target projection mapping (i, g, j) to $(j, s(g))$ and $(i, t(g))$ respectively. The composition $(i, g, j)(j, h, k)$ is equal to (i, gh, k).

In the particular case where the groupoid consists only of its unit space T, then its localization over the open cover $\mathcal{U}$ of T is the étale groupoid $(\mathcal{T}_{\mathcal{U}}, T_{\mathcal{U}})$, where the elements of $\mathcal{T}_{\mathcal{U}}$ are the triples (i, x, j) where $x \in U_i \cap U_j$.

1.3 EQUIVALENCE OF ÉTALE GROUPOIDS. Two étale groupoids $(\mathcal{G}, T)$ and $(\mathcal{G}', T')$ are *equivalent* if there exists an open cover $\mathcal{U}$ of T and an open cover $\mathcal{U}'$ of T' such that $(\mathcal{G}_{\mathcal{U}}, T_{\mathcal{U}})$ is isomorphic to $(\mathcal{G}'_{\mathcal{U}'}, T_{\mathcal{U}'})$.

For étale groupoids this definition is equivalent to the usual definition of Morita equivalence (see [**Ren82**]).

A homomorphism $(\mathcal{G}, T) \to (\mathcal{G}', T')$ is a continuous map $\phi : \mathcal{G} \to \mathcal{G}'$ inducing a continuous map $\phi^{(0)} : \mathcal{G}^{(0)} = T \to \mathcal{G}'^{(0)} = T'$ such that $\phi(g_1 g_2) = \phi(g_1 g_2)$. Such a homomorphism is an equivalence if it is an étale map inducing a bijection $\mathcal{G}\backslash T \to \mathcal{G}'\backslash T'$ and if for each $x \in T$ the restriction of ϕ to $\mathcal{G}_x$ is an isomorphism onto $\mathcal{G}'_{\phi^{(0)}(x)}$. Two étale groupoids $(\mathcal{G}, T)$ and $(\mathcal{G}', T')$ are equivalent if there is an open cover $\mathcal{U}$ of T and an equivalence $\phi : (\mathcal{G}_{\mathcal{U}}, T_{\mathcal{U}}) \to (\mathcal{G}', T')$. For more details, see for instance [**BrHa99**], Chap.III.$\mathcal{G}$.

1.4 DEVELOPABILITY. An étale groupoid is *developable* if it is equivalent to an étale groupoid associated to an action of a discrete group Γ acting on a space through homeomorphisms (see 1.2(a)).

2 Some étale groupoids associated to foliations

2.1 FOLIATIONS. A foliation $\mathcal{F}$ of codimension q on a manifold M can be defined by a so called foliated 1-cocycle. Such a cocycle is given by an open cover $\{U_i\}_{i\in I}$ of M and submersions f_i from each U_i onto an open subset T_i of $\mathbb{R}^q$ satisfying the following compatibility conditions: for each $x \in U_i \cap U_j$ there is an open neighbourhood $V \subseteq U_i \cap U_j$ of x and a homeomorphism $g_{ji} : f_i(V) \to f_j(V)$ such that

$$f_j = g_{ji} f_i \quad on \quad V.$$

Note that the germ of g_{ji} at $f_i(x)$ is independent of the choice of the open set V.

Another foliated cocycle $\{f_k : U'_k \to \mathbb{R}^q\}$ defines the same foliation on M if the compatibility conditions are satisfied for each pair (f_i, f'_k).

The *leaves* of $\mathcal{F}$ are the connected components of M with respect to the topology whose basis of open sets consists of the open subsets of the fibers of the submersions f_i. They are immersed submanifolds of codimension q of M.

This definition of a foliation was interpreted as follows in my thesis ([**Ha58**]. Let $(\Gamma^q, \mathbb{R}^q)$ be the groupoid of germs of partial homeomorphisms of $\mathbb{R}^q$. The collections of maps $U_i \cap U_j \to \Gamma^q$ sending $x \in U_i \cap U_j$ to the germ $g_{ji}(x)$ of g_{ji} at $f_i(x)$ can be considered as a 1-cocycle $\in Z^1(\mathcal{U}, \Gamma^q)$. Another family of submersions defining the same foliation gives a cohomologous cocycle, so that the foliation $\mathcal{F}$ can be defined as an element of $H^1(M, \Gamma^q)$. The restriction of such an element to the leaf L through $x \in U_i$ is an element of $H^1(L, \Gamma^q)$, completely characterized by a homomorphism $\pi_1(L, x) \to \Gamma^q_{f_i(x)}$ well defined up to coujugation. The image of this homomorphim is the *holonomy group* of L.

2.2 THE HOLONOMY GROUPOID. We can assume that the fibers of the submersions f_i are connected. Let T be the disjoint union of the T_i. The *holonomy groupoid* $(\mathcal{H}, T)$ of $\mathcal{F}$ associated to $\{f_i\}$ is the subgroupoid of the groupoid of germs of partial homeomorphisms of T generated by the germs $g_{ji}(x)$ of the g_{ji} at $f_i(x)$, $x \in U_i \cap U_j$. The holonomy pseudogroup of $\mathcal{F}$ is the pseudogroup of partial homeomorphisms of T generated by the g_{ji}: as observed in 1.1, the notions of holonomy groupoid and holonomy pseudogroup are equivalent notions.

It is easy to see that the holonomy groupoids associated to cocycles defining the same foliation are equivalent. By the holonomy groupoid of $\mathcal{F}$ we mean any representative in this equivalence class. Therefore to a foliation is associated an étale groupoid well defined UP TO EQUIVALENCE. This is an object of dimension q reflecting the dynamic of $\mathcal{F}$.

For instance there is a bijection between the set of leaves and the set of orbits of the holonomy groupoid, the minimal closed sets of M which are union of leaves and the minimal closed sets of T which are invariant by the holonomy groupoid. The holonomy groups of the leaves correspond to the isotropy groups of the holonomy groupoid, etc.

As in our definition the holonomy groupoid is well defined only up to equivalence, we shall consider only properties of the holonomy groupoid which are invariant under equivalences.

2.3 TRANSVERSE STRUCTURE. $\mathcal{F}$ is said to be *transversally* differentiable of class C^r, or real analytic, or holomorphic or Riemannian, etc..., if there is such a structure on T invariant by $\mathcal{H}$. The existence of such a structure is of course independent of the choice of the holonomy groupoid in its equivalence class. More generally a *transverse structure* for $\mathcal{F}$ is a structure on T invariant by the holonomy groupoid.

Let me mention a few examples showing the close relationship between the properties of the foliation and of its holonomy groupoid.

2.4 EXAMPLES. The following examples show that the existence of a transverse structure may have very strong implications for the corresponding foliation.

(1) Suppose that $\mathcal{F}$ is a foliation of codimension one on M which is transversally analytic. Then any closed curve transversal to the leaves is non-homotopic to zero in M (cf.[**Ha58**]). This follows from the following more general theorem whose proof relies essentially on the proof of the above fact:

Any groupoid of germs of partial analytic homeomorphisms of a one-dimensional analytic manifold, which is connected in the sense of 3.1, is developable. This means precisely that this groupoid is equivalent to the étale groupoid associated to the action of a group Γ through real analytic homeomorphisms of a one dimensional connected real analytic manifold (non-Hausdorff in general). Moreover if an element $\gamma \in \Gamma$ restricted to an open set is the identity, then this element is the unit element of Γ.

(2) In 1966 R. Sacksteder [**Sac65**] proved the following remarkable fact: If a foliation of codimension one on a compact manifold is transversally of class C^2 and has an closed exceptional minimal set saturated by leaves, then there is an element in some holonomy group with a contracting differential. (A minimal set is exceptional if transversally it looks like a Cantor set.)

I mention this result because I think that this is the first time that a global property of the holonomy groupoid was used in an essential way. In fact Sacksteder deduces his theorem from a general theorem on pseudogroups (or groupoids of germs).

(3) Given a smooth foliation on a compact manifold M, one can ask about the existence of a Riemannian metric such that all the leaves of the foliation are minimal submanifolds of M in the sense of Riemannian geometry. It is surprising that this existence depends only on a property of the holonomy groupoid (see [**Ha80**]).

There is another étale groupoid attached to a foliation and which plays an important role.

2.5 THE FUNDAMENTAL GROUPOID OF A FOLIATION. Given a foliation $\mathcal{F}$ of codimension q on a manifold M, consider a submanifold T of M of dimension q meeting every leaf at least once and transversally. The *fundamental groupoid* $(\mathcal{G}, T)$ of $\mathcal{F}$ associated to T is defined as follows. The elements of $\mathcal{G}$ are the homotopy classes of the paths in the leaves joining a point of T (its source) to a point of T (its target). A basis of open sets for the topology on $\mathcal{G}$ is obtained as follows. Let U be an open set of T and let $c : U \times [0,1] \to M$ be a continuous map such that, for each $u \in U$, the map $c_u : [0,1] \to M$ sending t to $c(u,t)$ is a path in a leaf joining u to a point of T; then the set of homotopy classes in the leaves of the paths c_u is an open set of the basis.

Let $(\mathcal{H}, T)$ be the associated étale groupoid of germs (cf.1.2 (b)). It is equivalent to the holonomy groupoid of $\mathcal{F}$ and is called the holonomy groupoid of $\mathcal{F}$ associated to the transversal T. The kernel of the natural projection from $\mathcal{G}$ to $\mathcal{H}$

is a sheaf of groups on T. Its stalk above $x \in T$ is the fundamental group of the holonomy covering of the leaf through x.

The fundamental groupoid plays the leading role in the study of germs of deformations of foliations (cf. [**BoHa90**]).

2.6 THE UNIVERSAL GROUPOIDS Γ_q^r AND $\overline{\Gamma}_q^r$. $(\Gamma_q^r, \mathbb{R}^q)$ is the groupoid of germs of partial diffeomorphisms of $\mathbb{R}^q$ of class C^r, where $r = 0, 1, ..., \infty, \omega$ (C^ω means real analytic).

Let $F\mathbb{R}^q$ be the bundle of q-frames on $\mathbb{R}^q$. For $r > 0$ the groupoid $(\Gamma_q^r, \mathbb{R}^q)$ acts on $F\mathbb{R}^q$ through the differential. The groupoid $(\overline{\Gamma}_q^r, F\mathbb{R}^q)$ is the étale groupoid associated to this action (its elements are the pairs formed by a frame at x and an element of Γ_q^r with source x).

For a foliation which is transversally of class C^r, there is an obvious homomorphism from its holonomy groupoid $\mathcal{H}$ (as defined in 2.1) to Γ_q^r. A 1-cocycle with values in $\overline{\Gamma}_q^r$ over an open cover of M projecting to a foliated 1-cocycle with values in Γ_q^r (as defined in 1.1) gives a foliation of codimension q on M together with a trivialization of the vector bundle normal to the leaves.

3 Homotopy and Cohomology for étale groupoids

In this paragraph $(\mathcal{G}, T)$ denotes an étale groupoid.

3.1 $\mathcal{G}$-PATHS. A $\mathcal{G}$-path c joining $x \in T$ to $y \in T$, over a subdivision $0 = t_0 \leq t_1 \leq ... \leq t_k = 1$ of the intervall $[0, 1]$, is a sequence $c = (g_0, c_1, g_1, ..., c_k, g_k)$, where the $c_i : [t_{i-1}, t_i] \to T$ are continuous paths, the g_i are elements of $\mathcal{G}$ with $s(g_i) = c_i(t_{i-1})$ for $i > 0$, $t(g_i) = c_i(t_i)$ for $i \leq k$ and $t(g_0) = x$, $s(g_k) = y$. A $\mathcal{G}$-loop at $x_0 \in T$ is a $\mathcal{G}$-path joining x_0 to x_0.

By definition the groupoid $(\mathcal{G}, T)$ is (arcwise) *connected* if, for any $x, y \in T$, there is a $\mathcal{G}$-path joining x to y.

One of the most important invariants of the equivalence class of a connected étale groupoid is its fundamental group.

3.2 THE FUNDAMENTAL GROUP. Given a point $x_0 \in T$, the fundamental group $\pi_1((\mathcal{G}, T), x_0)$ is the group of homotopy classes of $\mathcal{G}$-loops based at x_0. (See [**Ha70**], [**Sa88**] or [**BrHa99**], p.604–608, for the precise definition of homotopy.)

A covering $(\hat{\mathcal{G}}, \hat{T})$ of an étale groupoid $(\mathcal{G}, T)$ is the étale groupoid associated to an action of $\mathcal{G}$ on a covering $p : \hat{T} \to T$. Namely the elements of $\hat{\mathcal{G}}$ are the pairs $(g, \hat{x}) \in \mathcal{G} \times \hat{T}$ with $s(g) = p(\hat{x})$.

The theory of coverings and of the fundamental group of an étale groupoid is quite analogous to the corresponding theory for topological spaces. For a connected étale groupoid $(\mathcal{G}, T)$ with T locally simply connected, there is a 1-1 correspondence between pointed equivalence classes of connected coverings and subgroups of the fundamental group.

More generally one can define [**Ha70**] the higher homotopy groups $\pi_i((\mathcal{G}, T), x_0)$ as the groups of homotopy classes respecting base points of morphisms (in a generalized sense) from the sphere $\mathbb{S}^i$ of dimension i (considered as a trivial groupoid) to $\mathcal{G}$.

3.3 EXAMPLES.

(a) If $\mathcal{G} = \Gamma \ltimes T$ is the groupoid associated to the action of a group Γ on a simply connected space T, then its fundamental group is isomorphic to Γ and its universal covering is equivalent to T considered as a trivial groupoid.

(b) Let $\mathcal{F}$ be a foliation on a manifold M, and let $(\mathcal{G}, T)$ (resp. $(\mathcal{H}, T)$) be the fundamental groupoid (resp. the holonomy groupoid) of $\mathcal{F}$ with respect to a transversal T. Let x_0 be a base point in T. Then $\pi_1(M, x_0)$ is naturally isomorphic to $\pi_1((\mathcal{G}, T), x_0)$ (see [**Ha82**]). As the natural homomorphism $(\mathcal{G}, T) \to (\mathcal{H}, T)$ induces a surjective homomorphism $\pi_1((\mathcal{G}, T), x_0) \to \pi_1((\mathcal{H}, T), x_0)$, we get a surjection $\pi_1(M, x_0) \to \pi_1((\mathcal{H}, T), x_0)$.

In particular if M is simply connected, then the fundamental group of the holonomy groupoid and of the fundamental groupoid of $\mathcal{F}$ is trivial.

Cohomology.

3.4 $\mathcal{G}$-SHEAVES. We consider sheaves of abelian groups $\mathcal{A}$ on T on which $\mathcal{G}$ acts continuously. The stalk of $\mathcal{A}$ at x is denoted $\mathcal{A}_x$ and the action of $g \in \mathcal{G}$ with $s(g) = x$ on $a \in \mathcal{A}_x$ is noted $g.a \in \mathcal{A}_{t(g)}$. Such sheaves are called $\mathcal{G}$-sheaves (see [**Ha76**] and [**Ku88**] for more details). They form an abelian category. If $(\mathcal{G}', T')$ is an étale groupoid equivalent to $(\mathcal{G}, T)$, there is obviously an equivalence between the category of $\mathcal{G}$-sheaves on T and the category of $\mathcal{G}'$-sheaves on T'.

Examples of $\mathcal{G}$-sheaves are the constant sheaves (with the trivial action), the sheaves of differential forms when T is a differentiable manifold, the action being induced by the differential of the elements of $\mathcal{G}$. When T is locally simply connected and $(\mathcal{G}, T)$ is connected (see 3.1), then a locally trivial $\mathcal{G}$-sheaf with stalk an abelian group A is determined up to isomorphism by a homomorphism from $\pi_1((\mathcal{G}, T), x_0)$ to $\mathrm{Aut}(A)$, the group of automorphisms of A.

3.5 The cohomology groups $H^i(\mathcal{G}, \mathcal{A})$.

One can define (cf.[**Ha76**]), for $i \geq 0$, the cohomology groups $H^i(\mathcal{G}, \mathcal{A})$ of $\mathcal{G}$ with coefficients in $\mathcal{A}$. The group $H^0(\mathcal{G}, \mathcal{A})$ is the group of continuous sections of $\mathcal{A}$ which are invariant by the action of $\mathcal{G}$. Equivalently $H^0(\mathcal{G}, \mathcal{A})$ is the group of $\mathcal{G}$-homomorphisms $Hom_{\mathcal{G}}(\mathbb{Z}, \mathcal{A})$ from the constant $\mathcal{G}$-sheaf $\mathbb{Z}$ with stalk $\mathbb{Z}$ to $\mathcal{A}$. A short exact sequence of $\mathcal{G}$-sheaves induces a long exact sequence of cohomology groups.

These groups can be defined directly [**Ha76**] as the cohomology of a double complex mixing the standard flabby Godement resolution[1] [**Go64**] of $\mathcal{A}$ and the standard resolution for groups, or (see [**Ku88**], [**Ha92**]) as the right derived functors of the functor $Hom_{\mathcal{G}}(Z, \ )$ using an injective resolution of $\mathcal{A}$ (see Grothendiek [**Gr57**]). Those two approaches lead to the same result (cf. [**Ha92**]). The cohomology groups depend only on the equivalence class of $(\mathcal{G}, T)$. Cultivated people, like Moerdijk [**Moe91**], prefer to interpret them in the framework of the theory of topos of Grothendieck. We give below some details which will be useful later.

3.6 THE COMPLEX $C^*(\mathcal{G}, \mathcal{A})$. Let $C^i(\mathcal{G}, \mathcal{A})$ (see [**Ha76**]) be the abelian group whose elements are continuous maps associating to a composable sequence $(g_1, ..., g_i)$ of elements of $\mathcal{G}$ an element of $\mathcal{A}_{t(g_1)}$. We define a differential $\delta : C^i(\mathcal{G}, \mathcal{A}) \to$

[1] The canonical flabby resolution of Godement of $\mathcal{A}$ is a resolution by flabby $\mathcal{G}$-sheaves.

$C^{i+1}(\mathcal{G},\mathcal{A})$ for $i>0$ by the formula

$$\delta f(g_0,\ldots,g_i)=g_0.f(g_1,\ldots,g_i)-\sum_{k=0}^{i-1}(-1)^k f(g_0,\ldots,g_kg_{k+1},\ldots,g_i) \\ +(-1)^{i+1}f(g_0,\ldots,g_{i-1})$$

and for $i=0$ by

$$\delta f(g)=g.f(s(g))-f(t(g)).$$

Let $\mathcal{G}^{(k)}$ be the topological space of sequences $(g_1,\ldots,g_k)$ of k elements of $\mathcal{G}$ which are composable in this order, and let $t^{(k)}:\mathcal{G}^{(k)}\to T$ be the projection associating to such a sequence the target of g_1. The group $C^k(\mathcal{G},\mathcal{A})$ is the group of sections of the sheaf $(t^{(k)})^*\mathcal{A}$ on $\mathcal{G}^{(k)}$.

3.7 THE DOUBLE COMPLEX. Given a $\mathcal{G}$-sheaf $\mathcal{A}$ and a resolution of $\mathcal{A}$ by $\mathcal{G}$-sheaves

$$0\to\mathcal{A}\to\mathcal{J}^0\to\mathcal{J}^1\to\ldots,$$

we can form as usual the double complex $C^*(\mathcal{G},\mathcal{J}^*)$. One differential is δ and the other one d is induced by the differential of the resolution. We have $d\delta=\delta d$.

3.8 PROPOSITION ([**Ha76**] AND [**Ha92**]). *The cohomology groups $H^*(\mathcal{G},\mathcal{A})$ are isomorphic to the cohomology groups of the simple complex associated to the double complex $C^*(\mathcal{G},\mathcal{J}^*)$ when one of the following condition is satisfied:*

1. *the sheaves $\mathcal{J}^*$ are flabby*
2. *the sheaves $\mathcal{J}^*$ are soft and $\mathcal{G}$ is Hausdorff and locally compact.*

The next corollary (when applied to a suitable localization $\mathcal{G}_\mathcal{U}$) is an analogue of Leray theorem.

3.9 COROLLARY. *If the cohomology of the topological spaces $\mathcal{G}^{(k)}$ with coefficients in the sheaf ${t^{(k)}}^*\mathcal{A}$ vanishes in degree >0, then the cohomology $H^*(\mathcal{G},\mathcal{A})$ is isomorphic to the cohomology of the complex $C^*(\mathcal{G},\mathcal{A})$.*

For instance if $\mathcal{G}$ is the étale groupoid associated to the action of a discrete group Γ on a contractible space T (see 1.2(a)) and $\mathcal{A}$ is the constant sheaf with stalk A defined by a homomorphism $\pi_1((\mathcal{G},T),x_0)=\Gamma\to A$ (see 3.4), then $H^*(\mathcal{G},\mathcal{A})$ is isomorphic to the group cohomology $H^*(\Gamma,A)$.

3.10 **Čech cohomology**. The i-th cohomology group of the complex $C^*(\mathcal{G},\mathcal{A})$ is denoted $\hat{H}^i(G,A)$. Those groups are not an invariant of the equivalence class of $\mathcal{G}$. To achieve that, one can define the analogue of the Čech cohomology.

Given an open cover $\mathcal{U}$ of T, let $\mathcal{A}_\mathcal{U}$ be the sheaf on $T_\mathcal{U}$ inverse image of $\mathcal{A}$ by the natural projection $T_\mathcal{U}\to T$. It is naturally a $\mathcal{G}_\mathcal{U}$-sheaf. If $\mathcal{V}$ is another open cover of T which is finer than $\mathcal{U}$ and if we choose for each element U_i of $\mathcal{V}$ an element $\tau(U_i)$ of $\mathcal{U}$ containing it, we get a homomorphism $\mathcal{G}_\mathcal{V}\to\mathcal{G}_\mathcal{U}$. It gives a map of complexes $C^*(\mathcal{G}_\mathcal{V},\mathcal{A}_\mathcal{V})\to C^*(\mathcal{G}_\mathcal{U},\mathcal{A}_\mathcal{U})$ inducing a homomorphism $\hat{H}^*(\mathcal{G}_\mathcal{V},\mathcal{A}_\mathcal{V})\to\hat{H}^*(\mathcal{G}_\mathcal{U},\mathcal{A}_\mathcal{U})$ independent of the choice of τ. We define the Čech cohomology as

$$\check{H}^*(\mathcal{G},\mathcal{A}):=\varinjlim\hat{H}^*(\mathcal{G}_\mathcal{U},\mathcal{A}_\mathcal{U}),$$

where the direct limit is taken over the open covers of T.

It is clear that those groups depend only of the equivalence class of $\mathcal{G}$. For an étale groupoid which consists only of its space of units T, the Čech cohomology of this groupoid is the usual Čech cohomology of the topological space T. Unfortunately in general a short exact sequence of $\mathcal{G}$-sheaves does not give a long cohomology exact sequence of the Čech cohomology groups. As in the case of topological spaces [**Gr57**] we have the following property.

3.11 PROPOSITION. *The natural homomorphism $\check{H}^*(\mathcal{G},\mathcal{A}) \to H^*(\mathcal{G},\mathcal{A})$ is an isomorphism for $* = 0,1$ and is injective for $* = 2$.*

For instance, the isomorphisms classes of $\mathcal{G}$-principal bundles with structural group the circle S^1 are by definition the elements of $\check{H}^1(\mathcal{G},\mathcal{S}^1)$, where $\mathcal{S}^1$ is the sheaf of germs of continuous maps from open sets of T to S^1. This group is isomorphic to $H^1(\mathcal{G},\mathcal{S}^1)$ and the short exact sequence

$$0 \to \mathbb{Z} \to \mathcal{R} \to \mathcal{S}^1 \to 0,$$

where $\mathcal{R}$ is the sheaf of germs of continuous maps from open sets of T to $\mathbb{R}$ and $\mathbb{Z}$ the trivial $\mathcal{G}$-sheaf with stalk $\mathbb{Z}$, gives a long exact sequence

$$H^1(\mathcal{G},\mathcal{R}) \to H^1(\mathcal{G},\mathcal{S}^1) \to H^2(\mathcal{G},\mathbb{Z}) \to H^2(\mathcal{G},\mathcal{R}) \to \dots .$$

In general $H^*(\mathcal{G},\mathcal{R})$ does not vanish. It vanishes for $* > 0$ when $\mathcal{G}$ is the groupoid associated to an orbifold: in that case $H^1(\mathcal{G},\mathcal{S}^1)$ is isomorphic to $H^2(\mathcal{G},\mathbb{Z})$.

4 The classifying space

4.1 DEFINITION OF THE CLASSIFYING SPACE $B\mathcal{G}$. The geometric realization of the nerve of the topological category $\mathcal{G}$ (cf.[**Seg68**]) is a model for $B\mathcal{G}$. The weak homotopy type of $B\mathcal{G}$ depends only on the equivalence class of $\mathcal{G}$.

It is immediate to prove that the homotopy groups of $(\mathcal{G},T)$ are isomorphic to the homotopy groups of $B\mathcal{G}$. One can also show easily that for a constant $\mathcal{G}$- sheaf $\mathcal{A}$ of abelian groups with stalk A, the cohomology groups $H^*(B\mathcal{G},A)$ are isomorphic to the groups $H^*(\mathcal{G},\mathcal{A})$. For a general $\mathcal{G}$-sheaf $\mathcal{A}$, it has been proved by I. Moerdijk [**Moe98**] that $H^i(\mathcal{G},\mathcal{A})$ is isomorphic to the ith cohomology group of $B\mathcal{G}$ with coefficient in a sheaf on $B\mathcal{G}$ canonically associated to $\mathcal{A}$ (this was conjectured in [**Ha76**]).

It is explained in [**Ha70**] and [**Ha82**] in what sense $B\mathcal{G}$ (or other models) can be considered as a classifying space.

In many cases there are simpler models. For instance, if a foliation $\mathcal{F}$ on a manifold M is such that the universal (resp. holonomy) covering of each leaf is contractible, then M is a model for the classifying space of the fundamental (resp. holonomy) groupoid of $\mathcal{F}$ (cf.[**Ha82**]). Moerdijk result [**Moe98**] should also be valid in this case.

For instance if $\mathcal{F}$ is a foliation on a manifold M by curves or surfaces not homeomorphic to a 2-sphere or a projective plane, then M is a classifying space for the fundamental groupoid of $\mathcal{F}$.

4.2 The classification theorem.

Let E be a field of planes of codimension q on a manifold M (equivalently E is a vector subbundle of codimension q of the tangent bundle TM of M). A fundamental problem in foliation theory is to determine under what conditions such a field is homotopic to the field of planes tangent to a foliation of codimension q on M. In 1968, Bott [**B068**] found obstructions in terms of Pontrjagin classes of the bundle $E^\perp := TM/E$ complementary to E. Namely:

4.3 BOTT THEOREM. *If E is homotopic to a subbundle of TM tangent to a C^∞ foliation of codimension q, then any element of the real cohomology groups $H^*(M,\mathbb{R})$ which is of degree $* > 2q$ and which is a polynomial in the real Pontrjagin classes of $E^\perp$ is zero.*

He also found examples where this condition is not satisfied. This theorem can be rephrased as follows in terms of classifying spaces.

The homomorphism $\Gamma_q^\infty \to GL_q$ associating to a germ of C^∞-diffeomorphism of $\mathbb{R}^q$ the (q,q)-matrix of its differential induces a map $B\Gamma_q^\infty \to BGL_q$ of the corresponding classifying spaces. Up to homotopy we can replace this map by a fibration: its fiber has the homotopy type of $B\overline{\Gamma}_q^\infty$. As the real cohomology of BGL_q is the polynomial algebra in the Pontrjagin classes of the universal vector bundle over BGL_q, Bott theorem amounts to say that the map $H^*(BGL_q,\mathbb{R}) \to H^*(B\Gamma_q^\infty,\mathbb{R})$ induced by the projection $B\Gamma_q^\infty \to BGL_q$ has a trivial image for $* > 2q$.

In 72-74, Thurston proved the following remarkable theorem (see [**Th74**] and the references therein). The much easier case, when the manifold M is open, was proved in [**Ha70**] where it was shown that $B\overline{\Gamma}_q^\infty$ is q-connected.

4.4 THURSTON THEOREM. *Let E be a field of planes of codimension q on a differentiable manifold M. Then it is homotopic to the plane field tangent to a C^∞ foliation on M if and only if the map $M \to BGL_q$ classifying the normal bundle $E^\perp$ can be lifted to a map $M \to B\Gamma_q^\infty$.*

This deep theorem has striking corollaries, for instance any field of planes E of codimension q, whose complementary bundle $E^\perp$ is trivial, is homotopic to a field tangent to a C^∞ foliation of codimension q. It shows that, for the problem of deforming a field of planes of codimension q to a field tangent to a foliation, there is an obstuction theory with values in the homotopy groups of the fiber $B\overline{\Gamma}_q^\infty$. Therefore it would be very important to determine the connectivity of that fiber. From Bott theorem it follows that it cannot be more than $2q$-connected. It follows from a deep theorem of Mather (see [**Ma74**] and the references therein and also Thurston [**Th74**]) that $B\overline{\Gamma}_q^\infty$ is $q+1$-connected (this is essentially equivalent to the simplicity of the group of diffeomorphisms with compact support of $\mathbb{R}^q$ which are isotopic to the identity).

5 Illustrations of the use of cohomology

In this paragraph we give two examples to illustrate the role of cohomology.

5.1 The Godbillon-Vey invariant.

To each transversally oriented foliation $\mathcal{F}$ of class C^2 of codimension 1 on a manifold M, Godbillon and Vey [**GV71**] associated a cohomology class in $H^3(M,\mathbb{R})$, since called the Godbillon-Vey invariant of $\mathcal{F}$ and noted $GV(\mathcal{F})$. They defined it as the de Rham cohomology class of a closed differential form of degree 3 on M constructed from a 1-form defining $\mathcal{F}$. They also indicated its relation with the generator of the third cohomology group of the Lie algebra of formal vector fields on $\mathbb{R}$ defined earlier by Gelfand and Fuchs [**GF70**]. In fact the Godbillon-Vey invariant of a foliation comes from a real cohomology class $H^3(B\Gamma_1^2.\mathbb{R}) = H^3(\Gamma_1^2,\mathbb{R})$: it is represented by a natural cocycle (Thurston cocycle, see [**Ha76**], Chap.IV,4) in the double complex $C^*(\Gamma^2,\Omega^*)$, where Ω^* denotes the resolution of the constant sheaf $\mathbb{R}$ by the sheaves of differential forms on $\mathbb{R}$. Therefore $GV(\mathcal{F})$ comes from a real cohomology class of the holonomy groupoid of $\mathcal{F}$.

The seminal paper of Godbillon and Vey was the starting point of the discovery of a close relationship between the Gelfand-Fuchs cohomology of the Lie algebra of formal vector fields on $\mathbb{R}^q$ and the real cohomology of $B\overline{\Gamma}_q^\infty$, the former mapping to the later, i.e functorially to the real cohomology of manifolds with a differentiable foliation of codimension q and a trivialization of its normal bundle. Many other relationships of this kind were also discovered, leading to the construction of many characteristic classes for foliations.

For a long time the geometric interpretation of the Godbillon-Vey invariant remained mysterious. Eventually Duminy proved in 1982 that the non-vanishing of the Godbillon-Vey invariant of a foliation on a compact manifold[2] implies the existence of a leaf with a loop in it whose holonomy attracts this leaf on itself. For a report on the Godbillon-Vey invariant, see Ghys [**Gh89**].

5.2 The classification of extensions with abelian kernel.

Let $(\mathcal{G}, T)$ be an étale groupoid and let $\mathcal{A}$ be a $\mathcal{G}$-sheaf on T of abelian groups. By definition an extension of $\mathcal{G}$ by the kernel $\mathcal{A}$ is given by an open cover $\mathcal{U}$ of T, an étale groupoid $(\tilde{\mathcal{G}_{\mathcal{U}}}, T_{\mathcal{U}})$ and a continuous homomorphism from $(\tilde{\mathcal{G}_{\mathcal{U}}}, T_{\mathcal{U}})$ onto $(\mathcal{G}_{\mathcal{U}}, T_{\mathcal{U}})$ projecting to the identity of $T_{\mathcal{U}}$ with kernel $\mathcal{A}_{\mathcal{U}}$ (we use the notations of 1.2(c)). Such an extension is called in [**Ku88**] a twist of $\mathcal{G}_{\mathcal{U}}$ by $\mathcal{A}_{\mathcal{U}}$.

Another such extension $(\tilde{\mathcal{G}_{\mathcal{U}'}}, T_{\mathcal{U}'})$, where $\mathcal{U}'$ is an open cover of T, is equivalent to the extension $(\tilde{\mathcal{G}_{\mathcal{U}}}, T_{\mathcal{U}})$ if there exists an extension $(\tilde{\mathcal{G}_{\mathcal{V}}}, T_{\mathcal{V}})$, where $\mathcal{V}$ is an open cover of T finer than $\mathcal{U}$ and $\mathcal{U}'$, and a commutative diagram of continuous homomorphisms

$$\begin{array}{ccccc}
\mathcal{A}_{\mathcal{U}} & \longrightarrow & \tilde{\mathcal{G}_{\mathcal{U}}} & \longrightarrow & \mathcal{G}_{\mathcal{U}} \\
\uparrow & & \uparrow & & \uparrow \\
\mathcal{A}_{\mathcal{V}} & \longrightarrow & \tilde{\mathcal{G}_{\mathcal{V}}} & \longrightarrow & \mathcal{G}_{\mathcal{V}} \\
\downarrow & & \downarrow & & \downarrow \\
\mathcal{A}_{\mathcal{U}'} & \longrightarrow & \tilde{\mathcal{G}_{\mathcal{U}'}} & \longrightarrow & \mathcal{G}_{\mathcal{U}'}
\end{array}$$

where the vertical arrows are equivalences, the left and right ones being the natural homomorphisms induced by a choice for each open set of $\mathcal{V}$ of an open set of $\mathcal{U}$ and $\mathcal{U}'$ containing it.

It is clear that the set of equivalence classes of extensions of $\mathcal{G}$ by the kernel $\mathcal{A}$ depends only on the equivalene class of $\mathcal{G}$. It follows easily from Kumjan ([**Ku88**] p.224) that for a $\mathcal{G}$-sheaf $\mathcal{A}$ of abelian groups, the set of equivalence classes of extensions of $\mathcal{G}$ by $\mathcal{A}$ is isomorphic to $H^2(\mathcal{G}, \mathcal{A})$.

As an example, consider the foliation $\mathcal{F}$ of the 3-sphere S^3 whose leaves are the fibers of the Hopf fibration. The base space of this fibration is the 2-sphere S^2 and the (transverse) holonomy groupoid of $\mathcal{F}$ is equivalent to the trivial groupoid S^2. The fundamental groupoid of $\mathcal{F}$ is an extension of S^2 by the constant sheaf with stalk $\mathbb{Z}$, but in this case we need to pass to a localization of the trivial groupoid S^2 (any twist of S^2 by the constant sheaf $\mathbb{Z}$ is trivial). For instance the fundamental groupoid of $\mathcal{F}$ is the extension of S^2 corresponding to a generator of $H^2(S^2, \mathbb{Z})$.

[2]In fact this hypothesis can be replaced by the non-vanishing of the Godbillon-Vey invariant in the cohomology of its holonomy groupoid

6 Two basic problems

As the equivalence class of the holonomy groupoid of a foliation is a basic invariant, it is natural to study directly the properties of étale groupoids verifying some particular restricted conditions invariant under equivalences and to try to classify such groupoids up to equivalence. So we first formulate a first problem in vague terms.

6.1 THE CLASSIFICATION PROBLEM. *Classify up to equivalence étale groupoids* $(\mathcal{G}, T)$ *preserving a given structure on* T*, under suitable extra conditions.*

Often those suitable restrictions (for instance some finiteness conditions) are fullfilled by the holonomy (or fundamental) groupoid of a foliation on a *compact* manifold. Such groupoids always satisfy the following property.

6.2 DEFINITION. An étale groupoid $(\mathcal{G}, T)$ (where T is Hausdorff) is **compactly generated** if there is a finite number of open sets V_i of T, $i = 1, \ldots, n$ and continuous sections $g_i : V_i \to \mathcal{G}$ of the source map such that:

1. The union V of the V_i is relatively compact and meets all the orbits of $\mathcal{G}$,
2. The images $g_i(V_i)$ generate the restriction $(\mathcal{G}_V, V)$ of $(\mathcal{G}, T)$ to V,
3. Each map g_i can be extended to a continuous map from the closure of V_i to $\mathcal{G}$.

These conditions are satisfied by any étale groupoid $(\mathcal{G}', T')$ equivalent to $(\mathcal{G}, T)$, assuming T' Hausdorff. Also the open set V can be replaced by any other relatively compact open set meeting all the orbits.

The fundamental or holonomy groupoid of any foliation on a closed manifold (i.e. compact and without boundary) is compactly generated. To see it (for the holonomy groupoid), just replace the open cover $\{U_i\}$ in 2.1 by a finite suitable finer open cover, relatively compact in the former one, and restrict the submersions f_i. Many dynamic properties of foliations on closed manifolds are shared by compactly generated étale groupoids of germs.

6.3 THE REALIZATION PROBLEM. *Given an étale groupoid* $(\mathcal{G}, T)$ *such that* T *is a Hausdorff manifold, when is it equivalent to the fundamental or the holonomy groupoid of a foliation on a closed manifold ?*

As just mentioned, a necessary condition is the compact generation. In general it is not easy to determine when an étale groupoid is compactly generated (unless it is known to be equivalent to the holonomy groupoid of a foliation on a closed manifold!). Nevertheless the following remarkable result of G. Meigniez [**Mei97**] shows that compact generation can involve arithmetic.

Let Aff_1 be the solvable Lie group of affine transformations $x \mapsto ax + b$ of $\mathbb{R}$, where $a, b \in \mathbb{R}$ and $a > 0$. Let $\mu : Aff_1 \to \mathbb{R}_+^*$ be the homomorphism associating to $x \mapsto ax + b$ the number a.

6.4 MEIGNIEZ THEOREM. *Let* Γ *be a dense subgroup of* Aff_1 *and let* $\mathcal{H} = \Gamma \ltimes Aff_1$ *be the étale groupoid associated to the action of* Γ *by left translations on* Aff_1*. Then* $\mathcal{H}$ *is compactly generated if and only if* Γ *is finitely generated and*

1. 1 *is an integral linear combination of* $\mu(\Gamma) \cap (1, \infty)$,
2. 1 *is an integral linear combination of* $\mu(\Gamma) \cap (0, 1)$.

Many of those compactly generated étale groupoids are equivalent to holonomy groupoids of foliations on closed manifolds, but the question remains open for some of them (see [**Mei97**]).

The analogue of Meigniez theorem when Aff_1 is replaced by a non-compact semi-simple Lie group, like $\mathrm{Sl}(2,\mathbb{R})$, is a complete mystery.

In the next paragraph we shall consider the classification problem for complete Riemannian groupoids. We just mention here an interesting non-trivial particular case where the classification problem and the realization problem are solved completely. Indeed V. Cavalier [**Ca94**] has given a complete classification up to equivalence for compactly generated groupoids of germs of partial holomorphic automorphisms of a complex 1-dimensional manifold T preserving a non-trivial meromorphic vector field. It is proved in [**CH97**] that each of them is equivalent to the holonomy groupoid of a transversally holomorphic foliation on some closed manifold.

7 Complete Riemannian Groupoids

This is an interesting class of groupoids connected with Riemannian foliations.

7.1 RIEMANNIAN FOLIATIONS. A *(complete) Riemannian foliation* $\mathcal{F}$ of codimension q on a (complete) Riemannian manifold M is a foliation $\mathcal{F}$ defined by a foliated 1-cocycle $f_i : U_i \to T_i$ like in 2.1, where T_i has a Riemannian metric such that f_i is a Riemannian submersion. Therefore there is on T a Riemannian metric invariant by the holonomy groupoid.

This notion was introduced by B. Reinhart [**Rei59**]. The main features of the structure of complete Riemannian foliations was eventually established by P. Molino (for a detailed account, see the book of Molino [**Mo88**]).

Any representative of the equivalence class of the fundamental (or holonomy groupoid) of a complete Riemannian foliation satisfies the following conditions.

7.2 DEFINITION. A *complete Riemannian groupoid* is an étale groupoid $(\mathcal{G}, T)$ such that T is endowed with an invariant Riemannian metric, and which satisfies the following conditions:

1. *Hausdorff condition*: As a topological space, $\mathcal{G}$ is Hausdorff and if $c : [0,1) \to \mathcal{G}$ is a continuous path such that $\lim_{\tau\to 1}(s \circ c)(\tau)$ and $\lim_{\tau\to 1}(t \circ c)(\tau)$ exist, then $\lim_{\tau\to 1} c(\tau)$ exists.
2. *Completeness condition*: The space of orbits $\mathcal{G}\backslash T$ is complete with respect to the pseudometric[3] defined as follows: the distance between the orbits of two points $x, y \in T$ is the infimum of the length of $\mathcal{G}$-paths joining x to y.

REMARK. Let $\mathcal{U} = \{U_i,\}$ be an open cover of T by geodesically convex balls. It follows from the definition above that, for the localization $\mathcal{G}_\mathcal{U}$ of $\mathcal{G}$ over $\mathcal{U}$, the topological spaces ${\mathcal{G}_\mathcal{U}}^{(k)}$ (see 3.6) are union of balls. For a locally constant $\mathcal{G}$-sheaf $\mathcal{A}$ of abelian groups, this implies that the cohomology groups $H^*(\mathcal{G}, \mathcal{A})$ are isomorphic to the cohomology groups of the complex $C^*(\mathcal{G}_\mathcal{U}, \mathcal{A}_\mathcal{U})$ (see 3.9). In particular the natural homomorphism $\check{H}^*(\mathcal{G}, \mathcal{A}) \to H^*(\mathcal{G}, \mathcal{A})$ is an isomorphism.

[3] A pseudometric on a space X is given by a distance function $d : X \times X \to \mathbb{R}^+ \cup \infty$ which is symmetric, satisfies the triangle inequality and $d(x,y) = 0$ if $x = y$.

7.3 Basic properties of complete Riemannian foliations.

Let $\mathcal{F}$ be a Riemannian foliation on a complete connected Riemannian manifold M. Then

(1) The leaves have a common holonomy covering L [**Rei59**] and the homotopic fiber of the map from M to the classifying space $B\mathcal{H}$ of the holonomy groupoid of $\mathcal{F}$ has the homotopy type of L [**Ha82**].

(2) The closures of the leaves of $\mathcal{F}$ are properly embedded submanifolds of M [**Mo88**]

(3) There is a locally constant sheaf of germs of local transversal Killing vector fields whose stalk is a Lie algebra $\mathfrak{g}$, called the structural Lie algebra of $\mathcal{F}$. The orbits of the integration of these local transversal fields are the closures of the leaves (see [**Mo88**], p.161).

7.4 Corresponding properties for complete Riemannian groupoids.

Let $(\mathcal{G}, T)$ be a connected complete Riemannian groupoid and let $(\mathcal{H}, T)$ be the associated groupoid of germs. Then

(1) The kernel of the canonical homomorphism $\mathcal{G} \to \mathcal{H}$ is a locally constant sheaf of groups.

(2) The closures of the $\mathcal{H}$-orbits are properly embedded submanifolds of T.

(3) Let $\overline{\mathcal{H}}$ be the étale groupoid which is the closure of $\mathcal{H}$ in the C^1-topology ([**Sa88**]). It is also a complete Riemannian groupoid. To $(\overline{\mathcal{H}}, T)$ is attached a locally constant $\mathcal{H}$-sheaf $\underline{\mathfrak{g}}$ of germs of Killing vector fields characterized by the property that the elements of $\overline{\mathcal{H}}$ close to the identity in the C^1-topology are those of the form $\exp \xi$, where ξ is a local section of $\underline{\mathfrak{g}}$ close to 0. The stalk of $\underline{\mathfrak{g}}$ over a point $x_0 \in T$ is a finite dimensional Lie algebra $\mathfrak{g}$, called the *structural Lie algebra* of $\mathcal{H}$. The $\mathcal{H}$-sheaf $\underline{\mathfrak{g}}$ is determined by a representation $\pi_1((\mathcal{H}, T), x_0) \to \mathrm{Aut}(\mathfrak{g})$; the closure of its image contains $\mathrm{Ad}(\mathfrak{g})$ (the image by the exponential of the adjoint representation of $\mathfrak{g}$).

In particular, if $(\mathcal{H}, T)$ is simply connected, then the structural Lie algebra $\mathfrak{g}$ is abelian. If $(\mathcal{G}, T)$ is the groupoid associated to an action through isometries of a group Γ on a connected complete Riemannian manifold T, then the structural Lie algebra is the Lie algebra of the closure of Γ in the Lie group of isometries of T.

7.5 Remarks on developability.

(1) The holonomy groupoid of a complete Riemannian foliation $\mathcal{F}$ on a connected manifold M is developable in the sense of 1.4 if and only if the leaves of the lifting of $\mathcal{F}$ to the universal covering $\tilde{M}$ of M are the fibers of a fibration of $\tilde{M}$ with base space a complete (Hausdorff) Riemannian manifold.

In general for a foliation on a manifold M which is not complete Riemannian, the developability of its holonomy groupoid implies that the leaves of the foliation lifted to a suitable covering $\tilde{M}$ of M are the fibers of a submersion from $\tilde{M}$ onto a manifold (which might be non-Hausdorff). In general this last property does not imply the developability of the holonomy groupoid of the foliation.

As an example consider the free abelian subgroup Γ of rank 2 of the group of diffeomorphisms of $\mathbb{R}$ generated by two elements g_1 and g_2, where g_1 (resp. g_2) is the identity on the half-line $\{t \in \mathbb{R} : t \leq 0\}$ (resp. $\{t \in \mathbb{R} : t \geq 0\}$) and a contraction on the complementary half-line. The group Γ acts freely on $\mathbb{R}^2 \times \mathbb{R}$, the generator g_1 (resp. g_2) mapping (x_1, x_2, t) to $(x_1 + 1, x_2, g_1(t))$ (resp. to $(x_1, x_2 + 1, g_2(t))$. This action preserves the foliation $\tilde{\mathcal{F}}$ by the lines $t =$ constant and gives on the quotient M of $\mathbb{R}^2 \times \mathbb{R}$ a foliation $\mathcal{F}$ whose holonomy groupoid is equivalent to the holonomy groupoid of a Reeb foliation on S^3, which is of course non-developable. (The groupoid of germs of the elements of Γ is not the groupoid associated to the action of Γ on $\mathbb{R}$.)

(2) *Developability and curvature.* A complete Riemannian groupoid $(\mathcal{G}, T)$ is always developable if the sectional curvature of the Riemannian metric on T is non-positive (this was first proved by Hebda [**He86**] for the holonomy groupoid of a foliation). This theorem is in fact true in a much more general case. Let T be a space with a length metric (i.e. a metric such that the distance between two points is the infimum of the length of curves joining them: the distance is infinite if there are no rectifiable curves joining them). Assume that the metric space T is of non-positive curvature in the sense of Alexandrov and is locally complete. Let $(\mathcal{G}, T)$ be an étale groupoid verifying the two conditions in 7.2. Then $(\mathcal{G}, T)$ is developable. (For a proof of this theorem and its applications to complexes of groups, see [**BrHa99**], Chapter III $\mathcal{G}$).

Some classification results.

7.6 Description of complete Riemannian groupoids of germs with a dense orbit. This is done up to differentiable equivalence in [**Ha88**] (even for tubular neighbourhoods of orbits). For such a groupoid $(\mathcal{H}, T)$, let $\mathfrak{g}$ be its structural Lie algebra. If $(\mathcal{H}, T)$ is developable (it is the case in particular if $\mathfrak{g}$ is solvable, or if dim $T \leq 4$), then it is equivalent to the groupoid associated to the action of a dense subgroup Γ of G acting by left translations on G/K, where G is a Lie group with Lie algebra $\mathfrak{g}$ and K a compact subgroup of G. In the non-developable case (the lowest dimensional case arise for dim $T = 5$ and $\mathfrak{g}$ the Lie algebra of $SU_2 \times SU_2$), then $(\mathcal{G}, T)$ is Morita equivalent in the sense of Jean Renault (see [**Ren82**] and [**Ha82**]) to a non-étale groupoid associated to the action of a non-closed subgroup of $G \times K$ acting on G (see [**Ha88**], p.25).

7.7 The simply connected case. It is shown in [**HaSa92**] that the set of differentiable equivalence classes of simply connected complete Riemannian groupoids of germs whose restriction to a generic orbit is equivalent to the action of a dense discrete subgroup Λ of rank N of $\mathbb{R}^k$ is in bijection with the set of equivalence classes of smooth effective actions of a torus T^N on simply connected orbifolds. Those groupoids are developable only if they are equivalent to a trivial groupoid which is a simply connected manifold. Nevertherless they are always Morita equivalent to a non-étale groupoid associated to an action of a Lie group on a manifold. Simply connected complete Riemannian groupoids which are compactly generated are always equivalent to the holonomy groupoid of a Riemannian foliation on a closed manifold (problem 6.3).

7.8 The case dim $T = 2$.

As a sample of classification of connected complete Riemannian groupoids of germs $(\mathcal{H}, T)$, we consider the case where dim $T = 2$. The structural Lie algebra $\mathfrak{g}$ can be

1. the trivial Lie algebra,
2. the Lie algebra $\mathfrak{r}$ of $\mathbb{R}$,
3. the Lie algebra $\mathfrak{r}^2$ of $\mathbb{R}^2$,
4. the Lie algebra $\mathfrak{aff}_1$ of the affine group of transformations of the line,
5. the Lie algebra of the group of isometries of $\mathbb{R}^2$,

6. the Lie algebra $\mathfrak{so}_3$ of the group of isometries of $\mathbb{S}^2$,
7. the Lie algebra of the group of isometries of the hyperbolic plane.

If $\mathfrak{g}$ is non trivial or non equal to $\mathfrak{r}$, then $(\mathcal{H}, T)$ is always developable, equivalent to the groupoid $\Gamma \ltimes S$ associated to the action of a subgroup Γ of the group of isometries of a complete Riemannian manifold S of dimension 2.

If $\mathfrak{g} = 0$, then it is the groupoid of germs of changes of charts of an orbifold of dimension 2; those orbifolds are developable except in a few exceptional cases.

If $\mathfrak{g} = \mathfrak{r}$, then the space of orbit closures $\overline{\mathcal{H}}\backslash T$ is a circle or a closed interval (compact or not). The groupoid is developable, except possibly when this interval is compact. An example of a non-developable one is the holonomy groupoid of a foliation of $\mathbb{S}^3$ whose leaves are the orbits of a diagonal non-closed subgroup of dimension 1 of U_2 acting on $\mathbb{S}^3$.

In general if $(\mathcal{G}, T)$ is a connected complete Riemannian groupoid and if $(\mathcal{H}, T)$ is the associated groupoid of germs, then the kernel of the natural projection is a locally constant sheaf of groups with stalk a group N (see the next paragraph for more details on such extensions). The natural projection $\mathcal{G} \to \mathcal{H}$ induces a map $B\mathcal{G} \to B\mathcal{H}$ which can be considered up to homotopy as a fibration with fiber BN. Consider the homotopy sequence of this fibration:

$$\cdots \to \pi_2(B\mathcal{H}) \to N \to \pi_1(B\mathcal{G}) \to \pi_1(B\mathcal{H}) \to 1.$$

Assume that $(\mathcal{H}, T)$ is developable. Then $(\mathcal{G}, T)$ is also developable if and only, in this homotopy sequence, the image of $\pi_2(B\mathcal{H})$ in N is zero (this image is always contained in the center C of N).

If $\dim T = 2$, we see that $\pi_2(B\mathcal{H})$ is trivial when $\mathfrak{g}$ is not equal to 0, $\mathfrak{r}$, $\mathfrak{so}_3$, so that $(\mathcal{G}, T)$ is always developable. In the other cases, $(\mathcal{G}, T)$ can be non-developable, for instance if $\mathfrak{g} = \mathfrak{so}_3$, due to the fact that $\pi_2(B\mathcal{H}) = \pi_2(S^2) = \mathbb{Z}$.

8 More on extensions

In this paragraph we consider two particular extension problems related to the classification of complete Riemanniaan groupoids.

We have seen in 7.4(1) that, for a complete Riemannian groupoid $(\mathcal{G}, T)$, the kernel of the natural projection to the associated groupoid of germs is a locally constant $\mathcal{G}$-sheaf of groups. Moreover, replacing if necessary $(\mathcal{G}, T)$ by its localization over a cover by simply connected sets, we can assume that the extension splits topologically, i.e. there is a continuous map $\sigma : \mathcal{H} \to \mathcal{G}$ whose composition with the projection $\mathcal{G} \to \mathcal{H}$ is the identity of $\mathcal{H}$. So it is natural to study such extensions along the lines of 5.2.

8.1 Topologically split extensions with non-abelian kernel.
Let $(\mathcal{G}, T)$ be an étale groupoid and let N be a group. A *topologically split extension* of $(\mathcal{G}, T)$ by N is given by an open cover $\mathcal{U}$ of T, an étale groupoid $(\widetilde{\mathcal{G}_\mathcal{U}}, T_\mathcal{U})$ and a homomorphism $\phi : (\widetilde{\mathcal{G}_\mathcal{U}}, T_\mathcal{U}) \to (\mathcal{G}_\mathcal{U}, T_\mathcal{U})$ projecting to the identity of $T_\mathcal{U}$ with kernel a $\widetilde{\mathcal{G}_\mathcal{U}}$-sheaf $T_\mathcal{U} \times N$ (i.e this sheaf is trivial with stalk N as a sheaf on the topological space $T_\mathcal{U}$). Moreover we assume that there is a continuous map $\sigma : \widetilde{\mathcal{G}_\mathcal{U}} \to \mathcal{G}_\mathcal{U}$ such that $\phi \circ \sigma$ is the identity of $\mathcal{G}_\mathcal{U}$.

Let C be the center of N and let Out (N) be the quotient of the group Aut (N) of automorphisms of N by the subgroup of inner automorphisms. As usual we have the exact sequence

$$0 \to C \to N \to \text{Aut }(N) \to \text{Out }(N) \to 1.$$

The kernel $T_{\mathcal{U}} \times N$ of ϕ is a $\widetilde{\mathcal{G}_{\mathcal{U}}}$-sheaf of groups: the action of $\tilde{g} \in \widetilde{\mathcal{G}_{\mathcal{U}}}$ on $(s(\tilde{g}), n) \in T_{\mathcal{U}} \times N$ is of the form $((t(\tilde{g}), \nu(\tilde{g}).n)$, where ν is a (continuous) homomorphism from $\widetilde{\mathcal{G}_{\mathcal{U}}}$ to Aut (N). Passing to the quotient we get a homomorphism $\overline{\nu} : \widetilde{\mathcal{G}_{\mathcal{U}}} \to \text{Out }(N)$ which determines as above a $\widetilde{\mathcal{G}_{\mathcal{U}}}$-sheaf structure on $T_{\mathcal{U}} \times \text{Out }(N)$. Let $\underline{\text{Out }}(N)$ be the corresponding locally constant $\mathcal{G}$-sheaf of groups on T with stalk Out (N) and let $\underline{C}$ be the corresponding locally constant $\mathcal{H}$-sheaf of abelian groups with stalk C.

Two such extensions are equivalent if there is a diagram like in 5.2 (where all the extensions split topologically). The isomorphism class of the $\mathcal{G}$-sheaves $\underline{\text{Out }}(N)$ and $\underline{C}$ depends only of the equivalence class of the extension. It follows from this definition that the set of equivalence classes of such extensions depends only on the equivalence class of $(\mathcal{G}, T)$. It is expressed in terms of the Čech cohomology of $\mathcal{G}$. In the case of a complete Riemannian groupoid, those groups are isomorphic to the usual cohomology groups (see 7.2).

The following classification theorem is the generalisation of a theorem of Eilenberg-Mac Lane. Its proof is essentially the same.

8.2 THEOREM. *Let $\underline{Out}(N)$ be a locally constant $\mathcal{G}$-sheaf of groups with stalk $Out(N)$ and let $\underline{C}$ be the corresponding $\mathcal{G}$ -sheaf of abelian groups with stalk the center C of N. There is an extension of $\mathcal{G}$ by N with associated $\mathcal{G}$-sheaf $\underline{Out\ }(N)$ if and only if an obstruction in $\check{H}^3(\mathcal{G}, \underline{C})$ vanishes. If this is the case, the set of equivalence classes of extensions with associated sheaf $\underline{Out}(N)$ is in bijection with the set $\check{H}^2(\mathcal{G}, \underline{C})$.*

8.3 COROLLARY. *If the center of N is trivial, given a locally constant $\mathcal{G}$-sheaf $\underline{Out}(N)$ with stalk Out (N), then there always exists a topologically split extension of $\mathcal{G}$ by N with associated sheaf $\underline{Out}(N)$, and this extension is unique up to equivalence.*

This extension is constructed as follows: choose an open cover $\mathcal{U}$ of T such that the sheaf on $T_{\mathcal{U}}$ corresponding to $\underline{\text{Out }}(N)$ is trivial, isomorphic to $T_{\mathcal{U}} \times \text{Out }(N)$. The action of $\mathcal{G}_{\mathcal{U}}$ is given by a homomorphism $\overline{\nu} : \mathcal{G}_{\mathcal{U}} \to \text{Out }(N)$. Then $\widetilde{\mathcal{G}_{\mathcal{U}}}$ is the fiber product Aut $(N) \times_{\text{Out}(N)} \mathcal{G}_{\mathcal{U}}$ formed by pairs $(\alpha, g) \in \text{Aut }(N) \times \mathcal{G}_{\mathcal{U}}$ such that the image $\overline{\alpha}$ of α in Out (N) is equal to $\overline{\nu}(g)$. The composition $(\alpha_1, g_1)(\alpha_2, g_2)$ is defined if $g_1 g_2$ is defined and is equal to $(\alpha_1\alpha_2, \alpha_1(g_1)g_2)$. The projection $\phi : \widetilde{\mathcal{G}_{\mathcal{U}}} \to \mathcal{G}_{\mathcal{U}}$ maps (α, g) to g. To construct a topological splitting $\sigma : \mathcal{G}_{\mathcal{U}} \to \widetilde{\mathcal{G}_{\mathcal{U}}}$ we choose a lifting $\lambda : \text{Out }(N) \to \text{Aut }(N)$ of the natural projection and define $\sigma(g) = (\lambda\overline{\nu}(g), g)$.

8.4 Principal (G, Γ)-groupoids.

These groupoids arise naturally as holonomy groupoids of complete Riemannian foliations which are locally transversally parallelisable. In that case the restrictions of the foliation to each leaf closure are isomorphic and the space of leaf closures is a manifold W. The holonomy groupoid of the foliation restricted to the closure of a leaf is equivalent to the groupoid associated to the action by right translations of a dense subgroup Γ of a Lie group G.

8.5 DEFINITION. Let G be a simply connected Lie group, Γ be a dense subgroup of G and let $\mathcal{N} = G \rtimes \Gamma$ be the groupoid associated to the action of Γ on G by right translations. Let W be a manifold considered as a trivial groupoid.

A *locally trivial extension* of W by $\mathcal{N}$ (or a *principal* (G,Γ)*-groupoid over* W [**Ha85**], [**Si92**]) is given by an open cover $\mathcal{U} = \{U_i\}$ of W and an étale groupoid $(\mathcal{H}, T)$ such that

(a) T is the disjoint union of the $G \times U_i$ and the natural projection $T \to W$ induces an isomorphism from the space of orbit closures to W,

(b) the restriction of $\mathcal{H}$ to $G \times U_i$ is the groupoid associated to the action of Γ acting by right translations on the first factor and by the identity on the second one.

Let $(\mathcal{W}_{\mathcal{U}}, W_{\mathcal{U}})$ be the localization of the groupoid W over the cover $\mathcal{U}$ (we use the notations of 1.2(c)). There is a natural projection homomorphism $\mathcal{H} \to \mathcal{W}_{\mathcal{U}}$ whose kernel is the groupoid $\mathcal{N} \times T$. The notion of equivalence among such extensions is defined along the lines of 5.2 and 8.1.

Like for usual principal bundles, a principal (G,Γ)-bundle can be described by a 1-cocycle over an open cover $\mathcal{U}$ of W with values in a sheaf of groups $\mathcal{E}$. The group $\mathcal{E}(U)$ of sections of this sheaf over an open set U of W is the group of auto-equivalences of $(\mathcal{N} \times U, G \times U)$ projecting to the identity of U. When U is simply connected $\mathcal{E}(U)$ can be described as follows. Let Aut (G,Γ) be the group of automorphisms of G whose restriction to Γ is an automorphism of Γ. Let $G(U)$ be the group of continuous maps from U to G. Then one has the exact sequence:

$$1 \to \Gamma \to G(U) \rtimes \text{Aut } (G,\Gamma) \to \mathcal{E}(U) \to 1,$$

where $\gamma \in \Gamma$ is mapped to the pair $(\gamma^{-1}, \text{ad}\gamma)$ (here γ^{-1} stands for the constant map from U to γ^{-1}).

Note that, when $G = \mathbb{R}$ for instance, the group Aut $(\mathbb{R},\Gamma)$ depends on the arithmetic properties of Γ.

A remarkable fact proved in the thesis of Ana Maria da Silva [**Si92**] is that equivalence classes of étale groupoids which are locally principal (G,Γ)-bundles with base W are in bijection with homotopy classes of maps from W to a classifying space $B_{G,\Gamma}$ which is the base space of a universal (G,Γ)-principal bundle. The homotopy groups $\pi_i(B_{G,\Gamma})$ are isomorphic to $\pi_{i-1}(G)$ for $i > 2$, to the center Γ_0 of Γ for $i = 2$ and, for $i = 1$, to the group Out (G,Γ), quotient of Aut (G,Γ) by the group of inner automorphisms by the elements of Γ.

One has a homomorphism $\pi_2(W) \to \Gamma$, which is part of a long homotopy exact sequence and whose image is contained in the center of Γ. The groupoid $\mathcal{H}$ is developable if and only if this homomorphism is trivial.

References

[Bo68] R. Bott, *On a topological obstruction to integrability*, Global Analysis, Berkeley 1968, Proc. Sympos. Pure Math., vol. 16, 1970, pp. 127–131.

[BoHa90] C. Bonatti and A. Haefliger, *Déformations de feuilletages*, Topology **29** (1990), 205-229.

[BrHa99] M. Bridson and A. Haefliger, *Metric spaces of non-positive curvature*, Grundlehren der mathematischen Wissenschaften, vol. 319, 1999.

[Ca94] V. Cavalier, *Pseudogroupes complexes quasi parallélisés de dimension un*, Ann. Inst. Fourier (Grenoble) **44** (1994), 1539–1565.

[CH97] V. Cavalier et A. Haefliger, *Réalisations feuilletées de quelques pseudogroupes*, Integrable systems and foliations, Progress in mathematics, vol. 145, Bikhäuser, 1997, pp. 33–48.

[GF70] I.M. Gelfand and D.B. Fuchs, *Cohomologies of the Lie algebra of formal vector fields*, Izv. Acad. Nauk SSSR Ser.Mat **34** (1970), 322–337.

[Gh89] E. Ghys, *L'invariant de Godbillon-Vey*, Séminaire Bourbaki 1988-89, Astérisque, vol.177- 178, 1989, pp. 155–181.

[Go64] R. Godement, *Théorie des faisceaux*, Hermann, Paris, 1964.

[Gr57] A. Grothendiek, *Sur quelques points d'algèbre homologique*, Tohoku Math. J. **9** (1957), 119–221.

[GV71] C. Godbillon and J. Vey, *Un invariant des feuilletages de codimension* 1, C. R. Acad. Sc. Paris **273** (1971), 92–95.

[Ha58] A. Haefliger, *Structures feuilletées et cohomologie à valeur dans un faisceau de groupoides*, Comm. Math. Helv. **32** (1958), 248-329.

[Ha70] ———, *Homotopy and integrability*, Manifolds, Amsterdam, 1970, Springer Lecture Notes in Math. 197, 1971, pp. 133-163.

[Ha76] ———, *Differential cohomology*, Differential topology (Varenna 76), Liguori, Naples, 1979, pp. 19-70.

[Ha80] ———, *Some remarks on foliations with minimal leaves*, J. Differential Geometry **15** (198), 269-284.

[Ha82] ———, *Groupoïdes d'holonomie et classifiants*, Structure transverse des feuilletages, Toulouse 1982, Astérisque, vol. 116, 1984, pp. 70-97.

[Ha85] ———, *Pseudogroups of local isometries*, Differential geometry, Santiago di Compostela, 1984, Pitman Res. Notes in Math., vol. 131, 1985, pp. 174-197.

[Ha88] ———, *Leaf closures in Riemannian foliations*, A fête of topology, papers dedicated to I. Tamura, Academic Press, 1988, pp. 3–32.

[Hae92] ———, *Cohomology theory for étales topological groupoids,1992 (unpublished notes).*

[HaSa90] ——— and E. Salem, *Riemannian foliations on simply connected manifolds and actions of tori on orbifolds*, Illinois J. of Math. **34** (1990), 706–730.

[He6] J. Hebda, *Curvature and focal points in Riemannian foliations*, Indiana University Math. J. **35** (1986), 321–331.

[Ku88] A. Kumjian, *On equivariant sheaf cohomology and elementary C^*-bundles*, J. of Operator Algebra **20** (1988), 207–240.

[Ma74] J.N. Mather, *Foliations and local homology of groups of diffeomorphisms*, Proc. ICM Vancouver, 1974, vol. 2, pp. 35–37.

[Mei97] G. Meigniez, *Holonomy groups of solvable foliations*, Integrable systems and foliations, Progress in mathematics, vol. 145, Birkhäuser, 1997, pp. 107–146.

[Mo88] P. Molino, *Riemannian foliations*, Progress in Mathematics, vol. 73, Birkhäuser, 1988.

[Moe91] I. Moerdijk, *Classifying topos and foliations*, Ann. Inst. Fourier, Grenoble **41** (1991), 189–209.

[Moe98] ———, *Proof of a conjecture of A. Haefliger*, Topology **37** (1998), 735–741.

[Re52] G. Reeb, *Sur certaines propriétés topologiques des variétés feuilletées*, Hermann, Paris, 1952.

[Rei59] B. L. Reinhart, *Foliated manifolds with bundle like metric*, Ann. of Math **69** (1959), 119-132.

[Ren82] Jean Renault, *C^*-algebras of groupoids and foliations*, Proc. Sympos. Pure Math. **38** (1982), Amer. Math. Soc., Providence R. I., 339–350.

[Sa88] E. Salem, *Riemannian foliations and pseudogroups*, in P. Molino, Riemannian foliations, Progress in Mathematics, vol. 73, 1988, pp. 265–296.

[Seg68] G. Segal, *Classifying spaces and spectral sequences*, Publ. Math. IHES (1968), 105–112.

[Si92] Ana Maria da Silva, *Classification des pseudogroupes (G,Γ)-principaux*, C. R. Acad. Sc. Paris, Serie I. Math. **314** (1992), 463–466.

[Sac65] R. Sacksteder, *Foliations and pseudogroups*, Amer. J. Math. **87** (1965), 79-102.

[Th74] W.P. Thurston, *On the construction and classification of foliations*, Proc. ICM, Vancouver,1974, vol. 1, pp. 547–549.

SECTION DE MATHÉMATIQUES, 2–4 RUE DU LIÈVRE, GENÈVE
E-mail address: andre.haefliger@math.unige.ch

Contemporary Mathematics
Volume **282**, 2001

Etale Groupoids, Derived Categories, and Operations

Ieke Moerdijk

Etale groupoids form an important class of topological groupoids, which includes (at least up to equivalence) the groupoids arising from foliations and from orbifolds, and groupoids associated to actions of discrete groups or infinitesimally free actions of Lie groups. Many constructions for spaces or manifolds extend to étale groupoids without much difficulty, while their extension to arbitrary groupoids becomes complicated or impossible.

The purpose of this paper is to describe how sheaf theory extends to étale groupoids. More specifically, we discuss the construction of the "derived category" of an étale groupoid, and show how the six operations of Grothendieck (namely, tensor, hom, f^*, f_*, $f_!$ and $f^!$) extend from spaces to étale groupoids. Thus, our purpose is similar to that of Bernstein and Lunts [**BL**], who discuss the derived category and the operations for groupoids associated to actions of compact Lie groups. However, it will become apparent that the case of étale groupoids is somewhat different in nature.

This paper makes no claim to originality and is purely expository, its only virtue perhaps being the presentation from the unifying viewpoint of derived categories. After an introductory section which provides the basic definitions and notation concerning étale groupoids, I describe the derived category of sheaves on a given étale groupoid. In section 3 the operations f^* and f_*, and the related cohomology of étale groupoids are discussed. These are well known, and in a sense simply a special case of that for sheaves on an arbitrary site [**SGA IV**]. The case of groupoids has been discussed in different degrees of generality and from different points of view in many sources, including [**H2, H3, K, M2, M4**]. There is an explicit discussion of Morita equivalence of groupoids (in section 4), and invariance of the operations under this type of equivalence. This invariance is of crucial importance, because in many cases the étale groupoid associated to a given geometric object is defined only up to Morita equivalence: this is the case, for example, for the holonomy groupoid of a foliation. In section 5 I discuss the operation $f_!$, which is related to compactly supported cohomology and homology of étale groupoids. This homology was first discussed in [**CM1**] (but without any reference to the derived category), and is closely related to the cyclic homology of the convolution algebra of an étale

Key words and phrases. etale groupoid, foliation, classifying space, derived category, cohomology.

2000 Mathematics Subject Classification. 58H05, 57R32.

Research partially supported by the Dutch Science Foundation (NWO).

groupoid [**BN, C1**]. The construction of $f_!$ — and of its right adjoint $f^!$ — is somewhat involved, partly because of complications arising from the fact that we cannot assume all étale groupoids to be Hausdorff. In fact, the main examples arising in the theory of foliations aren't Hausdorff. To overcome this difficulty, I have relied on [**CM2**]. In section 6 I give a combinatorial way of describing the derived category of an étale groupoid, which makes use of a small (discrete) category associated to the étale groupoid, called the embedding category. This category has the same cohomology and homology as the étale groupoid itself. Section 7 discusses the relation between the derived category of an étale groupoid and its classifying space.

Acknowledgements: I would like to thank the organisers of the conference at Boulder for inviting me, and for giving me the opportunity to contribute this paper to the proceedings. It is clear from the text (section 5) that I have made considerable use of joint work with Marius Crainic, all still to appear. I am indebted to many people for helpful discussions over the years about étale groupoids and related subjects. Besides Crainic, I would in particular like to mention W.T. van Est, A. Haefliger, and J. Mrcun. This research is partially supported by the Dutch Science Foundation (NWO).

1. Etale groupoids.

1.1. Groupoids (notation). A groupoid is a category in which every arrow is invertible [**CWM**]. Thus, a groupoid G is given by a set of objects G_0 and a set of arrows G_1, together with structure maps satisfying the usual equations,

$$G_1 \times_{G_0} G_1 \xrightarrow{m} G_1 \underset{d_1}{\overset{d_0}{\rightrightarrows}} G_0 \xrightarrow{u} G_1 \xrightarrow{i} G_1. \tag{1.1}$$

Here d_0 and d_1 are the source (domain) and target (codomain), u is the unit, i is the inverse, and m the multiplication defined on the fibered product $\{(h,g)|d_0(h) = d_1(g)\}$. We will write $g : x \to y$ to indicate that $g \in G_1$ is an arrow with $d_0(g) = x$ and $d_1(g) = y$. Furthermore, we write hg for $m(h,g)$, 1_x or id_x for $u(x)$ and g^{-1} for $i(g)$, all as usual.

1.2. Etale groupoids. A topological groupoid is a groupoid G, as above, in which G_0 and G_1 are each equipped with a topology for which all the structure maps in (1.1) are continuous. Such a groupoid is *étale* if $d_0 : G_1 \to G_0$ is a local homeomorphism (this implies that all other structure maps in (1.1) are local homeomorphisms as well). In many examples of étale groupoids, G_0 and G_1 are smooth manifolds and d_0 is a local diffeomorphism. One refers to G as a *smooth* étale groupoid in this case.

1.3. Examples.

(a) Every topological space X can be viewed as an étale groupoid, with $X = G_0 = G_1$ and all structure maps in (1.1) identities.

(b) If Γ is a groupoid acting continuously on a space X, the translation groupoid $\Gamma \ltimes X$ is the groupoid with X as space of objects and $\Gamma \times X$ as space of arrows, an arrow $x \to y$ being a pair (γ, x) with $\gamma \cdot x = y$. This groupoid is étale if Γ is a discrete group.

(c) Many examples of étale groupoids arise in the context of foliations. Haefliger's groupoid Γ^q which classifies codimension q foliations [**H4**] is an étale groupoid. If $(M, \mathcal{F})$ is a foliated manifold, its holonomy and monodromy groupoids aren't étale, but they become so if the space of objects is reduced to a complete transversal T. Different choices of transversals give Morita equivalent (§4) étale groupoids. Thus, to each foliation one can associate two étale groupoids, each well-defined up to Morita equivalence [**H1, P**].

(d) Orbifolds give rise to étale groupoids [**H1**]. In fact, in the C^∞-context orbifolds are essentially equivalent to étale groupoids which are proper, in the sense that $(d_0, d_1) : G_1 \to G_0 \times G_0$ is a proper map (see [**MP**]).

1.4. Homomorphisms. A (continuous) homomorphism $f : H \to G$ between étale groupoids consists of two continuous maps, $f_0 : H_0 \to G_0$ and $f_1 : H_1 \to G_1$, commuting with all the structure maps in (1.1) (e.g. $f_1(hk) = f_1(h)f_1(k)$, $d_0 f_1(h) = f_0 d_0(h)$, etc.). Thus, f is just a *continuous functor*. In a similar fashion, one can define continuous natural transformations τ, from one homomorphism $H \to G$ to another, as suitable maps $\tau : H_0 \to G_1$.

2. Sheaves.

Is this section we introduce the category of G-sheaves for an étale groupoid G. We assume familiarity with the theory of sheaves on topological spaces, as exposed in, e.g., [**B, Go, I**]. Throughout this paper, the sheaves will be sheaves over a fixed field k (in practice $\mathbb{R}$ or $\mathbb{C}$), and we will not explicitly refer to k in our notation. We remark, however, that in this section and the next it would have been sufficient to assume that k is a commutative ring, while in §4 this ring would have to be Noetherian and of finite cohomological dimension, exactly as for spaces [**B**].

2.1. The category $\mathrm{Sh}(G)$**.** A G-sheaf (of k-modules) is a sheaf A on the space G_0, equipped with a continuous right G-action. This action gives for each arrow $g : x \to y$ a (k-module) map

$$A_y \to A_x, \quad a \mapsto a \cdot g$$

between the stalks of A, and the usual conditions for an action ($(a \cdot g) \cdot h = a \cdot (gh)$ and $a \cdot 1 = a$) should be satisfied. Continuity of the action can most easily be expressed in terms of the map $A \times_{G_0} G_1 \to A$, $(a, y) \mapsto a \cdot g$, defined on the fibred product of $d_1 : G_1 \to G_0$ and the étale space $A \to G_0$ of the sheaf. A map between G-sheaves is a map $\varphi : A \to B$ of sheaves (of k-modules) on the space G_0 (in the usual sense), which in addition respects the action,

$$\varphi(a \cdot g) = \varphi(a) \cdot g.$$

In this way one obtains a well-defined category $\mathrm{Sh}(G)$.

This category is the category of k-modules in the topos of G-sheaves of sets. In particular, it follows from the general properties of such categories that $\mathrm{Sh}(G)$ is an abelian category with enough injectives.

There is an obvious functor, "forget the action", which we denote

$$F : \mathrm{Sh}(G) \to \mathrm{Sh}(G_0), \tag{2.1}$$

from $\mathrm{Sh}(G)$ to the category of sheaves on the space G_0 of objects of the groupoid G. This functor is exact, and preserves injectives (the latter because G is étale).

2.2. Remark. For an étale groupoid G, there are always lots of G-sheaves. In fact, the sheaves on G_0 naturally constructed using the geometry of G_0 usually carry a G-action. For example, if G_0 is a smooth manifold, the sheaves Ω^q of differential q-forms on G_0 are G-sheaves. Many more examples of this kind can be found in the references at the end of this paper. (Notice, however, that this situation is completely different if G is not assumed étale. For instance, if G is a connected topological group, so that G_0 is a point, the functor F in (2.1) above is an equivalence of categories.)

For an étale groupoid G and an open set U in G_0, one can define a sheaf $k[U]$ whose stalk at a point x is the free k-module on the set of arrows in G from x into some point of U. This sheaf is free on U, in the sense that morphisms of G-sheaves from $k[U]$ to a given G-sheaf A correspond to sections of A over U.

2.3. Complexes. We write $\mathrm{Ch}(G)$, $\mathrm{Ch}^+(G)$, $\mathrm{Ch}^-(G)$ for the categories of cochain complexes (respectively bounded below and above such complexes) of G-sheaves of k-modules, and we use the obvious notions of G-equivariant chain maps and chain homotopies. Such a map $\varphi : A^\bullet \to B^\bullet$ is said to be a quasi-isomorphism (q.i.) if it induces isomorphisms between the cohomology G-sheaves. Since the functor F in (2.1) is exact, a map $A^\bullet \to B^\bullet$ is a q.i. between G-sheaves iff $F(A^\bullet) \to F(B^\bullet)$ is a q.i. of sheaves on the space G_0 in the usual sense.

2.4. Derived categories. The derived categories

$$\mathcal{D}(G),\ \mathcal{D}^+(G),\ \mathcal{D}^-(G)$$

of G-sheaves of k-modules are obtained, in the usual way, from the categories $\mathrm{Ch}(G)$, $\mathrm{Ch}^+(G)$ and $\mathrm{Ch}^-(G)$, by inverting the q.i.'s. These categories can be described in a more concrete way as complexes of injective sheaves, and chain homotopy classes of maps, as usual.

2.5. Tensor and Hom. For two G-sheaves A and B, the tensor product sheaf $A \otimes B$ on G_0 carries a natural G-action, and defines the tensor product in the category $\mathrm{Sh}(G)$. This tensor product extends to cochain complexes with the usual grading convention and, since we assume that k is a field, it induces a well-defined tensor product on derived categories.

Next, for G-sheaves A and B, the usual sheaf $\underline{\mathrm{Hom}}(A, B)$ on G_0 (more precisely, this is the sheaf $\underline{\mathrm{Hom}}(FA, FB)$) carries a natural G-action; to describe it explicitly, one uses again that G is étale. Thus $\underline{\mathrm{Hom}}(A, B)$ is an object of $\mathrm{Sh}(G)$. When $A^\bullet$ and $B^\bullet$ are cochain complexes of G-sheaves, one obtains a cochain complex, graded as usual, $\underline{\mathrm{Hom}}(A, B)^n = \prod_{p-q=n} \underline{\mathrm{Hom}}(A^q, B^q)$.

$\underline{\mathrm{Hom}}$ and tensor are related by the usual adjunction formula, which is an isomorphism between G-sheaves,

$$\underline{\mathrm{Hom}}(A \otimes B, C) = \underline{\mathrm{Hom}}(A, \underline{\mathrm{Hom}}(B, C)). \tag{2.2}$$

Applying $\underline{\mathrm{Hom}}$ to injective resolutions gives a functor

$$\mathrm{R}\underline{\mathrm{Hom}} : \mathcal{D}^-(G) \times \mathcal{D}^+(G) \to \mathcal{D}^+(G),$$

and the formula (2.2) holds when $=$ is interpreted as "canonical q.i.".

3. The operations f^* and f_*, and cohomology.

Let $f : H \to G$ be a homomorphism between étale groupoids (q.i.), fixed throughout this section.

3.1. The operation f^*. For a G-sheaf A, the pullback of A along $f_0 : H_0 \to G_0$ yields a sheaf $f^*(A)$ on H_0 with stalk $f^*(A)_y = A_{f(y)}$. This sheaf carries a natural action by the groupoid H. For an arrow $h : z \to y$ in H and $a \in f^*(A)_y$, this action is simply given in terms of the G-action on A by,

$$a \cdot h = a \cdot f(h).$$

In this way, we obtain a functor f^* making the following diagram commute (up to canonical isomorphism).

$$\begin{array}{ccc} \mathrm{Sh}(G) & \xrightarrow{f^*} & \mathrm{Sh}(H) \\ {\scriptstyle F}\downarrow & & \downarrow{\scriptstyle F} \\ \mathrm{Sh}(G_0) & \xrightarrow{f_0^*} & \mathrm{Sh}(H_0) \end{array}$$

Here F is the forgetful functor, as before, and f_0^* denotes the usual pullback operation for sheaves on spaces. The functor f^* is exact (since f_0^* is), hence preserves q.i. between cochain complexes. So it induces an evident functor

$$f^* : \mathcal{D}^{(\pm)}(G) \to \mathcal{D}^{(\pm)}(H)$$

between derived categories.

3.2. The operation Rf_*. The functor f^* above has a right adjoint $f_* : \mathrm{Sh}(H) \to \mathrm{Sh}(G)$, uniquely determined by the isomorphism

$$\mathrm{Hom}_H(f^*(A), B) = \mathrm{Hom}_G(A, f_*(B))$$

for a G-sheaf A and an H-sheaf B. A more explicit description of $f_*(B)$ can be obtained from this isomorphism by taking for A a sheaf of the form $k[U]$ where $U \subseteq G_0$ is an open set (cf. 2.2), yielding

$$\Gamma(U, f_*(B)) = \mathrm{Hom}_G(k[U], B),$$

where $\Gamma(U, -)$ denotes the k-module of sections over U. Unlike f^*, this functor f_* does not commute with the forgetful functor (i.e. $F \circ f_* \neq (f_0)_* \circ F$). However, f_* does preserve quasi-isomorphisms between injectives, hence induces a well-defined functor between derived categories

$$Rf_* : \mathcal{D}(H) \to \mathcal{D}(G),$$

adjoint to $f^* : \mathcal{D}(G) \to \mathcal{D}(H)$. The same holds for $\mathcal{D}^+$ (but not for $\mathcal{D}^-$ since $Rf_* : \mathcal{D}(H) \to \mathcal{D}(G)$ doesn't map $\mathcal{D}^-$ into $\mathcal{D}^-$, as is clear from 3.3 below).

3.3. Invariant sections and cohomology. Let A be a G-sheaf. A section $s \in \Gamma(G_0, A)$ is called invariant if, for any $g : x \to y$ in G,

$$s(y) \cdot g = s(x).$$

Write $\Gamma_{\mathrm{inv}}(G, A)$ for the k-module of invariant sections. The functor $\Gamma_{\mathrm{inv}}(G, -)$ from G-sheaves to k-modules is a functor of the form f_*, where $f : G \to \mathrm{pt}$ is the unique homomorphism to a point. The cohomology of G with coefficients in A is defined in terms of Γ_{inv}, as

$$H^i(G,A) = R(\Gamma_{\text{inv}}(G,-)^i)(A) = H^i(\Gamma_{\text{inv}}(G,I^\bullet))$$

where $I^\bullet$ is any injective resolution of A.

In the same way, one defines the hypercohomology $\mathbb{H}^i(G,A)$ for a cochain complex A.

This cohomology is the ordinary cohomology of the topos of G-sheaves, and has all the standard properties. If G is a space X (i.e. $G_0 = X = G_1$, see 1.3.(a)), it yields the usual sheaf cohomology of X, if G is a group it gives group cohomology and if G is an action groupoid it defines the standard equivariant sheaf cohomology [**Gr**].

4. Morita invariance.

In this section we explain that the category of G-sheaves only depends on the groupoid G up to Morita equivalence.

4.1. Essential equivalences. A homomorphism $f : H \to G$ between étale groupoids is called an *essential equivalence* if the following two conditions are satisfied.

(a) f is "fully faithful", in the sense that

$$\begin{array}{ccc} H_1 & \xrightarrow{f_1} & G_1 \\ {\scriptstyle (d_0,d_1)}\downarrow & & \downarrow{\scriptstyle (d_0,d_1)} \\ H_0 \times H_0 & \xrightarrow{f_0 \times f_0} & G_0 \times G_0 \end{array}$$

is a fibered product.

(b) the map $d_1\pi_1 : G_1 \times_{G_0} H_0 \to G_0$, sending a pair $(g : x \to x', y)$ with $f_0(y) = x$ to x', is an open surjection.

4.2. Morita equivalence. Two étale groupoids G and G' are said to be Morita equivalent if there are two essential equivalences

$$G \leftarrow H \rightarrow G'$$

from a third groupoid H. This defines an equivalence relation.

The main property of Morita equivalence in the present context is the following (see [**M1**]).

4.3. Theorem. *If $f : H \to G$ is an essential equivalence then $f^* : \mathrm{Sh}(G) \to \mathrm{Sh}(H)$ is an equivalence of categories. Hence, if G and G' are Morita equivalent étale groupoids, then $\mathrm{Sh}(G)$ and $\mathrm{Sh}(G')$ are equivalent categories.*

4.4. Remark. The converse is also true, at least for sheaves of sets: if G and G' define equivalent toposes of sheaves of sets, then G and G' are Morita equivalent (provided the spaces G_0 and G'_0 are 'sober' – this is a mild separation condition implied by Hausdorffness).

4.5. Corollary. *If $f : H \to G$ is an essential equivalence then $f^* : \mathcal{D}^{(\pm)}(G) \to \mathcal{D}^{(\pm)}(H)$ is an equivalence of categories. Hence, Morita equivalent étale groupoids have equivalent derived categories.*

4.6. Remark. It is possible to consider more general morphisms $f : H \to G$ between étale groupoids, given by equivalence classes of diagrams of homomorphism of the form,

$$H \xleftarrow{\sim} K \longrightarrow G, \tag{4.1}$$

where $K \xrightarrow{\sim} H$ is an essential equivalence. Two such diagrams $H \xleftarrow{\sim} K \longrightarrow G$ and $H \xleftarrow{\sim} L \longrightarrow G$ represent the same morphism iff there is a diagram of homomorphisms

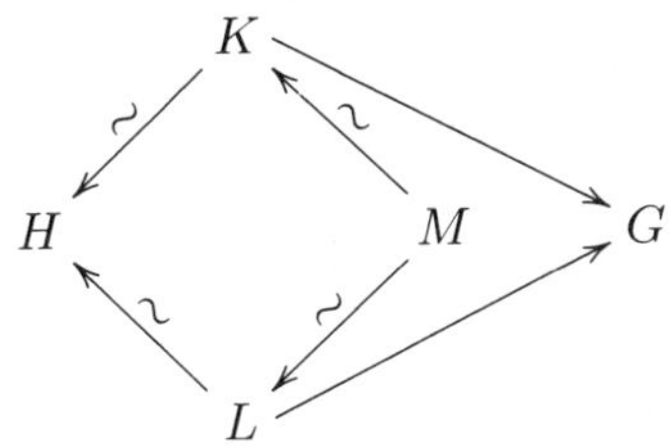

commuting up to natural transformations, where arrows marked "$\xrightarrow{\sim}$" are essential equivalences.

Two groupoids are Morita equivalent iff they become isomorphic in this category of general morphisms. This category can be described in various equivalent ways, e.g. in terms of principal bundles or in terms of topos morphisms, and plays a central role in the subject. See e.g. [**HS, MRW, Pra, Pro, Mr**].

A construction $\mathcal{C}$ on groupoids is functorial on this general category iff it is functorial on ordinary homomorphisms and has the following two properties: first, if $\mathcal{C}(f) = \mathcal{C}(g)$ if there is a natural transformation between f and g; and second, it turns essential equivalences into isomorphisms. (In fact, the first requirement is a consequence of the second.)

If $\mathcal{C}$ takes categories as values, then it is functorial on general morphisms as soon as it is functorial on ordinary homomorphisms up to (coherent) natural isomorphism, and it turns essential equivalences into equivalences of categories (see [**Pro**]). For example, by this universal property, one can construct p^* and Rp_* for a generalised morphism $p : H \to G$. If p is represented by an essential equivalence $e : K \xrightarrow{\sim} H$ and a homomorphism $f : K \to G$, then $e^* : \mathcal{D}(H) \to \mathcal{D}(K)$ is an equivalence with inverse $e_* = Re_*$, and one defines $p^* = f^* e_*$ and $Rp_* = Rf_* \circ e^*$.

5. The operations $f_!$ and $f^!$.

The operations discussed so far — $\otimes$, Hom, f^* and Rf_* — are standard, and exist for any topos map, not just ones of the form $\mathrm{Sh}(H) \to \mathrm{Sh}(G)$ induced by a groupoid homomorphism. The construction of the operations $f_!$ and $f^!$ is more special, and we need some specific assumptions.

5.1. Overall assumptions. From now on we will assume that all topological spaces X have a basis of locally compact Hausdorff open sets $U \subseteq X$ of cohomological dimension bounded by a number d depending on X but not on U. In particular, if G is an étale groupoid, we will assume that G_0 and (hence) G_1 have this property. As the examples of holonomy groupoids of foliations show, it is not possible to assume that G_1 is Hausdorff (but one can always assume that G_0 is).

5.2. c-Soft sheaves. For a space X satisfying the conditions above, we call a sheaf A on X c-soft if for any Hausdorff open $U \subseteq X$, the restriction $A|U$ is a c-soft sheaf on U (in the usual sense, [**B**]). Any sheaf has a finite c-soft resolution. The property of being c-soft is a local property. In particular, the pullback of a c-soft sheaf along an étale map is again c-soft. For an étale groupoid G, a G-sheaf is called c-soft if it is c-soft as a sheaf on G_0.

5.3. Compact supports. There are some subtleties involved in extending the usual treatment of compact supports [**B**] to non-Hausdorff spaces. The reader will find a detailed discussion of this in [**CM1**]. Here we summarize the main points.

For such a c-soft sheaf, define $\Gamma_c(X, A)$ to be the subgroup of the group of all (set-theoretic, i.e. not necessarily continuous) sections of A, generated by sections σ which are continuous on a neighbourhood of their support, and have the property that this support is a compact subset of a Hausdorff open set $U \subseteq X$. This definition gives a functor

$$\Gamma_c : \mathrm{Sh}(X) \to (k\text{-modules}),$$

extending the usual one for Hausdorff spaces, and still having all the standard properties.

5.4. $f_!$ for spaces ([CM1]). Let $f : X \to Y$ be a map between topological spaces. These spaces may be non-Hausdorff but satisfy the general assumptions of (5.1). There is an exact functor $f_!$ from sheaves on Y to sheaves on X, mapping c-soft sheaves to c-soft ones, and with the following properties, for any c-soft sheaf on Y, any $x \in X$ and any open $U \subseteq X$:

(a) $\Gamma_c(U, f_!B) = \Gamma_c(f^{-1}(U), B)$.

(b) $f_!(B)_x = \Gamma_c(f^{-1}(x), B)$.

Moreover, in a pullback square of the form

$$\begin{array}{ccc} Y' & \xrightarrow{q} & Y \\ {\scriptstyle f'}\downarrow & & \downarrow{\scriptstyle f} \\ X' & \xrightarrow{p} & X \end{array}$$

where p is étale (i.e. a local homeomorphism), the canonical map

$$f'_! q^*(B) \to p^* f_!(B)$$

is an isomorphism. (If p is not étale this property holds up to q.i., see [**CM1**]). Furthermore, if f is itself étale, and A is a c-soft sheaf on X, there is a natural morphism of sheaves

$$f_* : f_! f^*(A) \to A,$$

defined by summation along the fiber.

5.5. $f_!$ for groupoids. Now suppose $f : H \to G$ is a homomorphism between étale groupoids. Consider for each $n \geq 0$ the space $(G/f)_n = \bigcup_{x \in G_0} (G/f)_{n,x}$ of strings

$$(y_0 \xleftarrow{h_1} y_1 \longleftarrow \cdots \xleftarrow{h_n} y_n, f(y_n) \xleftarrow{g} x) \tag{5.1}$$

where $h_1, \ldots, h_n$ are arrows in H and g is one in G. Note that an arrow $u : x' \to x$ in G induces a map $(G/f)_{x,n} \to (G/f)_{n,x'}$, sending the string (5.1) to $(h_1, \ldots, h_n, gu)$.

In other words, the groupoid G acts on the space $(G/f)_n$. Write $\pi_n : (G/f)_n \to H_0$ for the map sending (5.1) to y_n

$$\pi_n(h_1, \dots, h_n, g) = \mathrm{domain}(h_n),$$

and observe that π_n is a local homeomorphism since G and H are étale groupoids. Furthermore, define $\sigma_n : (G/f)_n \to G_0$ by

$$\sigma_n(h_1, \dots, h_n, g) = x = \mathrm{domain}(g).$$

If B is a c-soft H-sheaf, then $\pi_n^*(B)$ is a c-soft sheaf on $(G/f)_n$, and one obtains a c-soft sheaf

$$f_!(B)_n := \sigma_{n!}\pi_n^*(B)$$

on G_0. The action of G on $(G/f)_n$ induces an action of G on $f_!(B)_n$, so that $f_!(B)_n$ is a c-soft G-sheaf.

Next, the usual simplicial face maps

$$d_i : (G/f)_n \to (G/f)_{n-1} \quad (0 \le i \le n)$$

defined by $d_i(h_1, \dots, h_n, g) = (h_1, \dots, h_i h_{i+1}, \dots, h_n, g)$ for $0 < i < n$, and by $(h_2, \dots, h_n, g)$ for $i = 0$, and $(h_1, \dots, h_{n-1}, f(h_n)g)$ for $i = n$, make $(G/f)_n$ into a simplicial G-space. Notice that each face map d_i is étale. We observe that

$$\sigma_{n-1} \circ d_i = \sigma_n \quad (0 \le i \le n).$$

Furthermore,

$$\pi_{n-1} \circ d_i = \pi_n \quad (0 \le i < n),$$

so that for an H-sheaf B as above there is a canonical isomorphism

$$d_i^* \pi_{n-1}^*(B) \cong \pi_n^*(B) \quad (0 \le i < n).$$

For $i = n$, the action by H (of h_n on the stalk at $(h_1, \dots, h_n, g)$) gives an isomorphism

$$d_i^* \pi_{n-1}^*(B) \cong \pi_n^*(B)$$

for $i = n$ also. Thus, for each $i = 0, \dots, n$, summation along the fiber d_{i*} defines a map

$$d_{i!}\pi_n^*(B) \cong d_{i!}d_i^*\pi_{n-1}^*(B) \xrightarrow{d_{i*}} \pi_{n-1}^*(B).$$

Applying the functor $(\sigma_{n-1})_!$ to this map, one obtains a map of c-soft G-sheaves

$$\sigma_{n!}\pi_n^*(B) = (\sigma_{n-1})_! d_{i!}\pi_n^*(B) \xrightarrow{(\sigma_{n-1})_!(d_{i*})} (\sigma_{n-1})_!\pi_{n-1}^*(B),$$

or, in other words, a map

$$\delta_i : f_!(B)_n \to f_!(B)_{n-1} \quad (0 \le i \le n).$$

Thus $f_!(B)$ has the structure of a simplicial G-sheaf of k-modules. By taking alternating sums of the δ_i in the usual way, we obtain a chain complex of G-sheaves, still denoted $f_!(B)$.

Finally, if we start with a bounded above cochain complex of c-soft H-sheaves $B^\bullet$, we obtain such a complex $f_!(B^\bullet)$ by setting

$$f_!(B^\bullet)^n = \bigoplus_{p-q=n} f_!(B^p)_q.$$

In [**CM1**] it is proved that this construction preserves quasi-isomorphisms, hence gives a well-defined functor

$$f_! : \mathcal{D}^-(H) \to \mathcal{D}^-(G).$$

(Here, it is convenient to represent objects of $\mathcal{D}^-(H)$ by bounded below complexes of c-soft sheaves. This is possible because each bounded below complex is q.i. to such a complex.)

5.6. Cohomology with Compact supports. For the special case of the unique morphism $p : G \to 1$ from any étale groupoid to the one-point groupoid, the construction above yields a functor $p_!$ from $\mathcal{D}^-(G)$ into bounded above cochain complexes of k-modules. We write

$$\mathbb{H}^i_c(G, B^\bullet) = H^i(p_!(B^\bullet)).$$

If B is (the c-soft resolution of) a single G-sheaf B concentrated in degree zero, this defines the cohomology of B with compact supports, denoted $H^i_c(G, B)$. (If G is a space, this is the usual cohomology with compact supports.)

We can now state the following version of "Verdier duality", essentially proved in [**CM1**].

5.7. Theorem. *Let $f : H \to G$ be a homomorphism between étale groupoids, as before. There exists a natural operation*

$$f^! : \mathcal{D}^+(G) \to \mathcal{D}^+(H),$$

uniquely determined by the property that there is a natural q.i.

$$\mathrm{R}\underline{\mathrm{Hom}}(f_! B, A) \cong Rf_*(\mathrm{R}\underline{\mathrm{Hom}}(B, f^! A)) \tag{5.2}$$

for any two objects A of $\mathcal{D}^+(G)$ and B of $\mathcal{D}^-(H)$.

5.8. Remark. The isomorphism (5.2) is an isomorphism in $\mathcal{D}^+(G)$. Applying Rp_* for $p : G \to 1$ and taking cohomology in degree zero yields the adjunction formula

$$\mathrm{Hom}_{\mathcal{D}^+(H)}(f_! B, A) = \mathrm{Hom}_{\mathcal{D}^+(G)}(B, f^! A), \tag{5.3}$$

as usual. Conversely, it is possible to deduce the "internal" isomorphism (5.3) from the natural isomorphism (5.3) applied to localisations of G and H involved in the explicit description of Rf_*.

6. The embedding category of an étale groupoid.

The purpose of this section is to show how the derived category of an étale groupoid can be described purely combinatorially, in terms of a small category associated to the étale groupoid, first introduced in [**M3**] and called the embedding category there. First, we recall some basic constructions related to small categories.

6.1. Small categories. [**CWM**] These are 'discrete' categories, without a topology, and 'small' in the sense that their objects and arrows are sets rather than classes (relative to some ambient set theory). Let $\mathcal{C}$ be such a category. A presheaf (of k-modules) on $\mathcal{C}$ is a contravariant functor A from $\mathcal{C}$ into k-modules. We write $A(C)$ for the value at an object $C \in \mathcal{C}$, and $\alpha^* : A(D) \to A(C)$ for the effect of an arrow $\alpha : C \to D$. A morphism between such presheaves is a natural transformation. This defines the category $\mathrm{PSh}(\mathcal{C})$ of presheaves of k-modules on $\mathcal{C}$. With cochain complexes of presheaves defined in the obvious way, one defines a map $A^\bullet \to B^\bullet$ between such complexes to be a q.i. if $A^\bullet(C) \to B^\bullet(C)$ is a q.i. of cochain complexes of k-modules, for every object $C \in \mathcal{C}$. The associated derived categories are denoted

$$\mathcal{D}(\mathcal{C}),\ \mathcal{D}^+(\mathcal{C}),\ \mathcal{D}^-(\mathcal{C}).$$

The nerve $N(\mathcal{C})$ of $\mathcal{C}$ is the simplicial set whose n-simplices are composable strings of the form $(C_0 \xleftarrow{\alpha_1} \cdots \xleftarrow{\alpha_n} C_n)$. The face maps d_i map such a string to $(C_0 \xleftarrow{\alpha_1} \cdots \xleftarrow{\alpha_i\alpha_i+1} \cdots \xleftarrow{\alpha_n} C_n)$ for $0 < i < n$ and to $(C_1 \xleftarrow{\alpha_2} \cdots \xleftarrow{\alpha_n} C_n)$ for $i = 0$ and $(C_0 \xleftarrow{\alpha_1} \cdots \xleftarrow{\alpha_{n-1}} C_{n-1})$ for $i = n$. The classifying space $B\mathcal{C}$ of $\mathcal{C}$ is defines as the geometric realization of $N(\mathcal{C})$. The cohomology groups $H^n(\mathcal{C}, A)$ with coefficients in a presheaf A are constructed from the complex

$$C^n(\mathcal{C}, A) = \prod_{C_0 \xleftarrow{\alpha_1} \cdots \xleftarrow{\alpha_n} C_n} A(C_n),$$

where the product ranges over $N_n(\mathcal{C})$. For $f \in C^{n-1}(\mathcal{C}, A)$ and $\alpha = (\alpha_1, \dots, \alpha_n) \in N_n(\mathcal{C})$, the coboundary is defined by

$$(df)(\alpha) = \sum_{i=0}^{n-1} (-1)^i f(d_i\alpha) + (-1)^n \alpha_n^* f(d_n\alpha).$$

If the presheaf A is constant, then this cohomology $H^*(\mathcal{C}, A)$ is isomorphic to the cohomology $H^*(B\mathcal{C}, A)$ of the classifying space [**Q**].

6.2. The category $\mathrm{Emb}(G)$**.** [**M3**] Let G be an étale groupoid. We construct a small category, called the embedding category of G and denoted $\mathrm{Emb}(G)$. The construction depends on the choice of a basis of open sets for the topology on G_0. Assume such a basis fixed from now on. (In practice there is often a natural choice for such a basis, e.g. consisting of (contractible) charts for a manifold structure on G_0). The objects of the category $\mathrm{Emb}(G)$ will be these open sets in G_0 belonging to the chosen basis. An arrow $U \to V$ is given by a section $s : U \to G_1$ of the domain map $d_0 : G_1 \to G_0$, with the property that $d_1 \circ s$ defines an embedding $U \to V$. For emphasis, we denote by $\bar{s} : U \to V$ the arrow of $\mathrm{Emb}(G)$ given by such a section s. One can think of $\bar{s}$ as a U-parametrised family of arrows in the groupoid G, namely $\{s(x) : x \to d_1 s(x)\}_{x \in U}$. Composition of two arrows $\bar{s} : U \to V$ and $\bar{t} : V \to W$ in $\mathrm{Emb}(G)$ is given by the formula

$$\bar{t} \circ \bar{s} = \overline{t * s}$$

where $(t * s)(x) = t(d_1 s(x)) \cdot s(x)$ (composition in G).

6.3. The functor Γ. Each G-sheaf A defines a presheaf $\Gamma(A)$ on $\mathrm{Emb}(G)$, by

$$\Gamma(A)(U) = \Gamma(U, A),$$

the set of sections of A over U. If $\overline{s} : U \to V$ in $\mathrm{Emb}(G)$ then $\overline{s}^* : \Gamma(V, A) \to \Gamma(U, A)$ is defined using the action of G, as $\overline{s}^*(a)(x) = a(d_1 s(x)) \cdot s(x)$. In this way one obtains a left exact functor

$$\Gamma : \mathrm{Sh}(G) \to \mathrm{PSh}(\mathrm{Emb}(G)),$$

into the category of presheaves on the embedding category. Although not explicitly stated in this way, the arguments in [**M3**] show the following.

6.4. Theorem. *The functor Γ induces a full and faithful embedding of derived categories*

$$R\Gamma : \mathcal{D}^+(G) \to \mathcal{D}^+(\mathrm{Emb}(G)).$$

(For the proof of this theorem, one uses that the functor Γ is itself full and faithful, and has a left adjoint Δ. For a presheaf B on $\mathrm{Emb}(G)$, the G-sheaf $\Delta(B)$ has for its stalk at $x \in G_0$,

$$\Delta(B)_x = \varinjlim_{x \in U} B(U),$$

where U ranges over all basic neighbourhoods of x. Then at the level of derived categories, Δ is left adjoint to $R\Gamma$, and the canonical map $\Delta(R\Gamma(A)) \to A$ is a q.i. for each cochain complex of presheaves A.)

6.5. Cohomology. For a G-sheaf A, the complex $R\Gamma(A)(U)$ computes the cohomology $H^*(U, A)$. In particular, if this cohomology vanishes in positive degrees for each open set U in the chosen basis, then the previous theorem implies that there is an isomorphism

$$H^*(G, A) = H^*(\mathrm{Emb}(G), \Gamma(A)).$$

This is the case, for example, if A is (locally) constant while each open set U in the chosen basis is contractible. (In [**M3**], only this case is discussed.)

6.6. Remark. There are analogous statements concerning a functor $R\Gamma_c : \mathcal{D}^-(G) \to \mathcal{D}^-(\mathrm{Emb}(G)^{\mathrm{op}})$, and the relation between the compactly supported cohomology of G and the homology of the category $\mathrm{Emb}(G)$. The category $\mathrm{Emb}(G)$ is also a natural and convenient geometric object for the construction of characteristic classes [**CM3, C2**].

7. Relation to classifying spaces.

To conclude this paper, we briefly discuss the relation with the classifying space BG of an étale groupoid.

7.1. The classifying space BG. Let G be an étale groupoid. The nerve of G is the simplicial space $N(G)$ whose space $N_n(G)$ of n-simplices is the space of strings of arrows of the form

$$x_0 \xleftarrow{g_1} x_1 \longleftarrow \cdots \xleftarrow{g_n} x_n,$$

equipped with the fibered product topology,

$$N_n(G) = G_1 \times_{G_0} G_1 \times \cdots \times_{G_0} G_1.$$

The face maps $d_i : N_n(G) \to N_{n-1}(G)$ are defined as usual (see 6.1). The classifying space BG is the geometric realization of $N(G)$. Here we take the "thick" realization in the sense of [**S**].

7.2. Sheaves. Each G-sheaf A induces in a natural way a sheaf $\tilde{A}$ on the space BG. This construction is explicitly described in [**M4**], where it is proved that there is a natural isomorphism in cohomology,

$$H^n(G, A) \xrightarrow{\sim} H^n(BG, \tilde{A}).$$

The same argument in fact shows the following somewhat stronger statement.

7.3. Theorem. *The construction $A \mapsto \tilde{A}$ defines a full embedding of derived categories*

$$\mathcal{D}^+(G) \hookrightarrow \mathcal{D}^+(BG).$$

7.4. Remark. The space BG is in general hard to deal with. It is an infinite dimensional space, and in the natural examples where G is not Hausdorff, neither is the space BG. The classifying space $B\,\mathrm{Emb}(G)$ is better behaved in this respect — it is always a CW-complex. The main theorem in [**M3**] asserts that these two classifying spaces have the same weak homotopy type,

$$BG \cong B\,\mathrm{Emb}(G).$$

However, I do not know whether Theorem 7.3 holds with $B\,\mathrm{Emb}(G)$ instead of BG. I also do not know how to deal with compactly supported cohomology of G (and the related operations $f_!$ and $f^!$ in terms of BG).

References

[B] A. Borel, *Intersection cohomology*, Birkhauser, 1984.

[BL] J. Bernstein, V. Lunts, *Equivariant sheaves and functors*, Springer Lecture Notes 1578 (1994).

[BN] J.-L. Brylinski, V. Nistor, *Cyclic cohomology of étale groupoids*, K-theory **8** (1994), 341–365.

[C1] M. Crainic, *Cyclic cohomology of étale groupoids: The general case*, K-Theory **17** (1999)319-362.

[C2] M. Crainic, PhD thesis, Utrecht University, April 2000.

[CM1] M. Crainic, I. Moerdijk, *A homology theory for étale groupoids*, to appear in Crelle's J. f. reine und angewandte Mathematik.

[CM2] M. Crainic, I. Moerdijk, *A remark on sheaf theory for non-Hausdorff spaces*, Utrecht preprint **1119** (1999), available at http://www.math.uu.nl/publications/Preprints/

[CM3] M. Crainic, I. Moerdijk, *Cech-De Rham theory for leaf spaces of foliations*, in prep.

[CWM] S. Mac Lane, *Categories for the Working Mathematician*, Springer-Verlag, 1971.

[Go] R. Godement, *Topologie algébrique et théorie des faisceaux*, Hermann, 1958.

[Gr] A. Grothendieck, *Sur quelques points d'homologie algébrique*, Tohoku Math. J. **9** (1957), 119-221.

[H1] A. Haefliger, *Groupoïdes d'holonomie et espaces classifiants*, Astérisque **116** (1984), 70-97.

[H2] A. Haefliger, *Differentiable cohomology*, CIME lectures, Varenna, 1976.

[H3] A. Haefliger, *Cohomology theory for étale groupoids*, unpublished ms, 1992.

[H4] A. Haefliger, *Feuilletages sur les variétes ouvertes*, Topology **9** (1970), 183-194.

[HS] M. Hilsum, G. Skandalis, *Morphismes K-orientés d'espaces de feuilles et fonctorialité en théorie de Kasparov*, Ann. Sc. E. N. S. **20** (1987), 325-390.

[I] B. Iversen, *Sheaf Cohomology*, Springer-Verlag, 1986.

[K] A. Kumjian, *On equivariant sheaf cohomology and elementary C*-bundles*, J. Operator Theory **20** (1988), 207-240.

[M1] I. Moerdijk, *The classifying topos of a continuous groupoid*, Trans. AMS **310** (1988), 629-668.

[M2] I. Moerdijk, *Classifying toposes and foliations*, Ann. Inst. Fourier **41** (1991),189-209.

[M3] I. Moerdijk, *On the weak homotopy type of étale groupoids*, in "Integrable Systems and Foliations", Birkhauser (1997), 147-156.

[M4] I. Moerdijk, *Proof of a Conjecture of A. Haefliger*, Topology **37** (1998), 735-741.

[MP] I. Moerdijk, D. Pronk, *Orbifolds, Sheaves and Groupoids*, K-Theory **12** (1997), 3-21.

[Mr] J. Mrcun, *Functoriality of the bimodule associated to a Hilsum-Skandalis map*, K-Theory **18** (1999), 235-253.

[MRW] P. Muhly, J. Renault, D. Williams, *Equivalence and isomorphism for groupoid C*-algebras*, J. Operator Theoy **17** (1987), 3-22.

[P] J. Philips, *The holonomic imperative and the homotopy groupoid of a foliated manifold*, Rocky Mountains J. Math. **17** (1987), 151-165.

[Pra] J. Pradines, *Morphisms between spaces of leaves viewed as fractions*, Cahiers de Top. et Geom. Diff. **30** (1989), 229-246.

[Pro] D. Pronk, *Etendues and stacks as bicategories of fractions*, Compositio Math. **102** (1996), 243-303.

[Q] D. Quillen, *Higher K-Theory I*, in: Springer Lecture Notes 341 (1973), 85-147.

[S] G. Segal, *Categories and cohomology theories*, Topology **13** (1974), 293-312.

[SGA IV] M. Artin, A. Grothendieck, J.-L. Verdier, *Théorie de topos et cohomologie étale des schémas ("SGA IV")*, Springer Lecture Notes 269, 270 (1972).

MATHEMATICAL INSTITUTE, UNIVERSITY OF UTRECHT, PO BOX 80.010, 3508 TA UTRECHT, THE NETHERLANDS

E-mail address: moerdijk@math.uu.nl

Contemporary Mathematics
Volume **282**, 2001

The analytic index for proper, Lie groupoid actions

Alan L. T. Paterson

ABSTRACT. Many index theorems (both classical and in noncommutative geometry) can be interpreted in terms of a Lie groupoid acting properly on a manifold and leaving an elliptic family of pseudodifferential operators invariant. Alain Connes in his book raised the question of an index theorem in this general context. In this paper, an analytic index for many such situations is constructed. The approach is inspired by the classical families theorem of Atiyah and Singer, and the proof generalizes, to the case of proper Lie groupoid actions, some of the results proved for proper locally compact group actions by N. C. Phillips.

1. Introduction

The objective of this paper is to prove the following theorem.

THEOREM 1.1. *Let G be a Lie groupoid for which the unit space G^0 is a proper, G-compact G-space. Let (X, p) be a G-manifold which is a fiber bundle over G^0 with compact smooth manifold Z as fiber and with structure group Diff(Z). Let $\tilde{E}, \tilde{F}$ be smooth G-vector bundles over X and $D = \{D^u\} : C^{0,\infty}(X, \tilde{E}) \to C^{0,\infty}(X, \tilde{F})$ be an invariant, continuous family of elliptic pseudodifferential operators on the fibers X^u of X. Then D determines a $(\mathbb{C}, C^*_{red}(G))$-Kasparov module whose $K_0(C^*_{red}(G))$-class is the analytic index of D.*

The motivation for this result comes from noncommutative geometry. In his book ([**6**, p.151]), Alain Connes comments that a number of index theorems both in classical and in noncommutative geometry are all special cases of the same index theorem for G-invariant elliptic operators D on a proper G-manifold X, G being a Lie groupoid. In this situation, the index of D belongs to $K_0(C^*(G))$. The present paper will establish this under certain (commonly satisfied) conditions. The approach to the analytic index of D in this paper is based on an adaptation of the equivariant Atiyah-Singer index theorem for families. There are, of course, other approaches to the analytic index in the context of groupoids, notably those using quasi-isomorphisms and the "deformation" approach using the tangent groupoid (e.g. [**6, 13, 14, 16, 17, 21, 22**]). However, these approaches apply (at the present time) only to the case where $X = G$, and in particular, do not cover the

1991 *Mathematics Subject Classification.* Primary: 19K35, 19K56, 22A22, 46L80, 46L87, 58B34, 58J40; Secondary: 19L47, 55N15, 58B15, 58H05.

classical families index theorem. Further, the action of G on itself is rather special: for example, that action is free. In the present paper, no freeness assumptions of the action of G on X are made.

Another merit in the approach here is that it relates more readily to pseudodifferential operators and Fredholm theory and the well-established, detailed, classical theory. It also gives rise to a groupoid equivariant K-theory for TX and the prospect of defining a topological index in the same spirit as that of the classical families index theorem. This will be discussed elsewhere.

The author believes that the proof of this paper can be adapted to give the analytic index in the full generality of [**6**, p.151] and plans to discuss this elsewhere. The restrictions in the present paper are largely a matter of convenience, since they ennable one to use existing theories to shorten and simplify the proof. In particular, the space X is assumed, as in the work of Atiyah and Singer ([**2**]), to be a manifold with compact fiber over the unit space G^0 of G. (The base space G^0 is not, however, assumed to be compact, but rather just G-compact and proper as a G-space.) This ennables one to appeal, for example, to the results on pseudodifferential operators used in [**2**]. Another advantage of this framework is that it gives rise to a Hilbert G-bundle, and this allows the techniques used by N. C. Phillips in his development of equivariant K-theory to be adapted to produce the required Kasparov module. Further, we have used the reduced C^*-algebra $C^*_{red}(G)$ rather than $C^*(G)$, thus avoiding having to use the disintegration theorem for locally compact groupoids. So the index of D constructed here lies in $K_0(C^*_{red}(G))$.

The second section collects, for the convenience of the reader, some facts and definitions about locally compact and Lie groupoids, while the third discusses G-spaces and G-manifolds. The fourth section discusses families of elliptic pseudodifferential operators invariant under the action of a Lie groupoid G. Such a family D gives an invariant Fredholm morphism on a certain Hilbert bundle, and the final section shows that associated with this bundle is a Hilbert module, whose bounded and compact module maps correspond exactly to the invariant bounded and compact morphisms on the Hilbert bundle. This section is modelled on the approach of N. C. Phillips in his book [**25**] – Phillips effectively deals with the case where G is a transformation group groupoid. So the Fredholm morphism D translates over to a Fredholm map on a Hilbert module over $C^*_{red}(G)$, which, by a standard process, gives a Kasparov module whose KK-class is the analytic index of D.

The author is grateful to the organizers for the opportunity to speak at the groupoid conference of the Colorado JSRC meeting of 1999, and for their warm hospitality. He would also like to express his thanks to N. C. Phillips for helpful conversations about equivariant K-theory.

2. Preliminaries on groupoids

If X is a locally compact Hausdorff space, then $\mathcal{C}(X)$ is the family of compact subsets of X, $C(X)$ is the algebra of bounded continuous complex-valued functions on X, and $C_c(X)$ is the subalgebra of $C(X)$ whose elements have compact support. For a smooth manifold X, we define $C^\infty(X)$ to be the algebra of $C^\infty-$ complex-valued functions on X, and $C_c^\infty(X)$ to be the algebra of functions in $C^\infty(X)$ with compact support.

A *groupoid* is most simply defined as a small category with inverses. Spelled out axiomatically, a groupoid is a set G together with a subset $G^2 \subset G \times G$, a product map $m : G^2 \to G$, where we write $m(a,b) = ab$, and an inverse map $i : G \to G$, where we write $i(a) = a^{-1}$ and where $(a^{-1})^{-1} = a$, such that:

1. if $(a,b), (b,c) \in G^2$, then $(ab,c), (a,bc) \in G^2$ and
$$(ab)c = a(bc);$$
2. $(b, b^{-1}) \in G^2$ for all $b \in G$, and if (a,b) belongs to G^2, then
$$a^{-1}(ab) = b \qquad (ab)b^{-1} = a.$$

We define the *range* and *source* maps $r : G \to G^0$, $s : G \to G^0$ by setting $r(x) = xx^{-1}, s(x) = x^{-1}x$. The *unit space* G^0 is defined to be $r(G)(= s(G))$, or equivalently, the set of idempotents u in G. The maps r, s fiber the groupoid G over G^0 with fibers $\{G^u\}, \{G_u\}$, so that $G^u = r^{-1}(\{u\})$ and $G_u = s^{-1}(\{u\})$. Note that $(x,y) \in G^2$ if and only if $s(x) = r(y)$.

For detailed discussions of groupoids (including locally compact and Lie groupoids as below), the reader is referred to the books [**13, 15, 22, 28**]. Important examples of groupoids are given by transformation group groupoids and equivalence relations.

For the purposes of this paper, a *locally compact groupoid* is a groupoid G which is also a second countable, locally compact Hausdorff space for which multiplication and inversion are continuous. More generally, there is a notion of *locally compact groupoid* which does not require the Hausdorff condition. These arise quite often in practice, and a detailed discussion of them is given in [**22**]. The details of the paper can be made to work in the general case since local arguments suffice. However, for convenience, the locally compact groupoids considered in the paper are assumed to be Hausdorff.

Let G be a locally compact groupoid. Note that G^2, G^0 are closed subsets of $G \times G, G$ respectively. Further, since r, s are continuous, every G^u, G_u is a closed subset of G.

For analysis on a locally compact groupoid G, it is essential to have available a *left Haar system*. This is the groupoid version of a *left Haar measure*, though unlike left Haar measure on a locally compact group, such a system may not exist and if it does, it will not usually be unique. However, in many cases, such as in the Lie groupoid case below, there is a natural choice of left Haar system.

A left Haar system on a locally compact groupoid G is a family of measures $\{\lambda^u\}$ $(u \in G^0)$, where each λ^u is a positive regular Borel measure on the locally compact Hausdorff space G^u, such that the following three axioms are satisfied:

1. the support of each λ^u is the whole of G^u;
2. for any $f \in C_c(G)$, the function f^0, where
$$f^0(u) = \int_{G^u} f \, d\lambda^u,$$
belongs to $C_c(G^0)$;
3. for any $g \in G$ and $f \in C_c(G)$,

$$\int_{G^{s(g)}} f(gh)\, d\lambda^{s(g)}(h) = \int_{G^{r(g)}} f(h)\, d\lambda^{r(g)}(h). \tag{2.1}$$

The existence of a left Haar system on G has topological consequences for G – it entails that both $r, s : G \to G^0$ are open maps ([**22**, p.36]). For each $u \in G^0$, the measure λ_u on $G_u = (G^u)^{-1}$ is given by: $\lambda_u(A) = \lambda^u(A^{-1})$.

Jean Renault showed ([**28**]) that $C_c(G)$, the space of continuous, complex-valued functions on G with compact support, is a convolution *-algebra with product given by:

$$f_1 * f_2(g) = \int_{G^{r(g)}} f_1(h) f_2(h^{-1}g)\, d\lambda^{r(g)}(h). \tag{2.2}$$

The involution on $C_c(G)$ is the map $f \to f^*$ where

$$f^*(g) = \overline{f(g^{-1})}. \tag{2.3}$$

As in the case of a locally compact group, there is a reduced C^*-algebra $C^*_{red}(G)$ (and a universal C^*-algebra $C^*(G)$) for the locally compact groupoid G. The reduced C^*-algebra can be defined as follows. For each $u \in G^0$, we first define a representation π_u of $C_c(G)$ on the Hilbert space $L^2(G, \lambda_u)$. To this end, regard $C_c(G)$ as a dense subspace of $L^2(G, \lambda_u)$ and define for $f \in C_c(G), \xi \in C_c(G)$,

$$\pi_u(f)(\xi)(g) = f * \xi \in C_c(G). \tag{2.4}$$

The reduced C^*-algebra-norm on $C_c(G)$ can then ([**22**, p.108]) be defined by:

$$\|f\|_{red} = \sup_{u \in G^0} \|\pi_u(f)\|.$$

The groupoid C^*-algebra $C^*(G)$ is the completion of $C_c(G)$ under its largest C^*-norm. In the paper, we will concentrate (for reasons of technical simplicity) on the reduced C^*-algebra of G.

A locally compact groupoid G is called a *Lie groupoid* ([**15, 22, 26, 27**]) if G is a manifold such that:

1. G^0 is a submanifold of G;
2. the maps $r, s : G \to G^0$ are submersions;
3. the product and inversion maps for G are smooth.

Note that G^2 is naturally a submanifold of $G \times G$ and every G^u, G_u is a submanifold of G. (See [**22**, pp.55-56].) Note that the maps r, s are open maps. The dimension of all of the G^u, G_u is the same, and will be denoted by l.

There is a basis for the topology of a Lie groupoid that proves useful. Since r is a submersion, it is locally equivalent to a projection. So for each $x \in G$, there exists an open neighborhood V of x and a diffeomorphism $\psi : V \to r(V) \times W$ where W is an open subset of $\mathbb{R}^l$, which is fiber preserving in the sense that we can write $\psi(x) = (r(x), \phi(x))$ for some function ϕ. So if $u \in r(V)$, then the restriction of ψ to $G^u \cap V$ is a diffeomorphism onto W. We will call such an open set V a *basic* open set and write $V \sim r(V) \times W$.

A. Connes ([**6**, p.101]) discusses convolution on $C_c^\infty(G)$ in terms of density bundles, and this is canonical. However, since the bundles involved are trivial, we can replace the densities by a left Haar system $\{\lambda^u\}$ which is *smooth* in the sense that in terms of a basic open set V, the map $(u, w) \to d\lambda^u/d\lambda_W(u, w)$ is C^∞ and strictly positive on $r(V) \times W$. (Here, λ_W is the restriction of Lebesgue measure on $\mathbb{R}^l$ to W.) For more details about this, see [**22**, 2.3]. All such smooth left Haar systems are equivalent in the obvious sense, and give the same $C^*_{red}(G)$ (and the same $C^*(G)$). For the rest of the paper, $\{\lambda^u\}$ will be a fixed smooth left Haar system on the Lie groupoid G.

3. G-spaces and G-manifolds

For the purposes of groupoid theory, we require the notions of proper G-spaces and G-manifolds (e.g. [**20**], [**6**, p.151]).

Let X, Y be locally compact, second countable topological spaces and $p : X \to Y$ be a continuous, open surjection. We will then say that (X, p) is *fibered over* Y. We will be particularly concerned with the case where $Y = G^0$, the unit space of a locally compact groupoid G. The space fibers of X are then the sets $X^u = p^{-1}(\{u\})$ $(u \in G^0)$.

If (X_i, p_i) $(i = 1, 2)$ are fibered over G^0, then the fibered product $(X_1 * X_2, \tilde{p})$ is also fibered over G^0, where

$$X_1 * X_2 = \{(x, y) \in X_1 \times X_2 : p_1(x) = p_2(y)\},$$

and $\tilde{p}(x, y) = p_1(x) = p_2(y)$.

Let $G*X$ be the fibered product of (G, s) with (X, p). The pair (X, p) is defined to be a *G-space* if there is given a continuous map $m : G * X \to X$, where we write gx for $m(g, x)$, such that for $(g, x) \in G * X$ and $(h, g) \in G^2(= G * G)$, we have:

1. $p(gx) = r(g)$;
2. $h(gx) = (hg)x$;
3. $g^{-1}(gx) = x$.

Of course, the pair (G, r) is a G-space under the groupoid multiplication. We note here that the unit space G^0 of G, as well as G itself, is a G-space. Here, the map p is just the identity map, and the action is given by:

$$g.s(g) = r(g).$$

If (X_i, p_i) $(i = 1, 2)$ are G-spaces, then the fibered product $(X_1 * X_2, \tilde{p})$ is also a G-space under the action: $(g, x, y) \to (gx, gy)$ $(g \in G^{p(x)})$. For a G-space X, let $G *_r X = \{(g, x) : r(g) = p(x)\}$ be the fibered product of the G-spaces (G, r) and (X, p).

Let (X, p) be a G-space. Then for each $g \in G$, the map $l_g : X^{s(g)} \to X^{r(g)}$, where $l_g(x) = gx$, is a homeomorphism, with inverse $l_{g^{-1}}$. Note also that if $u \in G^0$, then l_u is the identity map on X^u. Also, associated with the left action of G on X is a left action $g \to L_g$ of G on the bundle of algebras $C_c(X^u)$ $(u \in G^0)$. For this, we define $L_g : C_c(X^{s(g)}) \to C_c(X^{r(g)})$ by:

$$L_g f(x) = f(g^{-1}x) \quad (x \in X^{s(g)}).$$

Clearly, each L_g is an isomorphism of *-algebras, and both $(L_g)^{-1} = L_{g^{-1}}$ and $L_{gh} = L_g L_h$ for $(g, h) \in G^2$.

We now discuss properness and G-compactness for locally compact groupoid actions (on G-spaces). Proper actions for locally compact groups have been investigated by R. S. Palais ([**24**]) and N. C. Phillips ([**25**]). The basic results for proper locally compact group actions extend to the case of proper locally compact groupoid actions. Specifically, the action of a locally compact groupoid G on a G-space X gives the orbit equivalence relation on X. Further, if X is a proper G-space, then the space X/G is locally compact Hausdorff in the quotient topology, and the quotient map is open ([**19, 20**]).

The G-space X is called a *proper* G-space if the (continuous) map α, given by $(g, x) \to (gx, x)$, from $G*X$ into $X \times X$, is a proper map, i.e. if $\alpha^{-1}(C)$ is compact in $G*X$ whenever C is compact in $X \times X$. It is well-known, and easy to check, that

G itself as a G-space is proper. The G-space $X_1 * X_2$ is proper if both G-spaces X_1, X_2 are.

The G-space G^0 is not usually proper (even when G is a Lie group). However, if G^0 is a proper G-space – and this is an assumption of Theorem 1.1 – then *every* G-space is proper.

The G-space X is called *G-compact* (cf. [**25**, p.23]) if the quotient space X/G is compact. Following the proof of [**25**, Lemma, p.23-24], we obtain that *if X is a proper G-manifold, then X is G-compact if and only if there exists a compact subset B of X such that $X = GB$.*

One consequence of the existence of a G-compact, proper G-space X is that G^0 *as a G-space is also G-compact.* For then there is a canonical, surjective, continuous map from X/G onto G^0/G, so that the latter is compact if the former is. In this paper, we will assume that G^0 is G-compact. (This implies that the G-space X of the theorem of this paper is itself G-compact.)

We now discuss invariant averaging and G-partition of unities for a proper G-space. The discussion generalizes the approach of Phillips ([**25**, Chapter 2]). (For earlier special cases of this, see [**7, 11**].)

To deal with the varying fibers that arise in groupoid contexts and also with the vector bundle context, we use the following technical lemma. This is a variant of the lemma [**22**, Lemma C.0.2] (the main idea of which goes back to Renault and Connes). The proof of the lemma is the same *mutatis mutandis* as that of [**22**, Lemma C.0.2].

LEMMA 3.1. *Let (X, p) be a proper G-space and $(\mathcal{E}, \rho)$ be a vector bundle over X. Let $f' : G *_r X \to \mathcal{E}$ be continuous and such that $f'((g, x)) \in \mathcal{E}^x$ for each $x \in X$. Suppose further that for each $C \in \mathcal{C}(X)$ and with*

$$W_C = \{(g, x) \in G *_r X : x \in C\},$$

the restriction $f' \mid_{W_C} \in C_c(W_C, \mathcal{E})$. Then the fiber preserving map $F : X \to \mathcal{E}$, where

$$F(x) = \int f'(g, x)\, d\lambda^{p(x)}(g) \tag{3.1}$$

is continuous on X.

The scalar version of the above lemma is the case where $\mathcal{E}$ is the trivial bundle $X \times \mathbb{C}$: in that case, the function f' is scalar valued. If, in addition, $X = G$, then the result is equivalent to [**22**, Lemma C.0.2].

The scalar version of Lemma 3.1 ennables us, in the presence of properness, to average a $C_c(X)$-function into an invariant continuous function on X. The same argument applies more generally to sections of Banach G-bundles, cf. [**25**, Lemma 2.4].

A function $f : X \to \mathbb{C}$ is called *invariant* if for all $x \in X$ and all $h \in G^{p(x)}$, we have

$$f(h^{-1}x) = f(x).$$

PROPOSITION 3.2. *Let X be a proper G-space and $\xi \in C_c(X)$. Then the function $\eta : X \to \mathbb{C}$, where*

$$\eta(x) = \int_{G^{p(x)}} \xi(g^{-1}x)\, d\lambda^{p(x)}(g). \tag{3.2}$$

is invariant and continuous.

PROOF. Apply the scalar version of Lemma 3.1 with

$$f'(g,x) = \xi(g^{-1}x)$$

to obtain that η is well-defined and continuous. For invariance, we have, using (2.1):

$$\begin{aligned}\eta(h^{-1}x) &= \int \xi(g^{-1}h^{-1}x)\,d\lambda^{s(h)}(x)\\ &= \int \xi((hg)^{-1}x)\,d\lambda^{s(h)}(x)\\ &= \int \xi(k^{-1}x)\,d\lambda^{p(x)}(k)\\ &= \eta(x).\end{aligned}$$

□

Let X be a proper G-space and $\{U_\alpha\}$ be a collection of open subsets of X. A *G-partition of unity of X subordinate to* $\{U_\alpha\}$ (cf. [**25**, p.25]) is a collection of continuous non-negative functions f_γ with compact supports in the U_α's such that for every $x \in X$, we have

$$\sum_\gamma \int_{G^{p(x)}} f_\gamma(g^{-1}x)\,d\lambda^{p(x)}(g) = 1 \tag{3.3}$$

where the sum is locally finite.

PROPOSITION 3.3. *Let X be a proper G-space and $\{U_\alpha\}$ be a collection of open subsets of X such that the sets GU_α cover X. Then there exists a G-partition of unity subordinate to the collection $\{U_\alpha\}$.*

PROOF. We follow the proof of the group version of the result of Phillips in [**25**, Lemma 2.6]. Let $\pi : X \to X/G$ be the quotient map. Since X is a proper G-space, the space X/G is a locally compact Hausdorff, second countable space and π is open. The group argument of Phillips then goes through to give continuous functions $u_\gamma : X \to [0,1]$ $(\gamma \in S)$ with compact supports in the U_α's and the appropriate local finiteness properties, and for which the function f_1 on X given by

$$f_1(x) = \sum_{\gamma\in S} \int u_\gamma(g^{-1}x)\,d\lambda^{p(x)}(g)$$

is (by Proposition 3.2) strictly positive and continuous. One then takes $f_\gamma = u_\gamma/f_1$. □

PROPOSITION 3.4. *Let X be a proper, G-compact G-space. Then there exists a non-negative function $c \in C_c(X)$ such that for each $x \in X$, we have*

$$\int_{G^{p(x)}} c(g^{-1}x)\,d\lambda^{p(x)}(g) = 1. \tag{3.4}$$

PROOF. Let $Q : X \to X/G$ be the quotient map. Since Q is open and X/G is compact, there exists a compact subset C of X such that $Q(C) = X/G$. Let ξ be in $C_c(X)$ such that $0 \le \xi \le 1$ and $\xi(C) = \{1\}$. Define η as in (3.2) and take $c = \xi/\eta$. □

The smooth version of a G-space is a G-*manifold* (X, p). For this, G is now a Lie groupoid and X is a smooth manifold. The map $p : X \to G^0$ is required to be a submersion. Then each X^u is a submanifold of X of fixed dimension j. As in the case of G, there is a basis for the topology of X consisting of *basic* open sets $V \sim p(V) \times Y$ where Y is an open subset of $\mathbb{R}^j$.

Since the map $s \times p : G \times X \to G^0 \times G^0$ is also a submersion, the space $G * X = (s \times p)^{-1}(\Delta)$ is a submanifold of $G \times X$, where Δ is the diagonal $\{(u, u) : u \in G^0\}$. Under the above assumptions, the pair (X, p) is called a G-*manifold* if it is a G-space such that the multiplication map $m : G * X \to X$ is smooth and inversion is a diffeomorphism.

Note that the pairs $(G, r), (G^0, id)$ are G-manifolds. Also, if (X_i, p_i) $(i = 1, 2)$ are G-manifolds, then $(X_1 * X_2, \tilde{p})$ is also a G-manifold.

If (X, p) is a G-manifold, then for each $g \in G$, the map $l_g : X^{s(g)} \to X^{r(g)}$ is a diffeomorphism, with inverse $l_{g^{-1}}$. The (restricted) map $L_g : C_c^\infty(X^{r(g)}) \to C_c^\infty(X^{s(g)})$ is an isomorphism of *-algebras.

In this paper, hermitian metrics are assumed conjugate linear in the first variable and linear in the second. It will follow from Proposition 4.2, with $\tilde{E} = TX$, that there is a G-isometric hermitian metric $\langle , \rangle$ on the (complex) tangent bundle TX, the complexification of the vector bundle of vectors tangent along the fibers X^u's. (So in this paper, TX is *not* the (complex) tangent bundle to the manifold X.) This gives in the standard way a family of smooth measures μ^u on the X^u's. So in terms of local coordinates, we have $d\mu^u = \mid \det g_{ij}^u \mid^{1/2} dx$, where the hermitian metric $\langle , \rangle$ restricts over X^u to $\langle , \rangle^u$, and the latter is given locally by $\sum g_{ij}^u \, dx_i \otimes dx_j$. The G-isometric property of $\langle , \rangle$ just says that the differential $(\ell_g)_* : TX^{s(g)} \to TX^{r(g)}$ is an isometry. This in turn translates to the G-invariance of the μ^u's: for $f \in C_c(X^{r(g)})$, we have

$$\int_{X^{s(g)}} f(gx) \, d\mu^{s(g)}(x) = \int_{X^{r(g)}} f(y) \, d\mu^{r(g)}(y). \tag{3.5}$$

The counterparts to (i) and (ii) in the definition of a *left Haar system* (§2) also hold.

Note that the C^∞-versions of Lemma 3.1, Proposition 3.2, Proposition 3.3 and Proposition 3.4 hold for a G-manifold X – the proof modifications are simple.

4. Pseudodifferential operators on G-manifolds

For the rest of the paper, we will assume that *G is a Lie groupoid, G^0 is a proper, G-compact, G-manifold, and (X, p) is a G-manifold which is, in addition, a fiber bundle over G^0 with smooth compact fiber Z and structure group Diff(Z).* Consideration of this situation is motivated by the Atiyah-Singer families theorem, which can be interpreted as the case in which X is compact and $G = G^0$, a groupoid of units. (However, in the Atiyah-Singer situation, the base space G^0 is not assumed to be a manifold, and the fiber bundle is not assumed to be a manifold but rather a *manifold over* G^0. The "continuous family" version of the theorem of this paper requires continuous family groupoids ([**23**]) and will be considered elsewhere.)

Note that the existence of such a G-manifold imposes conditions on the Lie groupoid G. For example, since the G-action is proper and every X^u is compact, it follows that for every $u, v \in G^0$, the set $\{(g, x) \in G * X : (gx, x) \in X^v \times X^u\}$ is compact. So the sets $G_u \cap G^v$ are compact. In particular, the isotropy groups of

G are compact. It is left to the reader to check that X is G-compact if and only if G^0 is G-compact and X is proper if and only if G^0 is proper. (So in theorem of this paper, X is automatically proper and G-compact.) Note that by replacing μ^u by $\mu^u/\|\mu^u\|$, we can suppose that each μ^u is a probability measure.

We now discuss invariant families of pseudodifferential operators on X. It is natural to impose the condition on the symbols of such operators that they belong to one of the Hörmander class $L^m_{\rho,\delta}$ (as Kasparov does in [**11**]). However, as Atiyah and Singer comment ([**1**, p.508]), *for our purposes, any one of these classes would be equally good.* The reason is that for the purposes of the analytic index, the class of operators (associated with the class of symbols taken) gets closed up in a C^*-algebra, and this closure is the same for reasonable choices of starting class. We use the class of pseudodiferential operators in [**1**], and for the convenience of the reader, will now recall some of the details about these operators ([**1**],[**32**, Ch. 4]).

Let B be an open subset of some $\mathbb{R}^N$ and $m \in \mathbb{R}$. Then the space $\tilde{S}^m(B \times \mathbb{R}^N; B \times \mathbb{C}))$ of symbols is the set of smooth functions $a : B \times \mathbb{R}^N \to \mathbb{C}$ such that given multiindices α, β and a compact subset $K \subset B$, there exists a constant C such that

$$(4.1) \qquad \mid \partial_x^\beta \partial_\xi^\alpha a(x,\xi) \mid \leq C(1+ \mid \xi \mid)^{m-|\alpha|}$$

for all $x \in K, \xi \in \mathbb{R}^N$.

The symbols a that we will be considering are required to satisfy two other properties. Firstly, it is required that for all $x \in B$ and all $\xi \neq 0$ in $\mathbb{R}^N$, the principle symbol

$$(4.2) \qquad \sigma(a)(x,\xi) = \lim_{k\to\infty} \frac{a(x,k\xi)}{k^m}$$

exists. Clearly, $\sigma(a) : B \times (\mathbb{R}^N \sim \{0\}) \to \mathbb{C}$. Secondly, we also require that for any "cut-off" function $\psi \in C^\infty(\mathbb{R}^N)$, i.e. a C^∞-function on $\mathbb{R}^N$ which is 0 near $\xi = 0$ and 1 outside the unit ball, we have

$$a(x,\xi) - \psi(\xi)\sigma(a) \in \tilde{S}^{m-1}(B \times \mathbb{R}^N; B \times \mathbb{C}).$$

The set of symbols in $\tilde{S}^m(B \times \mathbb{R}^N; B \times \mathbb{C})$ which also satisfy these two properties is denoted by $S^m(B \times \mathbb{R}^N; B \times \mathbb{C})$.

More generally, the space $S^m(B \times \mathbb{R}^N; B \times \mathbb{C}^k)$ is defined in the obvious way by considering functions $a : B \times \mathbb{R}^N \to \mathbb{C}^k$ whose components are in $S^m(B \times \mathbb{R}^N; B \times \mathbb{C})$. The space $S^m(B \times \mathbb{R}^N; B \times \mathbb{C}^k)$ is a Fréchet space using as seminorms the $\|.\|_{m,B,\alpha,\beta,K}$'s, where $\|.\|_{m,B,\alpha,\beta,K}$ is the smallest constant C for which the inequality of (4.1) holds.

Any $a \in S^m(B \times \mathbb{R}^N; B \times M_k(\mathbb{C}))$ determines a pseudodifferential operator $a(x, D_x) : C_c^\infty(B; \mathbb{C}^k) \to C^\infty(B; \mathbb{C}^k)$, where for $f \in C_c^\infty(B; \mathbb{C}^k)$,

$$(4.3) \qquad a(x,D_x)f(x) = (2\pi)^{-N} \int_{\mathbb{R}^N} e^{ix.\xi} a(x,\xi)\hat{f}(\xi)\, d\xi.$$

Define the seminorm $\|,\|_{m,B,\alpha,\beta,K}$ on the operators $a(x, D_x)$ by:

$$\|a(x,D_x)\|_{m,B,\alpha,\beta,K} = \|a\|_{m,B,\alpha,\beta,K}.$$

Regarding $B \times \mathbb{R}^N$ as T^*B and replacing $B \times \mathbb{C}^k$ by a smooth complex vector bundle, the above definitions generalize in the standard way to the setting of smooth complex vector bundles E, F over a manifold M. (This is because the class of pseudodifferential operators above are invariant under diffeomorphisms.) We will take M to be compact but this is not essential. In this setting, an element of $S^m(M;E,F)$ is just a section of the bundle $Hom(E,F)$ pulled back over the cotangent bundle T^*M. Locally, $Hom(E,F)$ is of the form $B\times\mathbb{C}^k$ but $\mathbb{C}^k$ is realized as the space of $q\times p$-matrices, where p,q are the ranks of E and F. So with respect to a coordinate patch on M and bases in E,F, an element of $S^m(M;E,F)$ is just a matrix a_{ij} of symbols of the $S^m(B\times\mathbb{R}^N;B\times\mathbb{C}^k)$-kind earlier. A pseudodifferential operator

$$D: C^\infty(M,E)\to C^\infty(M,F)$$

is a map such that for every coordinate neighborhood U of M trivializing E,F, the map that sends $f\in C_c^\infty(U,E)$ to the restriction of $D\psi$ to U is a pseudodifferential operator of the form (4.3), or equivalently ([**9**, p.84]), that for any $\phi,\psi\in C_c^\infty(U)$, the operator $\phi D\psi$ is a pseudodifferential operator in the sense of (4.3).

The space of such pseudodifferential operators D is denoted by $\Psi^m(M;E,F)$ and is a Fréchet space under the seminorms $D\to \|(\phi D\psi)\|_{m,U,\alpha,\beta,K}$.

The principle symbol $\sigma(D)$ of D, defined locally in (4.2), is independent of coordinate choices, and is a section of the bundle $S^m(T_0^*M;Hom(\pi^*E,\pi^*F))$, where T_0^*M is T^*M with the zero section removed and π is the canonical map from T_0^*M to M. An explicit formula for $\sigma(D)$ is:

$$\sigma(D)(x,d\phi(x))f(x)=\lim_{k\to\infty}k^{-m}e^{-ik\phi(x)}D(e^{ik\phi}f)(x) \tag{4.4}$$

where $\phi\in C^\infty(M)$ and $f\in C^\infty(M,E)$ (e.g. [**1**, p.509], [**7**, p.298]).

It is obvious from (4.2) that the principle symbol is homogeneous of degree m, i.e. for $\xi\neq 0$ and $t>0$, we have

$$a_m(x,t\xi)=t^m a(x,\xi).$$

This homogeneity ennables one to regard the principle symbol as a section of the sphere bundle $Hom(\pi_S^*E,\pi_S^*F)$ over $S(M)$. Here, $S(M)$ is the unit sphere bundle of TM defined by some smooth Riemannian metric on M, $\pi_S: S(M)\to M$ is the canonical projection, and π_S^*E, π_S^*F are the associated pull-back bundles. Alternatively expressed, $\sigma(D)$ is an element of $\mathrm{Symb}^m(M;E,F)$, the space of smooth homomorphisms from π_S^*E to π_S^*F.

For convenience, from now on, we will only consider the case where $E=F$ and where the order of the pseudodifferential operators involved is O. For the first of these, a well-known argument ennables us to replace D by $D\oplus D^*$ on $E\oplus F$ (using hermitian metrics on E and F). So we can take $E=F$.

One can reduce to the case $m=0$ by giving E a hermitian metric and a connection. Then one just replaces D by by $(I+(D')^*D')^{-m/2}D$ where D' is the covariant derivative of the connection ([**1**, p.511]). We will write $\Psi(M;E)$ in place of $\Psi^0(M;E,E)$ and $\mathrm{Symb}(M;E)$ for $\mathrm{Symb}^0(M;E,E)$. An element $D\in\Psi(M;E)$ is called *elliptic* if the section $\sigma(D)$ is invertible.

Let E be given a hermitian metric (so that E becomes a finite-dimensional Hilbert bundle). Every $D\in\Psi(M;E)$ extends to a bounded linear operator on $L^2(M,E,\mu)$ for any smooth positive measure μ on M (e.g. [**31**, Theorem 6.5]). The closure of $\Psi(M;E)$ in $B(L^2(M,E,\mu))$ will be denoted by $\overline{\Psi}(M;E)$. For $D\in\Psi(M;E)$, let $\tilde{D}\in\overline{\Psi}(M;E)$ be the extension of D by continuity to $L^2(M,E,\mu)$. The following result is due to Atiyah and Singer ([**2**, p.124],[**1**, p.512]).

PROPOSITION 4.1. *The map* $D \to \tilde{D}$ *is continuous. Further the map* $\tilde{D} \to \sigma(D)$ *is continuous, where the norm on* $Symb(M;E)$ *is the* sup-*norm over the unit sphere bundle* SM *(treating* $\pi_S^* E$ *as a Hilbert bundle over* SM*).*

Let (X,p) be (as earlier) a G-manifold which is a smooth fiber bundle over G^0 with fiber Z. Let $(\tilde{E},q)$ be a *smooth complex vector bundle over* X in the sense of Atiyah and Singer ([**2**, p.123]). So $\tilde{E}$ is a vector bundle over X such that $(\tilde{E}, p \circ q)$ is a fiber bundle over G^0 with a fixed smooth complex vector bundle E over Z as fiber and structure group Diff(Z,E) (the topological group of diffeomorphisms of E that map fibers to fibers linearly). Locally on G^0, the restrictions $X\mid_U \sim U \times Z$ and $\tilde{E}\mid_U \sim U \times E$.

A natural example of a smooth vector bundle over X is TX.

For $u \in G^0, x \in X$, let $\tilde{E}^{(u)}$ be the restriction of $\tilde{E}$ to X^u and $\tilde{E}^x$ be the x-fiber of $\tilde{E}$. Note that $\tilde{E}^{(u)}$ is a vector bundle over X^u, while $\tilde{E}^x$ is a vector space. Let $C_c^\infty(X,\tilde{E})$ be the space of smooth sections of $\tilde{E}$ with compact supports. For $f \in C_c(X,\tilde{E})$, let $f^u \in C_c^\infty(X^u,\tilde{E}^{(u)})$ be the restriction of f to X^u.

Equivariance for the smooth fiber bundle $\tilde{E}$ is defined in a way similar to that for group actions (e.g. [**29**]). Precisely, we require that $\tilde{E}$ be a G-manifold with $q : \tilde{E} \to X$ a G-map. In addition, each $\ell_g : \tilde{E}^{(s(g))} \to \tilde{E}^{(r(g))}$, is a vector bundle isomorphism. If $\tilde{E}$ is equivariant, then $\tilde{E}$ is a proper G-space if X is proper.

The fibered product $X * X$ is a fiber bundle over G^0 with fiber $Z \times Z$ and structure group Diff$(Z \times Z)$. In addition, the external tensor product $\tilde{E} \boxtimes \tilde{E}$ is a smooth vector bundle over $X * X$ with the smooth vector bundle $E \boxtimes E$ as fiber.

PROPOSITION 4.2. *There exists a hermitian metric* $\langle , \rangle$ *on* $\tilde{E}$ *which is* G*-isometric.*

PROOF. (cf. [**25**, pp.40-41] for the group case.) Let $\langle , \rangle'$ be a hermitian metric on $\tilde{E}$ and c be as in Proposition 3.4. For $\xi, \eta \in \tilde{E}^x$, define

$$\langle \xi, \eta \rangle^x = \int c(g^{-1}x)\langle g^{-1}\xi, g^{-1}\eta\rangle' \, d\lambda^{p(x)}(g).$$

Applying the scalar version of Lemma 3.1 (to deal with continuity) with $\tilde{E} \boxtimes \tilde{E}$ in place of X and with $f'(g, \xi \otimes \eta) = c(g^{-1}x)\langle g^{-1}\xi, g^{-1}\eta\rangle'$, we obtain a hermitian metric $x \to \langle , \rangle^x$ for $\tilde{E}$. Arguing as in the proof of Proposition 3.2 gives that $\langle , \rangle$ is isometric. □

For the rest of the paper, $\tilde{E}$ (as well as TX) will be assumed to have G-isometric hermitian metrics. The smooth version of the following definition is given in [**14, 21**].

DEFINITION 4.3. A *pseudodifferential family* on X is a set $D = \{D^u\}$ $(u \in G^0)$ where:

1. each $D^u : C^\infty(X^u; \tilde{E}^{(u)}) \to C^\infty(X^u; \tilde{E}^{(u)})$ belongs to $\Psi(X^u; \tilde{E}^{(u)})$;
2. given a basic open subset $V \sim p(V) \times Y$ of X, trivializing $\tilde{E}$ (so that $\tilde{E}\mid_V \sim p(V) \times Y \times \mathbb{C}^k$), there exists a continuous function $a : p(V) \to S(T^*Y; Y \times M_k)$ such that for each $u \in p(V)$, and identifying $V \cap X^u$ with Y, we have

$$D_u \mid_{V \cap X^u} = a^u(x, D^x) \tag{4.5}$$

where $a^u(x,\xi) = a(u)(x,\xi)$.

Let $\Psi(X;\tilde{E})$ be the set of pseudodifferential families above. Definition 4.3 is equivalent to the corresponding definition of a pseudodifferential family given in [**2**]. There, Atiyah and Singer define H to be the closed subgroup of Diff$(Z,E)\times$ Diff(Z,E) consisting of pairs (Ψ,Φ) where Ψ,Φ determine the same element of Diff(Z). They show that $\Psi(X,\tilde{E})$ is a fiber bundle over G^0 with fiber $\Psi(Z;E)$ and structure group H. Then D is a pseudodifferential family if and only if the map $u\to D^u$ is a continuous section of $\Psi(X,\tilde{E})$.

DEFINITION 4.4. (cf. [**5**]) A function $f\in C(X,\tilde{E})$ is said to belong to $C^{0,\infty}(X,\tilde{E})$ if in local terms, the map $u\to f^u$ from G^0 into $C^\infty(Z,E)$ is continuous (where $C^\infty(Z,E)$ has its standard topology of uniform convergence on compacta for all derivatives). The space of functions f in $C^{0,\infty}(X,\tilde{E})$ that have compact support is denoted by $C_c^{0,\infty}(X,\tilde{E})$.

It is obvious that $C_c^{0,\infty}(X,\tilde{E})$ is a complex vector space which contains $C_c^\infty(X,\tilde{E})$. The space $C_c^{0,\infty}(X,\tilde{E})$ plays a similar role for the family $D=\{D_u\}$ as does $C_c^\infty(M,E)$ does for pseudodifferential operators on manifolds.

PROPOSITION 4.5. *Let D be a pseudodifferential family on X. Then $D: C^{0,\infty}(X,\tilde{E})\to C^{0,\infty}(X,\tilde{E})$, where for $x\in X^u$, we define*

$$Df(x)=D^u(f^u)(x).$$

PROOF. Using a partition of unity argument (cf. [**8**, p.23]), we can suppose that there is a basic open set $V\sim p(V)\times Y$ in X trivializing $\tilde{E}$ and functions $\phi,\psi\in C_c^\infty(V)$ such that $D=\phi D\psi$. Since $u\to(\psi f)^u$ is continuous from G^0 into $C_c^\infty(Y,\mathbb{C}^k)$ and D is a continuous family, it follows by the joint continuity of pseudodifferential operators on C_c^∞ functions ([**9**, Theorem 18.1.6]) that $Df\in C_c^{0,\infty}(X,\tilde{E})$. □

DEFINITION 4.6. A *parametrix* for $D\in\Psi(X;\tilde{E})$ is a pseudodifferential family $P=\{P^u\}$ such that for each u, P^u is a parametrix for D^u, i.e. each of D^uP^u-I, P^uD^u-I is a "smoothing" operator, an element $T\in\Psi(X^u;\tilde{E}^u)$ for which $\sigma(T)=0$.

PROPOSITION 4.7. *Let D be an elliptic, pseudodifferential family on X. Then there exists a parametrix P for D.*

PROOF. Note that SX is a manifold over G^0 with fiber SZ, and that $End(\pi_S^*(\tilde{E}))$ is a smooth vector bundle over SX. The vector bundle $End(\pi_S^*(\tilde{E}))$ restricts on each X^u to $\pi_{S,u}^*(X^u;End(\tilde{E}^{(u)}))$, where $\pi_{S,u}:SX^u\to X^u$ is the canonical projection. The "principle symbol" $\sigma(D)$ of D, where $\sigma(D)(u)=\sigma(D^u)$, is a section of the vector bundle $End(\pi_S^*(\tilde{E}))$. In fact, $\sigma(D)$ is continuous – this follows from the continuity of the maps $u\to D^u\to\sigma(D^u)$.

By ellipticity, the symbol $\sigma(D)^{-1}$ exists. In addition, straight-forward computations show that, locally, the matrices $b(u,x,\xi)$ belong to S^0, and that $u\to b^u$ is continuous. Following the argument of [**32**, Ch. IV,Theorem 3.15] (which deals with the case of a single pseudodifferential operator. i.e. where G^0 is a singleton), using basic open sets for an open cover, one constructs a pseudodifferential family P on X whose symbol is $\sigma(D)^{-1}$. Then ([**32**, Ch. IV, Theorem 4.4]) each P^u is a parametrix for D^u. So P is a parametrix for D. (It can be shown that a product of two pseudodifferential families is a pseudodifferential family (cf. [**32**, Theorem 3.16]) but we won't need this.) □

5. Construction of the analytic index

We now consider invariant pseudodifferential families on the G-manifold X. To this end, there is a section action $g \to L_g$ of G on $C_c^\infty(X, \tilde{E})$, where $L_g : C^\infty(X^{s(g)}, \tilde{E}^{(s(g))}) \to C^\infty(X^{r(g)}, \tilde{E}^{(r(g))})$ is the diffeomorphism given by:

$$L_g f(x) = g[f(g^{-1}x)] \quad (x \in X^{r(g)}). \tag{5.1}$$

There is a natural (algebraic) action of the groupoid G on $\Psi(X, \tilde{E})$ given by: $g \to L_g D^{s(g)} L_{g^{-1}}$. Note that $L_g D^{s(g)} L_{g^{-1}}$ belongs to $\Psi(X^{r(g)}; E^{r(g)})$ by [**10**].

DEFINITION 5.1. The pseudodifferential family D is called *invariant* if

$$L_g D^{s(g)} L_{g^{-1}} = D^{r(g)}. \tag{5.2}$$

The above notion is well-known in the literature (e.g. [**5, 6, 17, 21**]). We now discuss the symbol $\sigma(D)$ of an invariant pseudodifferential family D.

Since D is invariant then its symbol $\sigma(D)$ is also invariant under the natural action of G on the sections of $\pi_S^* End(\tilde{E})$). The action of G on the sections of this bundle is given by: $g \to g\alpha g^{-1}$. The map $D \to \sigma(D)$ is equivariant. One easy way to prove this is just to calculate the symbol of $g\sigma(D^{s(g)})g^{-1}$ using the explicit formula (4.4) and the fact that ℓ_g is a diffeomorphism from $X^{s(g)}$ onto $X^{r(g)}$. The details are left to the reader.

Suppose now that D is, in addition, elliptic, and let P be a parametrix for D (Proposition 4.7). There is actually an invariant parametrix P_1 for D. This can be proved by taking

$$P_1^u f(x) = \int_{G^u} c(g^{-1}x)(L_g P^{s(g)} L_{g^{-1}}) f(x)\, d\lambda^u(g)$$

where $p(x) = u$, $f \in C^\infty(X^u, \tilde{E}^{(u)})$, and c is smooth and as in Proposition 3.4. In this connection, cf. [**7, 11, 25**]. However, rather than giving the details of the proof of the existence of an invariant parametrix P_1, it is more convenient to deal with the corresponding question at the Fredholm level later (Proposition 5.4), where the corresponding proof in the group case by Phillips adapts easily.

For the rest of this paper, D is an invariant, elliptic pseudodifferential family on X as above. The analytic index of D will be constructed by adapting the approach of N. C. Phillips in [**25**] to equivariant K-theory for proper actions. (This in turn was motivated by the work of Segal ([**29, 30**].)

We first construct a Hilbert bundle $(\mathfrak{H}, \tilde{p})$ over G^0. By definition ([**25**, p.7]), a Hilbert bundle is a locally trivial fiber bundle with a Hilbert space H as fiber and with structure group $U(H)$, the unitary group of H with the strong operator topology. The fiber over $u \in G^0$ of this bundle is $L^2(X^u, \tilde{E}^{(u)}, \mu^u)$. The map $\tilde{p}$ just takes any $f \in L^2(X^u, \tilde{E}^{(u)}, \mu^u)$ to u.

PROPOSITION 5.2. *Let μ be a smooth measure on G^0 and give E a hermitian metric $\langle , \rangle$. Then, in a natural way, the bundle $\mathfrak{H}$ is a Hilbert bundle with fiber $L^2(Z, E, \mu)$, and it admits a canonical continuous unitary action of the Lie groupoid G.*

PROOF. We will establish the Hilbert bundle structure of $\mathfrak{H}$ by constructing a cocycle $\{g_{\alpha\beta}\}$ for that bundle (e.g. [**4**, p.48]). Let $\{(U_\alpha, \phi_\alpha)\}$ be an open cover of G^0 which trivializes the fiber bundle $(\tilde{E}, \pi \circ p)$. So each ϕ_α is a fiber preserving homeomorphism from $\tilde{E}\mid_{U_\alpha}$ onto $U_\alpha \times E$ which is a vector bundle isomorphism on fibers. For each u, let $\tau_\alpha^u : X^u \to Z$ be the diffeomorphism determined by ϕ_α. Let $x \in X$, and set $p(x) = u$, $\tau_\alpha^u(x) = z$. Let ϕ_α^u be the restriction of ϕ_α to $\tilde{E}^{(u)}$. Then there exists a positive definite, invertible element $A_\alpha^x \in End(E^z)$ such that for all $\xi \in \tilde{E}^x$, we have $\langle \xi, \xi \rangle^x = \langle A_\alpha^x \phi_\alpha(\xi), \phi_\alpha(\xi) \rangle^z$, and the map $x \to A_\alpha^x$ is continuous. Define $\chi_\alpha : \mathfrak{H} \mid_{U_\alpha} \to U_\alpha \times L^2(Z, E, \mu)$ by:

$$\chi_\alpha f^u(z) = r_\alpha^u(z)^{1/2}(A_\alpha^x)^{1/2}(f^u \circ (\phi_\alpha^u)^{-1})$$

where $r_\alpha^u = d((\tau_\alpha^u)^* \mu^u)/d\mu$. It is left to the reader to check that the map $\chi_\alpha \chi_\beta^{-1}$: $U_\alpha \cap U_\beta \to B(L^2(Z, E, \mu))$ is a cocycle with values in $U(L^2(Z, E, \mu))$. Then $\mathfrak{H}$ is the Hilbert bundle constructed in the standard way with transition functions $g_{\alpha\beta} = \chi_\alpha \chi_\beta^{-1}$. To check that $g_{\alpha\beta}$ is continuous into $U(L^2(Z, E, \mu))$ with the strong operator (=weak operator) topology, one uses elementary measure theory.

Turning to the groupoid action on $\mathfrak{H}$, for each $g \in G$, by the invariance of the μ^u's and the isometric action of G on $\tilde{E}$, the map L_g extends from $C(X^{s(g)}, E^{(s(g))})$ to give a unitary element of $B(L^2(X^{s(g)}, \tilde{E}^{(s(g))}, \mu^{s(g)}), L^2(X^{r(g)}, \tilde{E}^{(r(g))}, \mu^{r(g)}))$. Clearly, algebraically, $\mathfrak{H}$ is a G-space with unitary action $g \to L_g$ where L_g has the same formula as in (5.1). It remains to be shown that the product map $(g, f) \to L_g f$ from $G * \mathfrak{H}$ into $\mathfrak{H}$ is continuous. Suppose then that $g_n \to g$ in G, $f_n \to f$ in $\mathfrak{H}$ with $s(g_n) = \tilde{p}(f_n)$ and $s(g) = \tilde{p}(f)$. Translating this into local terms, we can suppose that U, V are open subsets of G^0, such that $s(g_n), s(g) \in U, r(g_n), r(g) \in V$ and $Z, \tilde{E}$ are trivial over U and V. In addition, we can suppose that $f_n, f \in L^2(Z, E, \mu)$, and, regarding the L_{g_n}, L_g as unitary on $L^2(Z, E, \mu)$, we have to show that $\|L_{g_n} f_n - L_g f\|_2 \to 0$. Given $\epsilon > 0$, there exists $F \in C_c(Z, E)$ such that $\|F - f\|_2 < \epsilon$. By the continuity of the action of G on X we have $\|L_{g_n} F - L_g F\|_\infty \to 0$, and so $\|L_{g_n} F - L_g F\|_2 \to 0$. An elementary triangular inequality argument then shows that $\|L_{g_n} f_n - L_g f\|_2 < \epsilon$ eventually, so that $\|L_{g_n} f_n - L_g f\|_2 \to 0$ as required. □

We now recall some facts about morphisms on Hilbert bundles ([**25**, Chapter 1]). A *morphism* on $\mathfrak{H}$ is a continuous map $T : \mathfrak{H} \to \mathfrak{H}$ which restricts to a linear map T^u on each fiber $\mathfrak{H}^u$ and is such that the adjoint map T^* on $\mathfrak{H}$, where $(T^*)^u = (T^u)^*$, is also continuous. The morphism T is called *equivariant* if for each $g \in G$, we have

$$L_g T^{s(g)} L_{g^{-1}} = T_{r(g)}. \tag{5.3}$$

In particular, if T is equivariant, then

$$\|T^{s(g)}\| = \|T^{r(g)}\| \tag{5.4}$$

for all g.

The morphism T is called *bounded* if

$$\|T\| = \sup_{u \in G^0} \|T^u\| < \infty. \tag{5.5}$$

The set of all bounded morphisms on $\mathfrak{H}$ is a unital C^*-algebra $\mathbf{B}(\mathfrak{H})$ in the obvious way under the norm of (5.5).

It follows by the G-compactness of G^0, the local boundedness of morphisms ([**25**, Cor. 1.8]) and (5.4) that every equivariant morphism is automatically bounded. The set of equivariant morphisms of $\mathfrak{H}$ is a unital C^*-subalgebra of $\mathbf{B}(\mathfrak{H})$, and will be denoted by $\mathbf{B}_G(\mathfrak{H})$.

An element $T \in \mathbf{B}(\mathfrak{H})$ is called *compact* if for every compact subset A of G^0, the set $\{T^u(\xi^u) : \xi^u \in \mathfrak{H}^u, \|\xi^u\| \leq 1, u \in A\}$ has compact closure in $\mathfrak{H}$. The set of bounded compact morphisms is a closed ideal of $\mathbf{B}(\mathfrak{H})$ and is denoted by $\mathbf{K}(\mathfrak{H})$ (cf. [**25**, Lemma 1.12]). Similarly, the set $\mathbf{K}_G(\mathfrak{H})$ of equivariant compact morphisms is a closed ideal of $\mathbf{B}_G(\mathfrak{H})$.

Next we introduce the equivariant compact morphisms which correspond to the "rank 1" operators on a Hilbert module. Let $e_1, e_2 \in C_c(X, \tilde{E})$ and define a section h_{e_1,e_2} of $\tilde{E} \boxtimes \tilde{E}$ by:

$$h_{e_1,e_2}(x, y) = \int L_g e_1(x) \boxtimes L_g e_2(y)\, d\lambda^{p(x)}(g). \tag{5.6}$$

Note that the section h_{e_1,e_2} in (5.6) has compact support and is continuous by the section version of Proposition 3.2.

Now define the operator $T(e_1, e_2)^u$ on $\mathfrak{H}^u$ by:

$$T(e_1, e_2)^u(\xi)(x) = \langle \xi, h_{e_1,e_2}(x, .)\rangle \tag{5.7}$$

where the inner product on the right-hand side of (5.7) is calculated in $\mathfrak{H}^u = L^2(X^u, \tilde{E}^{(u)}, \mu^u)$. Then $T(e_1, e_2)^u$ is a compact operator since it is a kernel operator whose kernel is continuous with compact support. The equation (5.7) can be usefully written:

$$T(e_1, e_2)^u(\xi) = \int L_g e_1 \overline{\langle L_g e_2, \xi\rangle}\, d\lambda^u(g). \tag{5.8}$$

PROPOSITION 5.3. *The map $T(e_1, e_2) \in \boldsymbol{K}_G(\mathfrak{H})$, and the span of morphisms of the form $T(e_1, e_2)$ is dense in $\boldsymbol{K}_G(\mathfrak{H})$.*

PROOF. We first show that $T(e_1, e_2)$ is well-defined and is continuous on $\mathfrak{H}$. We just need to show this on some trivialization $U \times L^2(Z, E, \mu)$ of $\mathfrak{H}$. Applying Lemma 3.1 with $X * X$ in place of X, $\mathcal{E} = \tilde{E} \boxtimes \tilde{E}$ and

$$f'(g, x, y) = L_g e_1(x) \boxtimes L_g e_2(y)$$

gives that h_{e_1,e_2} is continuous on $X * X$. It follows that $T(e_1, e_2)$ is continuous on $\mathfrak{H}$ and that for any compact subset A of U (and hence of G^0) the set

$$\{T(e_1, e_2)^u(\xi^u) : \xi^u \in \mathfrak{H}^u, \|\xi^u\| \leq 1, u \in A\}$$

has compact closure in $\mathfrak{H}$. Next, $T(e_1, e_2)$ is adjointable with adjoint $T(e_2, e_1)$. So $T(e_1, e_2) \in \mathbf{K}(\mathfrak{H})$. We now show that $T = T(e_1, e_2)$ is invariant. This follows from the argument below using (5.8) and (5.3):

$$\begin{aligned}
T^{r(g_0)} L_{g_0}\xi &= \int L_g e_1 \overline{\langle L_g e_2, L_{g_0}\xi\rangle}\, d\lambda^{r(g_0)}(g) \\
&= \int L_{g_0}[L_{g_0^{-1}g} e_1] \overline{\langle L_{g_0^{-1}g} e_2, \xi\rangle}\, d\lambda^{r(g_0)}(g) \\
&= \int L_{g_0}[L_h e_1] \overline{\langle L_h e_2, \xi\rangle}\, d\lambda^{s(g_0)}(h) \\
&= L_{g_0} T^{s(g_0}\xi.
\end{aligned}$$

For the last part of the proposition, we have to show that given $R \in \mathbf{K}_G(\mathfrak{H})$ and $\epsilon > 0$, then there exist ξ_i, η_i $(1 \leq i \leq n)$ in $C_c(X, \tilde{E})$ such that

$$\|R^u - \textstyle\sum_{i=1}^n T(\xi_i, \eta_i)^u\| \leq \epsilon \tag{5.9}$$

for all $u \in G^0$. To this end, we adapt the corresponding argument of Phillips in the locally compact group case ([**25**, pp.92-93]). Restricting R to trivializing subsets of G^0, there is then an open cover $\{U_u\}$ of G^0 such that for each u, there exist $\xi_{i,u}, \eta_{i,u} \in C_c(X, \tilde{E})$ such that

$$\|R^v - \sum_i (\xi_{i,u} \otimes \eta_{i,u})^v\| < \epsilon$$

for all $v \in U_u$. Here $(\xi_{i,u} \otimes \eta_{i,u})^v(w) = (\xi_{i,u})^v \overline{\langle (\eta_{i,u})^v, w\rangle}$ for $w \in L^2(X^v, \tilde{E}^{(v)}, \mu^v)$. Using Proposition 3.3 with G^0 in place of X, there exists a G-partition of unity $\{f_\gamma\}$ subordinate to the cover $\{U_u\}$ of G^0. By considering terms of the form $f_\gamma^{1/2}\xi_{i,u} \otimes f_\gamma^{1/2}\eta_{i,u}$ and using the G-compactness of G^0 and the invariance of the morphisms involved, there exist $\xi_j, \eta_j \in C_c(X, \tilde{E})$ (j in some finite index set J) such that for all $u \in G^0$,

$$\|h(u)R^u - \sum_j (\xi_j \otimes \eta_j)^u\| \leq \epsilon h(u)$$

where $h = \sum_\gamma f_\gamma$. Then for any $u \in G^0$, we have

$$\begin{aligned}
&\|R^u - \textstyle\int \sum_j L_g(\xi_j \otimes \eta_j)L_{g^{-1}}\, d\lambda^u(g)\| \\
&\quad = \|\textstyle\int [h(g^{-1}u)R^u - \sum_j L_g(\xi_j \otimes \eta_j)L_{g^{-1}}]\, d\lambda^u(g)\| \\
&\quad = \|\textstyle\int L_g[h(g^{-1}u)R^{g^{-1}u} - \sum_j (\xi_j \otimes \eta_j)^{g^{-1}u}]L_{g^{-1}}\, d\lambda^u(g)\| \\
&\quad \leq \int \|h(g^{-1}u)R^{g^{-1}u} - \textstyle\sum_j (\xi_j \otimes \eta_j)^{g^{-1}u}\|\, d\lambda^u(g) \\
&\quad \leq \int \epsilon h(g^{-1}u)\, d\lambda^u(g) = \epsilon.
\end{aligned}$$

Since $L_g(\xi_j \otimes \eta_j)L_{g^{-1}} = L_g\xi_j \otimes L_g\eta_j$, we obtain (5.9). $\square$

An element $T \in \mathbf{B}(\mathfrak{H})$ is called *Fredholm* if there exists $S \in \mathbf{B}(\mathfrak{H})$ such that both $ST - I, TS - I \in \mathbf{K}(\mathfrak{H})$.

PROPOSITION 5.4. (i) *Suppose that $T \in \boldsymbol{B}_G(\mathfrak{H})$ is Fredholm. Then there exists $S \in \boldsymbol{B}_G(\mathfrak{H})$ such that both $ST - I, TS - I \in \boldsymbol{K}_G(\mathfrak{H})$.*

(ii) *The elliptic pseudodifferential family $D = \{D^u\}$ defines, by extending each D^u to $\tilde{D}^u \in B(\mathfrak{H}^u)$ (Proposition 4.1) an invariant Fredholm morphism on $\mathfrak{H}$.*

PROOF. (i) The locally compact group version of this is given in [**25**, Lemma 3.7]. For each $u \in G^0$, define $A^u \in B(\mathfrak{H}^u)$ by: for $\xi \in \mathfrak{H}^u$, $x \in X^u$, set

$$A^u\xi(x) = \int_{G^u} c(g^{-1}x)L_g S^{s(g)} L_{g^{-1}} f(x)\, d\lambda^u(g) \tag{5.10}$$

where c is as in Proposition 3.4. The argument of Phillips adapts directly to show that $A \in \mathbf{B}_G(\mathfrak{H})$ and satisfies $TA - I, AT - I \in \mathbf{K}_G(\mathfrak{H})$.

(ii) By Proposition 4.1, in local terms, the map $u \to \tilde{D}^u$, regarded as a map into $B(L^2(Z, E, \mu))$, is strong operator – even norm – continuous, and has an adjoint $u \to (\tilde{D}^u)^*$. So D, identified with $\{\tilde{D}^u\}$, is a morphism. The invariance of this morphism follows from the invariance of each D^u on the dense subspace $C^\infty(X^u, E^{(u)})$ of $L^2(X^u, \tilde{E}^{(u)}, \mu^u)$. $\square$

We now construct the Fredholm module which will give the index of the elliptic pseudodifferential family D. The proof constructs a certain Kasparov $(\mathbb{C}, C^*_{red}(G))$

module. The proof in the locally compact group case is effectively given by Phillips ([**25**, Ch. 6]) in his discussion of the generalized Green-Rosenberg theorem. Actually, Phillips (*loc. cit.*) proves more than that, showing that his equivariant K-theory group $K^0_G(X)$ is isomorphic as an abelian group to $K_0(C^*(G,X))$. We will not consider the groupoid version of this in this paper but consider only the corresponding Kasparov $(\mathbb{C}, C^*_{red}(G))$-module that gives the analytic index.

Let $\Gamma_c(\mathfrak{H}) = C_c(G^0, \mathfrak{H})$. For $T \in \mathbf{B}_G(\mathfrak{H})$ and $f \in \Gamma_c(\mathfrak{H})$, define a section $\Phi(T)(f)$ of $\mathfrak{H}$ by:

$$\Phi(T)(f)(u) = T^u f \quad (= T^u(f(u))). \tag{5.11}$$

PROPOSITION 5.5. *Let $f \in \Gamma_c(\mathfrak{H})$. Then the section $\Phi(T)(f)$ belongs to $\Gamma_c(\mathfrak{H})$.*

PROOF. Trivially, the support of $\Phi(T)$ is compact in G^0. Let $u \in G^0$ and $u_n \to u$ in G^0. Trivializing in a neighborhood of u in G^0, we can regard $f(u_n), f(u) \in L^2(Z, E, \mu)$ (cf. Proposition 5.2), and by the continuity of f, $\|f(u_n) - f(u)\|_2 \to 0$. Since T is continuous on $\mathfrak{H}$, we have $T_{u_n} f \to T_u f$. So $\Phi(T)(f)$ belongs to $\Gamma_c(\mathfrak{H})$. $\square$

We will show that $\Gamma_c(\mathfrak{H})$ is a pre-Hilbert module over the pre-C^*-algebra $C_c(G)$, with $C_c(G)$-inner product and module operations given below, where $g \in G$, $x \in X$, $e, e_1, e_2 \in \Gamma_c(\mathfrak{H})$ and $f \in C_c(G)$:

$$\langle e_1, e_2\rangle(g) = \int_{X^{r(g)}} \overline{\langle e_1(x), L_g e_2(x)\rangle}\, d\mu^{r(g)}(x) \tag{5.12}$$

$$ef(x) = \int_{G^{p(x)}} L_g e(x) f(g^{-1})\, d\lambda^{p(x)}(g). \tag{5.13}$$

The preceding equality can alternatively be written as:

$$(ef)(u) = \int_{G^u} (L_g e) f(g^{-1})\, d\lambda^u(g). \tag{5.14}$$

The following equality is useful:

$$\langle e_1, L_g e_2\rangle = \langle L_{g^{-1}} e_1, e_2\rangle \tag{5.15}$$

where we have omitted the superscripts $r(g), s(g)$ respectively on the two preceding inner products. This follows from (3.5) and the isometric action of G on $\tilde{E}$.

Note that (5.12) can then be rewritten succinctly as:

$$\langle e_1, e_2\rangle(g) = \overline{\langle e_1, L_g e_2\rangle}. \tag{5.16}$$

PROPOSITION 5.6. *The space $\Gamma_c(\mathfrak{H})$ is a pre-Hilbert module over the pre-C^*-algebra $C_c(G) \subset C^*_{red}(G)$ with $C_c(G)$-inner product and module action given by (5.12) and (5.13) respectively.*

PROOF. We claim first that $\langle e_1, e_2\rangle \in C_c(G)$. That $\langle e_1, e_2\rangle$ is continuous follows using (5.16), the continuity of the sections e_1, e_2 and the strong operator continuity of the groupoid action $g \to L_g$ on $\mathfrak{H}$. Now let C_i be the (compact) support of the sections e_i. By the properness of the action of G on X, it follows that the set $\{(g,x) : (g^{-1}x, x) \in C_2 \times C_1\}$ is compact, and the support of $\langle e_1, e_2\rangle$ is contained in the projection of that set onto the first coordinate. A similar argument (using (5.14)) shows that $ef \in \Gamma_c(\mathfrak{H})$.

It remains to check ([**3**, p.126]) that (a) $\langle , \rangle$ is sesquilinear, and that for all $e, e_1, e_2 \in \Gamma_c(\mathfrak{H})$ and all $f \in C_c(G)$, we have (b) $\langle e_1, e_2 f\rangle = \langle e_1, e_2\rangle f$, (c) $\langle e_1, e_2\rangle^* =$

$\langle e_2, e_1\rangle$, and that (d) $\langle e,e\rangle \geq 0$, and is 0 if and only if $e = 0$. (a) is obvious. We prove the others in turn.

For (b), using (2.2):

$$\begin{aligned}
\langle e_1, e_2 f\rangle(g) &= \int \overline{\langle e_1(x), L_g(e_2 f)(x)\rangle}\, d\mu^{r(g)}(x) \\
&= \int \overline{\langle e_1(x), g[e_2 f(g^{-1}x)]\rangle}\, d\mu^{r(g)}(x) \\
&= \int\int \overline{\langle e_1(x), g[L_k(e_2)(g^{-1}x)]\rangle} f(k^{-1})\, d\lambda^{s(g)}(k) d\mu^{r(g)}(x) \\
&= \int\int \overline{\langle e_1(x), L_{gk}e_2)(x)\rangle} f(k^{-1})\, d\lambda^{s(g)}(k) d\mu^{r(g)}(x) \\
&= \int\int \overline{\langle e_1(x), L_h e_2(x)\rangle} f(h^{-1}g)\, d\lambda^{r(g)}(h) d\mu^{r(g)}(x) \\
&= \int \langle e_1, e_2\rangle(h) f(h^{-1}g)\, d\lambda^{r(g)}(h) \qquad (5.17)\\
&= \langle e_1, e_2\rangle f(g).
\end{aligned}$$

For (c), using (5.15) and (5.16), we have $\langle e_2, e_1\rangle^*(g) = \overline{\langle e_2, L_{g^{-1}}e_1\rangle} = \overline{\langle L_g e_2, e_1\rangle} = \langle e_1, e_2\rangle(g)$.

For (d), from the definition of $C^*_{red}(G)$ in §2, we just have to show that for $u \in G^0$, we have $\pi_u(\langle e,e\rangle) \geq 0$ where π_u was defined in (2.4). Taking $f = \langle e,e\rangle$, we get, by [**22**, (3.42)], that for $\xi \in C_c(G)$

$$\begin{aligned}
\langle \pi_u(f)\xi, \xi\rangle &= \int\int f(gh)\xi(h^{-1})\overline{\xi(g)}\, d\lambda^u(h)\lambda_u(g) \\
&= \int\int \overline{\langle e, L_{gh}e\rangle}\xi(h^{-1})\overline{\xi(g)}\, d\lambda^u(h) d\lambda_u(g) \\
&= \int\int \overline{\langle L_{g^{-1}}e, L_{h^{-1}}e\rangle}\xi(h)\overline{\xi(g)}\, d\lambda_u(h) d\lambda_u(g) \\
&= \langle \int \xi(g) L_{g^{-1}}e\, d\lambda_u(g), \int \xi(h) L_{h^{-1}}e\, d\lambda_u(h)\rangle_{L^2(X^u, \tilde{E}^{(u)}, \mu^u)} \qquad (5.18)\\
&\geq 0. \qquad (5.19)
\end{aligned}$$

Since for $u \in G^0$, $\langle e,e\rangle(u) = \int \|e(x)\|^2\, d\lambda^u(g)$, it follows that $\langle e,e\rangle = 0$ if and only if $e = 0$. □

It is elementary that the $C_c(G)$-valued inner product $\langle,\rangle$ on $\Gamma_c(\mathfrak{H})$ extends to a $C^*_{red}(G)$-valued inner product, also denoted by $\langle,\rangle$, on the completion $\Gamma(\mathfrak{H})$ of $\Gamma_c(\mathfrak{H})$ under the norm $e \to \|\langle e,e\rangle\|^{1/2}$, and is a Hilbert $C^*_{red}(G)$-module. The C^*-algebras of bounded and compact module maps on $\Gamma(\mathfrak{H})$ will be denoted by $\mathbf{B}(\Gamma(\mathfrak{H}))$ and $\mathbf{K}(\Gamma(\mathfrak{H}))$ respectively. In the following theorem, the map Φ actually determines an isomorphism from $\mathbf{B}_G(\mathfrak{H})$ onto $\mathbf{B}(\Gamma(\mathfrak{H}))$ but we don't include the proof since it is not needed for our purposes.

THEOREM 5.7. *The map Φ of Proposition 5.5 determines a C^*-isomorphism from $\mathbf{B}_G(\mathfrak{H})$ into $\mathbf{B}(\Gamma(\mathfrak{H}))$ that takes $\mathbf{K}_G(\mathfrak{H})$ onto $\mathbf{K}(\Gamma(\mathfrak{H}))$.*

PROOF. Trivially, $\Phi(T) : \Gamma_c(\mathfrak{H}) \to \Gamma_c(\mathfrak{H})$ is linear. We will show first that is continuous with norm $\leq \|T\|$. In particular, it extends by continuity to a bounded linear map on $\Gamma(\mathfrak{H})$.

Let $e \in \Gamma_c(\mathfrak{H})$ and $T \in \mathbf{B}_G(\mathfrak{H})$. It is sufficient to show that

$$\|\langle \Phi(T)e, \Phi(T)e\rangle\| \leq \|T\|^2 \|\langle e, e\rangle\|. \tag{5.20}$$

This will follow if we can show that for u, ξ as in the proof of (d) of Proposition 5.6,

$$\langle \pi_u(\langle \Phi(T)e, \Phi(T)e\rangle)\xi, \xi\rangle \leq \|T\|^2 \langle \pi_u(\langle e, e\rangle)\xi, \xi\rangle. \tag{5.21}$$

Using (5.18), the invariance of T and the fact that (in the proof) $s(g) = u$:

$$\begin{aligned} \langle \pi_u(\langle \Phi(T)e, \Phi(T)e\rangle)\xi, \xi\rangle &= \|\textstyle\int \xi(g) L_{g^{-1}} T^{r(g)} e \, d\lambda_u(g)\|_2^2 \\ &= \|\textstyle\int \xi(g) T^u L_{g^{-1}} e \, d\lambda_u(g)\|_2^2 \\ &\leq \|T^u\|^2 \|\textstyle\int \xi(g) L_{g^{-1}} e \, d\lambda_u(g)\|^2 \end{aligned}$$

and (5.21) follows.

So $\Phi(T)$ is a bounded linear operator on $\Gamma(\mathfrak{H})$ and $\|\Phi(T)\| \leq \|T\|$. To obtain $\Phi(T) \in \mathbf{B}(\Gamma(\mathfrak{H}))$, we show that $\Phi(T)$ is an adjointable module map on $\Gamma(\mathfrak{H})$ with adjoint $\Phi(T^*)$.

For $e_1, e_2 \in \Gamma(\mathfrak{H})$ and $g \in G$, we have using (5.16) and the invariance of T^*,

$$\begin{aligned} \langle \Phi(T)e_1, e_2\rangle(g) &= \overline{\langle \Phi(T)e_1, L_g e_2\rangle} \\ &= \overline{\langle T^{r(g)} e_1, L_g e_2\rangle} \\ &= \overline{\langle e_1, (T^{r(g)})^* L_g e_2\rangle} \\ &= \overline{\langle e_1, L_g (T^{s(g)})^* e_2\rangle} \\ &= \langle e_1, \Phi(T^*) e_2\rangle. \end{aligned}$$

So $\Phi(T^*)$ is an adjoint for $\Phi(T)$.

Next we have to show that $\Phi(T)$ is a module map. To this end, for $e \in \Gamma_c(\mathfrak{H}), f \in C_c(G)$, we have by (5.14),

$$\begin{aligned} [\Phi(T)(ef)](u) &= T^u(ef)(u) \\ &= T^u \int L_g(e) f(g^{-1}) \, d\lambda^u(g) \\ &= \int T^{r(g)} L_g(e) f(g^{-1}) \, d\lambda^u(g) \\ &= \int L_g T^{s(g)}(e) f(g^{-1}) \, d\lambda^u(g) \\ &= (Te)f(u). \end{aligned}$$

So $\Phi(T)$ is a module map. From (5.11), Φ is a *-homomorphism that is one-to-one. So Φ is a C^*-isomorphism from $\mathbf{B}_G(\mathfrak{H})$ into $\mathbf{B}(\Gamma(\mathfrak{H}))$.

For the last part of the theorem, from Proposition 5.3 and the density of $\Gamma_c(\mathfrak{H})$ in $\Gamma(\mathfrak{H})$, we just have to show that $\Phi(T(e_1, e_2)) \in \mathbf{K}(\Gamma(\mathfrak{H}))$. In fact, $\Phi(T(e_1, e_2))$ is just the "rank 1" operator θ_{e_1, e_2} ([**3**, p.128]). Indeed, for $e \in \Gamma_c(\mathfrak{H})$ and $u \in G^0$, using 5.8:

$$\begin{aligned}\Phi(T(e_1,e_2))e(u) &= T(e_1,e_2)^u e \\ &= \int L_g e_1 \overline{\langle L_g e_2, e\rangle}\, d\lambda^u(g) \\ &= \int L_g e_1 \overline{\langle e_2, L_{g^{-1}} e\rangle}\, d\lambda^u(g) \\ &= \int L_g e_1 \langle e_2, e\rangle(g^{-1})\, d\lambda^u(g) \\ &= e_1\langle e_2, e\rangle(u) \\ &= \theta_{e_1,e_2}(e)(u).\end{aligned}$$

□

We can now complete the construction of the analytic index of the pseudodifferential operator D. By Proposition 5.4, D is an invariant Fredholm morphism on $\mathfrak{H}$ and there exists an $S \in \mathbf{B}_G(\mathfrak{H})$ such that both $SD - I, DS - I \in \mathbf{K}_G(\mathfrak{H})$. Let $a = \Phi(D), b = \Phi(S)$. By Theorem 5.7, we have $ab - I, ba - I \in \mathbf{K}(\Gamma(\mathfrak{H}))$. We summarize the well-known construction of a Fredholm module from a. Let $\pi : \mathbf{B}(\Gamma(\mathfrak{H})) \to \mathbf{B}(\Gamma(\mathfrak{H}))/\mathbf{K}(\Gamma(\mathfrak{H}))$ be the quotient map. Then $\pi(a)$ is invertible. Let u be the unitary part of $\pi(a)$ given by the polar decomposition of $\pi(a)$: so $u = \pi(a)[\pi(a^*a)]^{-1/2}$. Let $c \in \mathbf{B}(\Gamma(\mathfrak{H}))$ be such that $\pi(c) = u$. Then $cc^* - I, c^*c - I \in \mathbf{K}(\Gamma(\mathfrak{H}))$. We then get a Kasparov $(\mathbb{C}, C^*_{red}(G))$-bimodule

$$(\Gamma(\mathfrak{H}) \oplus \Gamma(\mathfrak{H}), \begin{pmatrix} 0 & c^* \\ c & 0 \end{pmatrix})$$

which gives an element of $KK^0(\mathbb{C}, C^*_{red}(G)) = K_0(C^*_{red}(G))$. This element is independent of the choice of c by the invariance of Kasparov classes under compact perturbations. This element is the analytic index of D.

References

[1] M. F. Atiyah and I. Singer, *The index of elliptic operators, I*, Ann. of Math. 87(1968), 484-530.

[2] M. F. Atiyah and I. Singer, *The index of elliptic operators, IV*, Ann. of Math. 93(1971), 119-38.

[3] B. Blackadar, *K-theory for operator algebras*, MSRI Publications, Vol. 5, Springer-Verlag, New York, 1986.

[4] R. Bott and L. W. Tu, *Differential Forms in Algebraic Topology*, Graduate Texts in Mathematics, Springer-Verlag, New York, 1982.

[5] A. Connes, *Sur la théorie non commutative de l'intégration*, Lecture Notes in Mathematics, 725(1979), 19-143.

[6] A. Connes, *Noncommutative Geometry*, Academic Press, Inc., New York, 1994.

[7] A. Connes and H. Moscovici, *The L^2-index theorem for homogeneous spaces of Lie groups*, Ann. of Math. 115(1982), 291-330.

[8] Y. V. Egorov and M. A. Shubin (Eds.), *Partial Differential Equations II*, Encyclopedia of Mathematical Sciences, Vol. 31, Springer-Verlag, Berlin, 1994.

[9] L. Hörmander, *The Analysis of Linear Partial Differential Operators.*, Vol 3, New York, Springer-Verlag, 1985.

[10] L. Hörmander, *Pseudodifferential operators and hypoelliptic equations*, Amer. Math. Soc., Proc. Symposia Pure Math. 10(1967), 138-183.

[11] G. G. Kasparov, *An index for invariant elliptic operators, K-theory, and representations of Lie groups*, Soviet Math. Dokl. 27(1983), 105-109.

[12] G. G. Kasparov, *Equivariant KK-theory and the Novikov conjecture*, Invent. Math. 91(1988), 147-201.

[13] N. P. Landsman, *Mathematical Topics between Classical and Quantum Mechanics*, Springer-Verlag, New York, 1998.

[14] R. Lauter and V. Nistor, *Pseudodifferential analysis on groupoids and singular spaces*, preprint, 1999.

[15] K. C. H. Mackenzie, *Lie Groupoids and Lie algebroids in Differential Geometry*, London Mathematical Society Lecture Note Series, vol. 124, Cambridge University Press, Cambridge, 1987.

[16] B. Monthubert, *Pseudodifferential calculus on manifolds with corners and groupoids*, Preprint, 1997.

[17] B. Monthubert and F. Pierrot, *Indice analytique et groupoïdes de Lie*, C. R. Acad. Sci. Paris, Ser. I, 325(1997), 193-198.

[18] P. S. Muhly, *Coordinates in Operator Algebra*, to appear, CBMS Regional Conference Series in Mathematics, American Mathematical Society, Providence, 180pp..

[19] P. S. Muhly, J. N. Renault and D. P. Williams, *Equivalence and isomorphism for groupoid C^*-algebras*, J. Operator Theory 17(1987), 3-22.

[20] P. S. Muhly and D. P. Williams, *Groupoid cohomology and the Dixmier-Douady class.* Proc. London Math. Soc. 71(1995), 109-134.

[21] V. Nistor, A. Weinstein and P. Xu, *Pseudodifferential operators on differential groupoids*, Pacific J. Math. 189(1999), 117-152.

[22] A. L. T. Paterson, *Groupoids, inverse semigroups and their operator algebras*, Progress in Mathematics, Vol. 170, Birkhäuser, Boston, 1999.

[23] A. L. T. Paterson, *Continuous family groupoids*, preprint, 1999.

[24] R. S. Palais, *On the existence of slices for actions of non-compact Lie groups*, Ann. of Math. 73(1961), 295-323.

[25] N. C. Phillips, *Equivariant K-theory for proper actions*, Pitman Research Notes in Mathematics, Vol. 178, John Wiley, New York, 1988.

[26] J. Pradines, *Théorie de Lie pour les groupoïdes différentiables. Relations entre propriétés locales et globales.* C. R. Acad. Sci. Paris Sér. A-B 263(1966), A907-A910.

[27] J. Pradines, *Théorie de Lie pour les groupoïdes différentiables. Calcul différential dans la catégorie des groupoïdes infinitésimaux*, C. R. Acad. Sci. Paris Sér. A-B 264(1967), A 245-A 248.

[28] J. N. Renault, *A groupoid approach to C^*-algebras*, Lecture Notes in Mathematics, Vol. 793, Springer-Verlag, New York, 1980.

[29] G. Segal, *Equivariant K-theory*, Publ. Math. Inst. Hautes Études Sci. 34(1968), 129-151.

[30] G. Segal, *Fredholm Complexes*, Quart. J. Math. Oxford 21(1970), 385-402.

[31] M. A. Shubin, *Pseudodifferential Operators and Spectral Theory*, Springer-Verlag, New York, 1987.

[32] R. O. Wells, *Differential Analysis on Complex Manifolds*, Graduate Texts in Mathematics, Springer-Verlag, New York 1980.

DEPARTMENT OF MATHEMATICS, UNIVERSITY OF MISSISSIPPI, UNIVERSITY, MISSISSIPPI 38677
E-mail address: mmap@olemiss.edu

Contemporary Mathematics
Volume **282**, 2001

Groupoid C^*-algebras and Operator K-theory

Pierre-Yves Le Gall

ABSTRACT. Let $\mathcal{G}$ be a locally compact topological groupoid, A and B two C*-algebras endowed with continuous actions of $\mathcal{G}$. We define an operator K-theory group $KK_{\mathcal{G}}(A, B)$, generalizing Kasparov's construction. We describe the functorial properties of this theory: the existence of a "Kasparov product" and functoriality with respect to groupoid cocycles, and explain some applications to the computation of the K-theory of groupoid C*-algebras.

1. Introduction

Jean N. Renault has explained how many important examples of operator algebras can be defined as the (full or reduced) envelopping C*-algebra of the convolution algebra of a groupoid. This construction has found many important applications in Alain Connes' Non Commutative Geometry. In particular these ideas allows one to state and prove many powerful index theorems.

One of the main problems for these algebras is the computation of their K-theory. This problem is closely related to the Baum-Connes conjecture which states that for a groupoid $\mathcal{G}$ (under some mild assumptions) a certain map between a geometrically defined group $K_*^{top}(\mathcal{G})$ and $K_*(C_r^*(\mathcal{G}))$ is an isomorphism.

In the case of groups, the study of this conjecture relies mainly on Kasparov's equivariant (with respect to groups) theory. The purpose of this paper is to give the main definitions and properties of a Kasparov theory equivariant with respect to groupoids ([L1] [L2]) and to survey some applications (due to J. L. Tu) of this theory to the Baum-Connes conjecture.

2. Basic definitions

2.1. Morphisms of groupoids. Let $\mathcal{G}$ and $\mathcal{H}$ be two topological groupoids (by topological groupoids we mean, in this paper, Hausdorff locally compact and σ-compact groupoids). We shall call a continuous functor from $\mathcal{G}$ to $\mathcal{H}$ a *strict morphism*. There is another, more general class of morphisms of groupoids which is just a continuous analogue of C^∞ morphisms of foliations spaces (in the sense of [C1], [HS]):

1991 *Mathematics Subject Classification.* Primary 19K35; Secondary 22A22, 20L13.

Key words and phrases. Equivariant Kasparov theory, topological groupoids, equivalence of groupoids, Baum-Connes conjecture.

Recall that a continuous *right action* of $\mathcal{G}$ on a topological space Z is given by two continuous maps, $\sigma : Z \to \mathcal{G}^{(0)}$ and $(z,\gamma) \mapsto z\gamma$ from $Z \rtimes_r \mathcal{G} = \{(z,\gamma) \in Z \rtimes \mathcal{G}\,;\, \sigma(z) = r(\gamma)\}$ into Z such that : $\sigma(z\gamma) = s(\gamma)$ if x is a unit $zx = z$ and if $(\gamma, \gamma') \in \mathcal{G}^{(2)}$, $(z\gamma)\gamma' = z(\gamma \circ \gamma')$. One defines the left actions in the same way.

DEFINITION 2.1. A generalized morphism φ from $\mathcal{G}$ to $\mathcal{H}$ is a topological space Ω_φ with:

a) Two continuous applications $r : \Omega_\varphi \to \mathcal{G}^{(0)}$ and $s : \Omega_\varphi \to \mathcal{H}^{(0)}$

b) A continuous left action of $\mathcal{G}$ on Ω_φ (with respect to r) and a right action of $\mathcal{H}$ on Ω_φ (with respect to s) such that r is a principal fibration.

c) A continuous, open and surjective map p from a locally compact space Y to $\mathcal{G}^{(0)}$ admitting a factorization by a continuous map $\widetilde{p} : Y \to \Omega_\varphi$ such that $p = r \circ \widetilde{p}$.

To be more precise, b) means that the action of $\mathcal{H}$ on Ω_φ is proper, r is surjective and, forall $x, y \in \Omega_\varphi$ with $r(x) = r(y)$, there is a unique $h \in \mathcal{H}$ such that $xh = y$.

The generalized morphism φ is an *equivalence of groupoids* [MRW] when s is a fibration of strucural groupoid $\mathcal{G}$ and the maps r and s are open. Generalized morphisms can be composed in the same way as C^∞ morphisms of leaf spaces [HS].

Remarks: a) In many examples r is open or has local sections so the last condition is automatically satisfied.

b) There is an alternative description of generalized morphisms (in terms of "cocycles") [Ha], [HS]: For each continuous map h from a topological space Z to $\mathcal{G}^{(0)}$ one defines on $\mathcal{G}_Z = \{(z,\gamma,y) \in Z \times \mathcal{G} \times Z\,;\, s(\gamma) = h(y) \,\text{and}\, r(\gamma) = h(z)\}$ a groupoid structure by $(z,\gamma,y)\circ(y,\gamma',x) = (z,\gamma\gamma',x)$ and $(z,\gamma,y)^{-1} = (y,\gamma^{-1},z)$.

Since r is a fibration, to the generalized morphism φ is canonically associated a strict f morphism from the groupoid $\mathcal{G}_{\Omega_\varphi}$ to $\mathcal{H}$ in the following way: let (z,γ,y) be an element of $\mathcal{G}_{\Omega_\varphi}$. Since $r(z) = r(\gamma y)$, there exists an unique $g \in \mathcal{H}$ such that $zg = \gamma y$. Let $f((z,\gamma,y)) = g$. The application thus defined is a strict morphism. Composing with the obvious strict morphism $\mathcal{G}_Y \to \mathcal{G}_{\Omega_\varphi}$ one sees that φ determines a triple (that we shall call *premorphism*) (Y,p,f) where p is a continuous, open and surjective map from a locally compact space Y to $G^{(0)}$ and $f : \mathcal{G}_Y \to \mathcal{H}$ is a strict morphism.

Conversely it can be proved that a premorphism determines a generalized morphism, so that generalized morphisms are just equivalence classes (in an appropriate sense) of premorphisms.

2.2. Actions of groupoids on fields of C*-algebras.

2.2.1. $C_0(X)$-*algebras.* Let us recall the definition of $C_0(X)$-algebras ([DG], [Ka2], [Bl]):

DEFINITION 2.2. Let X be a locally compact topological space. A $C_0(X)$-algebra is a pair (A,θ) consisting of a C*-algebra A and an homomorphism $\theta : C_0(X) \to \mathbf{Z}\mathcal{M}(A)$ such that: $\theta(C_0(X))A = A$.

Such an algebra is given (using θ) a canonical $C_0(X)$-module structure and is identified with the algebra of continuous sections of a bundle having for *fiber*, at each $x \in X$ the algebra $A_x = A/C_xA$ where $C_x = \{h \in C_0(X)\,;\, h(x) = 0\}$. The class in A_x of an element $a \in A$ is noted a_x. Let us remark that a is completely determined by the family $(a_x)_{x\in X})$ and that the bundle A is *semi-continuous* in the sense that for all $a \in A$ the application $x \mapsto \|a_x\|$ is upper semi-continuous.

A $C_0(X)$-agebra A can be restricted to each open subset U of X as $A_U = C_0(U)A$ (which is a closed ideal of A) or closed subset F as $A_F = A/A_{X-F}$.

One sets: $A_b = \{a \in \mathcal{M}(A)\,;\, \forall \varphi \in C_0(X) : \varphi a \in A\}$ and $A_c = AC_c(X)$ the bounded and compactly supported sections of A.

Let us recall the construction of the *pullback* of A along a continuous map $p : Y \to X$. The C*-algebraic tensor product $A \otimes C_0(Y)$ is a $C_0(X \times Y)$-algebra. The pullback p^*A of A along p is its restriction to the closed subspace $\{(y, p(y))\,;\, y \in Y\}$ of $Y \times X$. It is in a natural way a $C_0(Y)$-algebra and, for each $y \in Y$, one has a canonical identification $(p^*A)_y = A_{p(y)}$.

2.2.2. *Action of a groupoid on a $C_0(X)$-algebra.* Let $\mathcal{G}$ a groupoid having for unit space a locally compact topological space X and A a $C_0(X)$-algebra. An action of $\mathcal{G}$ on A is given by a C*-algebra isomorphism $\alpha_\gamma : A_{s(\gamma)} \to A_{r(\gamma)}$ for all $\gamma \in \mathcal{G}$, such that, for all composable pair $(\gamma, \gamma') \in \mathcal{G}^{(2)}$, one has: $\alpha_{\gamma\circ\gamma'} = \alpha_\gamma \circ \alpha_{\gamma'}$.

When $\mathcal{G}$ is a locally compact topological groupoid, taking into account the canonical identifications: for all $\gamma \in \mathcal{G}$, $(s^*A)_\gamma = A_{s(\gamma)}$ and $(r^*A)_\gamma = A_{r(\gamma)}$, to a $C_0(\mathcal{G})$-algebra isomorphism α from s^*A into r^*A is associated, for all $\gamma \in \mathcal{G}$, a C*-algebra isomorphism $\alpha_\gamma : A_{s(\gamma)} \to A_{r(\gamma)}$.

DEFINITION 2.3. Let $\mathcal{G}$ be a locally compact topological groupoid of basis X. A continuous action of $\mathcal{G}$ on a $C_0(X)$-algebra A is an isomorphism of $C_0(\mathcal{G})$-algebra $\alpha : s^*A \to r^*A$ such that, for all composable pair $(\gamma, \gamma') \in \mathcal{G}^{(2)}$: $\alpha_{\gamma\circ\gamma'} = \alpha_\gamma \circ \alpha_{\gamma'}$.

2.2.3. *General constructions.* We will now survey some basic constructions for these algebras, namely crossed products and induction with respect to (strict and generalized) morphisms of groupoids.

Crossed products

Let us assume that $\mathcal{G}$ is endowed with a Haar system $\lambda = (\lambda_x)_{x\in X}$ (as defined by J. Renault) [Re]:

DEFINITION 2.4. A Haar system on $\mathcal{G}$ is a family of positive mesures $(\lambda^x)_{x\in X}$ with supp $(\lambda^x) \subset \mathcal{G}^x$ and such that:

a) $\forall x \in X$, $\forall \phi \in C_c(\mathcal{G})$:

$$x \mapsto \int_{\gamma\in\mathcal{G}^x} \phi(\gamma)d\lambda^x(\gamma) \in C_c(X).$$

b) $\forall x, y \in X$, $\forall \gamma \in \mathcal{G}_x^y$, $\forall \varphi \in C_c(\mathcal{G})$:

$$\int_{\gamma'\in\mathcal{G}^x} \varphi(\gamma\circ\gamma')d\lambda^x(\gamma') = \int_{\gamma'\in\mathcal{G}^y} \varphi(\gamma')d\lambda^y(\gamma').$$

As in the case of groups, one defines on $(r^*A)_c = C_c(\mathcal{G})r^*A$ an involution and a convolution product by the formulas: for all a, $b \in (r^*A)_c$ $\gamma \in \mathcal{G}$,

$$(a * b)_\gamma = \int_{\gamma'\in\mathcal{G}^{r(\gamma)}} a_{\gamma'}\alpha_{\gamma'}(b_{\gamma'^{-1}\circ\gamma})d\lambda^{r(\gamma)}(\gamma') \qquad (a^*)_\gamma = \alpha_\gamma((a_{\gamma^{-1}})^*).$$

One can prove that the families $((a*b)_\gamma)_{\gamma\in\mathcal{G}}$ and $((a^*)_\gamma)_{\gamma\in\mathcal{G}}$ define elements $a * b$ and a^* of $(r^*A)_c$.

Let $\mathcal{D}_A^1$ be the Banach involutive algebra obtained by completing $\mathcal{D}_A$ with respect to the norm: $\|a\|_1 = \max\{|a|_1, |a^*|_1\}$ and

$$|a|_1 = \sup_{x\in X} \int_{\gamma\in\mathcal{G}^x} \|a_\gamma\| d\lambda^x(\gamma).$$

For all $x \in X$ one has a representation λ_x of $\mathcal{D}_A^1$ in $\mathcal{L}(L^2(\mathcal{G}^x; A_x)$ (where $L^2(\mathcal{G}^x; A_x)$ is the Hilbert A_x-module $L^2(\mathcal{G}^x, d\lambda^x) \otimes A_x$) given by

$$(a * b)_\gamma = \int_{\gamma'\in\mathcal{G}^{r(\gamma)}} a_{\gamma'}\alpha_{\gamma'}(b_{\gamma'^{-1}\circ\gamma}) d\lambda^{r(\gamma)}(\gamma')$$

The family $(\lambda_x)_{x\in X}$ of representations thus obtained is separating for $\mathcal{D}_A^1$.

DEFINITION 2.5. a) The crossed product $A \rtimes \mathcal{G}$ of A by $\mathcal{G}$ is the envelopping C*-algebra of the Banach involutive algebra $\mathcal{D}_A^1$.

b) The completion of $\mathcal{D}_A^1$ for the C*-norm $\|f\| = \sup\limits_{x\in X} \|\lambda_x(f)\|$ is the reduced crossed product of A by $\mathcal{G}$ and will be designed by: $A \rtimes_r \mathcal{G}$.

c) The full and reduced C*-algebras of $\mathcal{G}$ are, respectively, $C^*(\mathcal{G}) = C_0(X) \rtimes \mathcal{G}$ and $C_r^*(\mathcal{G}) = C_0(X) \rtimes_r \mathcal{G}$.

Morphisms of groupoids

Let $\mathcal{G}$ and $\mathcal{H}$ be two topological groupoids and D be a C*-algebra endowed with a continuous action $\alpha : s^*D \to r^*D$ of $\mathcal{H}$.

a) Let $f : \mathcal{G} \to \mathcal{H}$ be a strict morphism. One defines on the pullback f^*D a continuous action of $\mathcal{G}$ by $\gamma \in \mathcal{G}$, $(f^*\alpha)_\gamma = \alpha_{f(\gamma)} : D_{s(f(\gamma))} \to D_{r(f(\gamma))}$.

b) This construction can be extended to general morphisms. Let (Y, p, f) a premorphism from $\mathcal{G}$ to $\mathcal{H}$. By a) one defines a $\mathcal{G}_Y$-algebra structure on f^*D. One has the following lemma:

LEMMA 2.6. *For all $\mathcal{G}_Y$-algebra B there exists a $\mathcal{G}$-algebra $p_!B$ such that $p^*p_!B = B$ (as $\mathcal{G}_Y$-algebras).*

Let us sketch the construction of $p_!B$. One defines as $Y \times_p Y$ the open subgroupoid of $\mathcal{G}_Y$: $\{(z, u_{p(z)}, y) \in \mathcal{G}_Y\}$ (here $u : \mathcal{G}^{(0)} \to \mathcal{G}$ is the unit map).

One sets $B^{Y\times_p Y} = \{d \in B_b\,;\, \forall\gamma \in Y \times_p Y\,,\, \alpha_\gamma(d_{s(\gamma)}) = d_{r(\gamma)}\}$ the C*-algebra of $Y \times_p Y$-invariant bounded sections of B. Since $C_b(X) = C_0(Y)^{Y\times_p Y}$, the algebra $B^{Y\times_p Y}$ is in a natural way a $C_b(X)$-module. One then defines on the $C_0(X)$-algebra $p_!B = C_0(X)B^{Y\times_p Y}$ an action $p_!\alpha$ of $\mathcal{G}$ by $p_!\alpha_g = \alpha_{y,g,z}$ for any $y, z \in Y$ such that $p(y) = r(g)$ and $p(z) = s(g)$. This definition is of course independent of the choice of y and z and the action thus defined is proved to be continuous. Since $p_!B$ is included in $\mathcal{M}(B)$ there is a natural morphism $p^*p_!B \to B$ which is proved to be an isomorphism.

Let us remark that the proof of this last assertion relies explicitly on the semi-continuity of the field of C*-algebras associated to the $C_0(X)$-algebra D.

DEFINITION 2.7. Let φ a generalized morphism from $\mathcal{G}$ to $\mathcal{H}$ given by a premorphism (Y, p, f). One defines as φ^*D the $\mathcal{G}$-algebra $p_!f^*D$. This algebra is independant of the choice of the premorphism.

(For an example of this construction see the proof of theorem 3.4).

It is not possible to recall here the precise definition of Hilbert modules in the sense of Kasparov (see [B]). Let us just mention that if B is a C*-algebra, a B-Hilbert module $\mathcal{E}$ is an analogue of a Hilbert space, but with scalar product taking values in B. The properties of these modules are very similar to Hilbert spaces'. In particular one can define the C*-algebra $\mathcal{L}(\mathcal{E})$ of "bounded" and $\mathcal{K}(\mathcal{E})$ of "compact" (in a proper sense) operators on $\mathcal{E}$.

Let D be a $\mathcal{G}$-algebra and $\mathcal{E}$ a Hilbert D-module. One can think of $\mathcal{E}$ as a "continuous field of Hilbert modules" over X, with fiber on $x \in X$ the D_x-module $\mathcal{E}_x = \mathcal{E} \otimes_D D_x$. A continuous action of $\mathcal{G}$ on $\mathcal{E}$ is a continuous family of unitaries (in a proper sense) $V_\gamma : \mathcal{E}_{s(\gamma)} \to \mathcal{E}_{r(\gamma)}$ such that, if $(\gamma, \gamma') \in \mathcal{G}^{(2)}$ then $V_\gamma V_{\gamma'} = V_{\gamma \circ \gamma'}$.

When $\mathcal{E}$ is $\mathbf{Z}/2\mathbf{Z}$-gradued (this means that $\mathcal{E}$ splits into a direct sum of two Hilbert modules) one assume that the unitaries $(V_\gamma)_{\gamma \in \mathcal{G}}$ are homogeneous of degree 0.

3. Kasparov's groups and functorial properties

In this section we give the definitions and main properties of the equivariant KK-groups, generalizing [Ka2].

Let A and B be two $\mathcal{G}$-algebras. A $\mathcal{G}$-equivariant A,B-bimodule is a pair $(\mathcal{E}, \pi)$, where $\mathcal{E}$ is a Hilbert B-module $\mathbf{Z}/2\mathbf{Z}$-gradued endowed with a contiuous action of $\mathcal{G}$, and π a $\mathcal{G}$-equivariant representation of A in $\mathcal{L}(\mathcal{E})$ which we shall often denote $a\xi = \pi(a)(\xi)$.

DEFINITION 3.1. Let A and B be two $\mathcal{G}$-algebras. A $\mathcal{G}$-equivariant A,B Kasparov bimodule is a triplet $(\mathcal{E}, \pi, F)$ composed by a $\mathcal{G}$-equivariant A,B-bimodule $(\mathcal{E}, \pi)$ and an homogeneous operator of degree one $F \in \mathcal{L}(\mathcal{E})$ such that $(\mathcal{E}, \pi\, F) \in E(A,B)$ and $\forall a \in r^*A$: $\left(a_\gamma(F_{r(\gamma)} - V_\gamma F_{s(\gamma)} V_\gamma^*)\right)_{\gamma \in \mathcal{G}}$ defines an element of $r^*\mathcal{K}(\mathcal{E})$.

One defines $E_{\mathcal{G}}(A,B)$ to be the set of (the unitary classes, in an obvious sense, of) $\mathcal{G}$-equivariant A,B-Kasparov bimodules, which is endowed with a semi-group structure by direct sum of bimodules. An homotopy in $E_{\mathcal{G}}(A,B)$ is given by an element of $E_{\mathcal{G}}(A, B[0,1])$.

DEFINITION 3.2. The set of homotopy classes of elements of $E_{\mathcal{G}}(A,B)$ is denoted $KK_{\mathcal{G}}(A,B)$.

As in the non equivariant case, $KK_{\mathcal{G}}(A,B)$ is an abelian group. When the groupoid is defined by a continuous action of a group G on a topological space X, one gets, for G-$C_0(X)$-algebras (in the sense of Kasparov) A and B : $KK_{X \rtimes G}(A,B) = \mathbf{R}KK_G(X; A, B)$ (the usual equivariant KK-group).

Let us explain the functorial properties of the theory. Since theorems 3.3 and 3.4 are direct generalizations of Kasparov's work we will concentrate on theorem 3.6 which is a groupoid statement. Let A, B and D be separable $\mathcal{G}$-algebras.

THEOREM 3.3. *There is a well defined bilinear coupling (a Kasparov product) $KK_{\mathcal{G}}(A,D) \times KK_{\mathcal{G}}(D,B) \to KK_{\mathcal{G}}(A,B)$. This product has the same properties as in the usual case ([Ka2], theorem 2.14) and is in particular associative.*

Remark : let us draw some easy consequences of this theorem.

a) The Kasparov product endows $KK_{\mathcal{G}}(C_0(X), C_0(X))$ with a unital ring structure which is proved to be commutative. This ring is the analogue for groupoids of Kasparov's representation ring for locally compact groups.

b) Let A and B be $\mathcal{G}$-algebras. For all $n \in \mathbf{N}$ one defines $KK^n_{\mathcal{G}}(A,B) = KK_{\mathcal{G}}(A \otimes C_0(\mathbf{R}^n), B)$. Then (Bott periodicity) $KK^{n+2}_{\mathcal{G}}(A,B) = KK^n_{\mathcal{G}}(A,B)$.

Futhermore, when $\mathcal{G}$ has a Haar system, one has the following generalization of [Ka2] theorem 3.11.

THEOREM 3.4. *There are well defined group homomorphisms* $j_{\mathcal{G}} : KK_{\mathcal{G}}(A,B) \to KK(A \rtimes \mathcal{G}, B \rtimes \mathcal{G})$ *and* $j_{\mathcal{G},r} : KK_{\mathcal{G}}(A,B) \to KK(A \rtimes_r \mathcal{G}, B \rtimes_r \mathcal{G})$ *natural with respect to Kasparov product, in the sense that for all* $\mathcal{G}$*-algebra* D, $x \in KK_{\mathcal{G}}(A,D)$ *and* $y \in KK_{\mathcal{G}}(D,B)$ *one has* $j_{\mathcal{G}}(x \otimes_D y) = j_{\mathcal{G}}(x) \otimes_{D \rtimes \mathcal{G}} j_{\mathcal{G}}(y)$ *and* $j_{\mathcal{G},r}(x \otimes_D y) = j_{\mathcal{G},r}(x) \otimes_{D \rtimes_r \mathcal{G}} j_{\mathcal{G},r}(y)$.

Moreover one can prove that the theory is functorial with respect to generalized morphisms of groupoids: Let $\mathcal{G}$ and $\mathcal{H}$ be two locally compact and σ-compact groupoids, f a strict morphism from $\mathcal{G}$ to $\mathcal{H}$ and $(\mathcal{E}, \pi, T)$ an element of $E_{\mathcal{H}}(A,B)$.

Let us define $\pi \otimes 1 : f^*A \to \mathcal{L}(f^*\mathcal{E})$ the representation $[\pi \otimes 1](a \otimes g) : \xi \otimes b \otimes h \mapsto \pi(a)\xi \otimes b \otimes gh$. The map $z \mapsto f^*z = (f^*\mathcal{E}, \pi \otimes 1, f^*T)$ defines a group homomorphism $f^* : KK_{\mathcal{H}}(A,B) \to KK_{\mathcal{G}}(f^*A, f^*B)$ natural with respect to Kasparov product.

This result can be extended to the case of generalized morphisms. Let (Y,p,f) be a premorphism from $\mathcal{G}$ to $\mathcal{H}$ and let us still denote by $p : \mathcal{G}_Y \to \mathcal{G}$ the strict morphim $(z, \gamma, y) \mapsto \gamma$.

LEMMA 3.5. *For all* $\mathcal{G}_Y$*-algebra* A *and* B *the application*

$$p^* : KK_{\mathcal{G}}(p_!A, p_!B) \to KK_{\mathcal{G}_Y}(A,B)$$

is a group isomorphism.

The proof of this lemma relies on the fact that every element $(\mathcal{E}, \pi, T) \in E_{\mathcal{G}_Y}(A,B)$ is homotopic (in Kasparov's sense) to an element $(\mathcal{E}, \pi, \widetilde{T})$ the operator $\widetilde{T}$ being $Y \times_p Y$-equivariant. Generalizing lemma 2.1 one constructs easily an element $p_!(\mathcal{E}, \pi, \widetilde{T}) \in E_{\mathcal{G}}(p_!A, p_!B)$. The construction $(\mathcal{E}, \pi, T) \mapsto p_!(\mathcal{E}, \pi, \widetilde{T})$ gives a well defined map $p_! : KK_{\mathcal{G}_Y}(A,B) KK_{\mathcal{G}}(p_!A, p_!B) \to KK_{\mathcal{G}}(p_!A, p_!B)$ wich is the inverse of p^*. As a corollary, one gets:

THEOREM 3.6. *Let* A *and* B *be two separable* $\mathcal{H}$*-algebras and* φ *a generalized morphism from* $\mathcal{G}$ *to* $\mathcal{H}$ *given by a premorphism* (Y,p,f).

The map $z \mapsto p_! f^* z$ *defines a group homomorphism (depending only of* φ*)*

$$\varphi* : KK_{\mathcal{H}}(A,B) \to KK_{\mathcal{G}}(\varphi^*A, \varphi^*B).$$

a) This morphism is natural with respect to Kasparov product.

b) The construction $\varphi \mapsto \varphi^*$ *is functorial with respect to composition of general morphism. In particular* φ^* *is an isomorphism if* φ *is an equivalence of groupoids.*

c) Let 1_A *be the unit of the ring* $KK_{\mathcal{G}}(A,A)$ *(defined as the class of* $(A, i_A, 0)$ *where* $\forall a \in A ; i_A(a) = a \in \mathcal{K}(A)$ *). Then* $\varphi^*(1_A) = 1_{\varphi^*A} \in KK_{\mathcal{G}}(\varphi^*A, \varphi^*A)$.

Property b) should be compared to the well known result of [MReW] that the envelopping C*-algebras of equivalent groupoids are Morita equivalent (and as a consequence K-equivalent).

Since many equivalences of groupoids appear in Kasparov's work, this result has many applications. For example it gives an easy proof of the descent isomorphism ([Ka2], theorem 3.5). In the case of foliations, it simplifies the constructions of [HS].

As an example of the way it may be used, we would like to give an easy proof of the Thom isomorphism in usual K-theory ([Ka1], 5, theorem 8): Let X be a locally compact topological space and E be a real fiber bundle over X of rank n and K-oriented. Let $\varphi : X \to \mathrm{Spin}^c(n)$ the groupoid morphism (i.e. the cocycle) wich defines E. One notes $\alpha_n \in KK^n_{Spin^c(n)}(\mathbf{C}, C_0(\mathbf{R}^n))$ and $\beta_n \in KK_{Spin^c(n)}(C_0(\mathbf{R}^n), \mathbf{C})$ the Bott and Bott "inverse" elements ([Ka1], 5, theorem 7).

THEOREM 3.7. *One has the following equalities:*

a) $\varphi^*\alpha_n \otimes_{C_0(E)} \varphi^*\beta_n = 1_{C_0(X)} \in KK^n_X(C_0(X), C_0(X))$.

b) $\varphi^*\beta_n \otimes_{C_0(X)} \varphi^*\alpha_n = 1_{C_0(E)} \in KK^n_X(C_0(E), C_0(E))$.

Proof : By construction $\varphi^*\mathbf{C} = C_0(X)$ and $\varphi^*C_0(\mathbf{R}^n) = C_0(E)$ algebra of sections of the bundle E going to zero at infinity. Moreover, $\alpha_n \otimes_{C_0(R^n)} \beta_n = 1_C$ and $\beta_n \otimes_C \alpha_n = 1_{C_0(R^n)}$ so $\varphi^*\alpha_n \otimes_{C_0(E)} \varphi^*\beta_n = \varphi^*1_{C_0(C)}$ and $\varphi^*\beta_n \otimes_{C_0(X)} \varphi^*\alpha_n = \varphi^*1_{C_0(R^n)}$. As a result, one gets $\varphi^*1_{C_0(C)} = 1_{\varphi^*C_0(C)}$ and $\varphi^*1_{C_0(R^n)} = 1_{\varphi^*C_0(R^n)}$.□

Remark: Most of the above constructions can be extended (with appropriate modifications of the definition of the KK-groups) to the case of a groupoid $\mathcal{G}$ endowed with a (non Hausdorff) manifold structure and having for unit space $\mathcal{G}^{(0)}$ and fibers $(\mathcal{G}_x)_{x\in\mathcal{G}^{(0)}}$) Hausdorff manifolds (this case covers the graph of any foliation).

4. The Baum-Connes conjecture for groupoids

This conjectures proposes a scheme to compute the K-theory groups of $C^*_r(\mathcal{G})$ (or more generally of reduced crossed-products $D\rtimes_r\mathcal{G}$). It states that a map (which we will not define) μ_r from geometrically defined groups $K^{top}_*(\mathcal{G})$ to $K_*(C^*_r(\mathcal{G}))$ is an isomorphism. If true this conjecture has many important consequences. Injectivity is in itself a very important property (in fact, in the case of discrete groups, it implies the well known Novikov conjecture on homotopy invariance of higher signatures).

In this section we shall review some results in this direction due to J. L. Tu ([Tu1], [Tu2], and the survey paper [Tu3]).

A topological space M is called a classifying space for proper actions of $\mathcal{G}$ if it is a a proper $\mathcal{G}$-space such that for every proper G-space Z there exists a unique (up to $\mathcal{G}$-equivariant homotopy) $\mathcal{G}$-equivariant map $Z \to M$.

Such a space always exists and is unique up to $\mathcal{G}$-equivariant homotopy. It may be choosen locally compact, σ-compact and Hausdorff. We shall denote by $E\mathcal{G}$ a classifying space for $\mathcal{G}$ having these properties.

One then defines $K^{top}_*(\mathcal{G})$ as $\varinjlim KK_{\mathcal{G}}(C_0(Y), C_0(X))$ the inductive limit being taken over all $\mathcal{G}$-invariant and $\mathcal{G}$-compact subspaces Y of $E\mathcal{G}$.

Using a Dirac-dual Dirac method inspired by the work of Kasparov [Ka2] and Higson-Kasparov [HK], J. L. Tu proved:

THEOREM 4.1. *If there exist a $\mathcal{G}$-proper algebra A and elements $\eta \in KK_{\mathcal{G}}(C_0(X), A)$ and $D \in KK_{\mathcal{G}}(A, C_0(X))$ such that there Kasparov product $\gamma = \eta \otimes_A D$ satisfies $p^*\gamma = 1 \in KK_{E\mathcal{G}\rtimes\mathcal{G}}(C_0(E\mathcal{G}), C_0(E\mathcal{G}))$ then:*

a) η, D *are uniquely defined by this condition.*

b) μ_r *is split injective.*

c) If moreover $\gamma = 1$ *then* μ_r *is bijective.*

In fact J. L. Tu proves that injectivity is true under weaker assumptions [Tu1] (this is what he calls a Dirac, Dual Dirac construction in limit) that applies to a large class of groupoids he calls "bolics". Moreover, generalizing works by Higson and Kasparov [HK], he gives a geometrical criterium for the equality $\gamma = 1$:

A σ-compact locally compact groupoid $\mathcal{G}$ is said to have the *Haagerup property* if there is a continuous field of real Hilbert spaces $(H_x)_{x\in\mathcal{G}^{(0)}}$ over its unit space with an action of $\mathcal{G}$ by affine isometries such that the map $\mathcal{G} \to \mathbf{R}_+ \times \mathcal{G}^{(0)} \times \mathcal{G}^{(0)}$ defined by $\gamma \mapsto (\|\gamma 0_{s(\gamma)} - 0_{r(\gamma)}\|, s(\gamma), r(\gamma))$ is proper. Note that (topologically) amenable groupoids satisfy the Haagerup property.

THEOREM 4.2. *If* $\mathcal{G}$ *has the Haagerup property it has a* γ *element and* $\gamma = 1 \in KK_{\mathcal{G}}(C_0(X), C_0(X))$.

(see [Hi2] for an original application of this last result to the Novikov conjecture for discrete groups).

Remark: There are some serious obstructions for $\gamma = 1$ (for example this is never the case when the groupoid is a (non-compact) group having property T.

To overcome this difficuly, V. Lafforgue has recently defined an analogue of the groupoid equivariant KK-theory that works for (fields of) Banach algebras [La]. It is not known wether his theory has a Kasparov product or not. But as in the case of groups, it's compatibility properties with the usual Kasparov theory may be used to prove several cases of the Baum-Connes conjecture.

References

[Hi] P. BAUM, A. CONNES, N. HIGSON *Classifying space for proper actions and K-theory of group C*-algebras*, Contempporary math. **167** (1994) p. 241-291.

[B] B. BLACKADAR *K*-theory for Operator Algebras, M.S.R.I. Publications **5**, Springer Verlag, New York (1986).

[Bl] E. BLANCHARD *Déformations de C*-algèbres de Hopf.* Bull. Soc. math. France, 124 (1996), p. 141-215.

[C1] A. CONNES *A survey of foliations and operator algebras*, in Operator algebras and applications part 1. Proc. Symp. Pure Math. A.M.S. **38** (1982), 521–628.

[C2] A. CONNES Non commutative geometry, Academic Press (1994).

[CS] A. CONNES et G. SKANDALIS *The longitudinal index theorem for foliations*, Publ. Res. Inst. Math. Sci. Kyoto Univ **20** n° 6 (1984), 1139–1183.

[DG] M.J. DUPRÉ and R.M. GILLETTE Banach bundles, Banach modules and automorphisms of C*-algebras, Pitman Research Notes in Math. **92** Pitman Publishing inc. Melbourne (1983).

[Ha] A. HAEFLIGER *Groupoïde d'holonomie et classifiant*, Asterisque **116** (1984), 70–97.

[Hi1] N. HIGSON *A primer on KK-theory*, Proc. Symp. in pure math. vol. 51 (1990) p. 239-283.

[Hi2] N. HIGSON *Bivariant K-theory and the Novikov conjecture*, preprint.

[HK] N. HIGSON and G. KASPAROV *Operator K-theory for groups wich act properly and isometrically on Hilbert space*, Electron. Res. Announc. Ame. Math. Soc. **3** (1997), 131-142.

[HS] M. HILSUM et G. SKANDALIS *Morphismes K-orientés d'espaces de feuilles et fonctorialité en théorie de Kasparov*, Ann. Sci. E.N.S. 4e série **20** (1987), 325–390.

[Ka1] G. G. KASPAROV *The operator K-functor and extensions of C*-algebras*, Izv. Akad. Nauk. S.S.S.R. Ser. Mat. **44** (1980) 571–636. Trad. angl. in: Math. U.S.S.R. Izv. **16** (1981), n°3, 513–572.

[Ka2] G. G. KASPAROV *Equivariant KK-theory and the Novikov conjecture*, Invent. Math. **91** (1988), 147–201.

[La] V. LAFFORGUE *K-théorie bivariante pour les algèbres de Banach, groupoïdes et conjecture de Baum-Connes*, in preparation.

[L2] P. Y. LE GALL *Théorie de Kasparov équivariante et groupoïdes I*, K-theory **16**, 361-390, 1999.

[L2] P. Y. LE GALL *Théorie de Kasparov équivariante et groupoïdes II*, in preparation.

[MReW] P. S. MUHLY, J.N.RENAULT and D.P. WILLIAMS *Equivalence and isomorphisms for groupoid C*-algebras*, Journal of Operator Theory. **17** (1987), 3–22.

[Re] J.N. RENAULT A Groupoid Approach to C*-algebras, Springer Lecture Note in Math. **793** (1980).

[Tu1] J. L. TU *La conjecture de Baum-Connes pour les feuilletages hyperboliques*, K-theory **16**, 126-184, 1999.

[Tu2] J. L. TU *La conjecture de Baum-Connes pour les feuilletages moyennables*, K-theory **16**, 215-264, 1999.

[Tu3] J. L. TU *The Baum-Connes for groupoids*, preprint.

UNIVERSITÉ PARIS SUD, BÂT 425, 91405 ORSAY CEDEX (FRANCE)
E-mail address: Pierre-Yves.Le-Gall@math.u-psud.fr

Contemporary Mathematics
Volume **282**, 2001

Groupoids of manifolds with corners and index theory

Bertrand Monthubert

ABSTRACT. This article is a survey on the relations between pseudodifferential calculus on manifolds with corners and groupoids, based on a contribution to the conference "Groupoids in Geometry, Analysis and Physics" held in Boulder, USA in 1999.

1. Introduction

In the spirit of noncommutative geometry, a pseudodifferential calculus is a groupoid. This approach is due to A. Connes, who made an extensive use of the holonomy groupoid of a foliation to study its index properties ([**2, 4**], and who gave a proof of Atiyah-Singer's index theorem through the use of groupoids ([**3**]): he showed that in the case of a smooth manifold M, the classical pseudodifferential calculus is given by the groupoid $M \times M$, and the analytic index is related to the tangent groupoid.

By replacing some analysis by some geometry, this approach allows one to use general techniques on groupoids and C^*-algebras in order to get results on pseudodifferential calculus and index theory.

In this survey we will see how Connes' ideas can be extended, especially to the case of manifolds with corners. First, we define the pseudodifferential calculus of a Lie groupoid, so that the construction of a pseudodifferential calculus in a specific context is reduced to the construction of a Lie groupoid. Then, in the context of manifolds with boundary, we associate a Lie groupoid to any such manifold, and compare it with Melrose's b-calculus. Last, we build a Lie groupoid for any manifold with corners, and we finally show how this helps to understand the b-calculus.

2. Pseudodifferential calculus on Lie groupoids

2.1. Pseudodifferential operators. In this section we define algebras of pseudodifferential operators on Lie groupoids, following [**12, 10**]. An equivalent definition has been independently obtained by V. Nistor, A. Weinstein and P. Xu ([**13**]).

1991 *Mathematics Subject Classification.* Primary 35S15, 47G30, 22A22; Secondary 46L80, 19K56, 58J42.

Key words and phrases. Pseudodifferential calculus, manifolds with corners, groupoids, C^*-algebras.

Let G be a Hausdorff, Lie groupoid with base $G^{(0)}$; we denote by s and r its source and range maps. We will not assume $G^{(0)}$ to be a smooth manifold without boundary; it can be any C^∞ manifold, but we require that each fiber $s^{-1}(y)$ (for any $y \in G^{(0)}$) is a manifold without boundary: G is thus a longitudinally smooth groupoid. In this framework, one can define a notion of pseudodifferential operators through their Schwartz kernels: let $I_c^m(G, G^{(0)}; \Omega^{\frac{1}{2}})$ be the space of compactly supported distributional sections K on G with values in the bundle of half-densities $\Omega^{\frac{1}{2}}$, which are of class C^∞ outside $G^{(0)}$, and given by an oscillatory integral in a neighborhood of $G^{(0)}$:

$$K(\gamma) = (2\pi)^{-n} \int_{\mathcal{A}_{r(\gamma)}(G)^*} e^{i\langle\varphi(\gamma),\xi\rangle} a(\gamma, \xi) d\xi.$$

Here, n is the dimension of $G^{(0)}$, a is a polyhomogeneous symbol of order m with values in $\Omega^{\frac{1}{2}}$, $\mathcal{A}(G)$ is the Lie algebroid associated to G, and φ is a phase function with values in $\mathcal{A}(G)$, such that the following diagram commutes:

$$\begin{array}{ccc} O & \xrightarrow{\varphi} & \mathcal{A}(G) \\ {\scriptstyle r}\downarrow & & \downarrow \\ G^{(0)} & \xrightarrow{=} & G^{(0)} \end{array}$$

It is a diffeomorphism between a tubular neighborhood O of $G^{(0)}$ in G and $\mathcal{A}(G)$, such that $\varphi(\gamma) = 0 \iff \gamma \in G^{(0)}$, and its differential is the isomorphism $T_x G^x \simeq N_x G^{(0)}$.

Such a Schwartz kernel defines a pseudodifferential operator by

$$Tf(\gamma) = \int_{G_{s(\gamma)}} K(\gamma'\gamma^{-1}) f(\gamma').$$

We obtain thus an algebra of pseudodifferential operators denoted by $\Psi_c^\infty(G) = \cup_{m\in\mathbb{Z}} \Psi_c^m(G)$.

REMARK 2.1. If G is non-Hausdorff, $\Psi_c^\infty(G)$ can also be defined.

REMARK 2.2. If $G = M \times M$, with M a smooth manifold without boundary, we recover precisely the definition of pseudodifferential operators of L. Hörmander ([**6**]).

If we restrict to order 0 operators, we can define $\overline{\Psi_0}(G)$ as the norm closure (in the multiplier algebra of $C_c^\infty(G)$) of the involutive algebra $\Psi_c^0(G)$.

THEOREM 2.3. *There exists a symbol map* $\sigma : \overline{\Psi_0}(G) \to C_0(S^* N^G_{G^{(0)}})$, *whose kernel is* $C^*(G)$. *The following sequence is thus exact:*

$$0 \to C^*(G) \to \overline{\Psi_0}(G) \xrightarrow{\ \sigma\ } C_0(S^* N^G_{G^{(0)}}) \to 0,$$

where $S^* N^G_{G^{(0)}}$ *is the cosphere bundle of the Lie algebroid of* G, $\mathcal{A}(G) = N^G_{G^{(0)}}$.

This extends the classical theorem of manifolds without boundary M: if $G = M \times M$, then $C^*(G) = \mathcal{K}$, and $S^* N^G_{G^{(0)}} = S^* M$. We will thus call it the Atiyah-Singer's exact sequence of G, even though it does not give conditions for an operator to be Fredholm.

2.2. Analytic index and tangent groupoid. As in the classical case, one can define the analytic index as the boundary map of the exact sequence of the theorem above. We prefer to give a somewhat improved definition, which does not involve $\overline{\Psi_0}(G)$.

In [**5**], M. Hilsum and G. Skandalis defined the normal cone deformation of a groupoid, which generalizes Connes' tangent groupoid; for this reason we will call it the tangent groupoid of G. As a set, it is defined as $G_T = G\times]0,1]\cup\mathcal{A}(G)\times\{0\}$. Its groupoid structure is induced by that of G and of $\mathcal{A}(G)$, and it has a Lie groupoid structure so that the following sequence is exact:

$$0 \to C^*(G)\otimes C_0(]0,1]) \to C^*(G_T) \to C_0(\mathcal{A}^*(G)) \to 0.$$

As $K_*(C^*(G)\otimes C_0(]0,1]))$, one gets an isomorphism

$$ev_0 : K_*(C^*(G_T)) \tilde{\to} K_*(C_0(\mathcal{A}^*(G)).$$

Moreover, the evaluation at 1 induces a morphism

$$ev_1 : K_*(C^*(G_T)) \to K_*(C^*(G)).$$

DEFINITION 2.4. The analytic index of G is

$$Ind_a = ev_1 \circ ev_0^{-1} : K_*(C_0(\mathcal{A}^*(G)) \to K_*(C^*(G)).$$

REMARK 2.5. The analytic index takes its values in $K_*(C^*(G))$, which is not equal to $\mathbb{Z}$ in general. This is not a Fredholm index. A Fredholm condition is developed in section 5.2

3. Pseudodifferential calculus on manifolds with boundary

3.1. Melrose's *b*-calculus. In this section, we briefly recall the definition of a particular pseudodifferential calculus on manifolds with boundary, Melrose's *b*-calculus ([**9**]).

Let X be a manifold with boundary. The *b*-stretched product of X is a C^∞ manifold defined as the gluing

$$X_b^2 = X^2\backslash(\partial X)^2 \cup S_+N,$$

where $S_+N = (N_{(\partial X)^2}^{X^2} - \{0\})/\mathbb{R}_+^*$ can be trivialized as $(\partial X)^2\times[-1,1]$ if one chooses a defining function ρ of ∂X (a differentiable function which is zero only on ∂X and whose differential is non-zero on ∂X). Under this choice, X_b^2 is a submanifold of $X^2\times[-1,1]$:

$$X_b^2 = \Big\{(x,y,t) \in X\times X\times\mathbb{R}, (1-t)\rho(x) = (1+t)\rho(y)\Big\}.$$

It has three boundary components,

$$lb = \Big\{(x,y,-1) \in \partial X\times X\times\mathbb{R}\Big\} \simeq \partial X\times X, rb = \Big\{(x,y,1) \in X\times\partial X\times\mathbb{R}\Big\} \simeq X\times\partial X,$$

as well as S_+N. It has a "diagonal", denoted by

$$\Delta_b = \Big\{(x,x,0) \in X\times X\times\mathbb{R}\Big\} \simeq X$$

A good picture, due to R. Melrose, is the following, with $X = \mathbb{R}_+$:

This allows R. Melrose to define a b-pseudodifferential operator as an operator whose Schwartz kernel K lives in $I^\infty(X_b^2, \Delta_b; \Omega^{\frac{1}{2}})$, and such that all the derivatives of K vanish at $lb \cup rb$ (as $lb \cup rb$ is away from Δ_b, K is a C^∞ function on a neighborhood of $lb \cup rb$).

3.2. The groupoid of a manifold with boundary. The definition of the b-calculus is close to that of the pseudodifferential calculus on a groupoid. There are nevertheless two differences: first, X_b^2 is not a Lie groupoid, second, the Schwartz kernels of b-pseudodifferential operators need to fulfill an extra condition, being zero in Taylor series on $lb \cup rb$ (*i.e.* all the derivatives vanish on $lb \cup rb$). But recall for instance that the space of C^∞ functions on the circle, which are zero in Taylor series at 0 are in a one-to-one correspondence with the Schwartz space of $\mathbb{R}$, the C^∞ functions on $\mathbb{R}$ which are rapidly decreasing as well as all their derivatives. We will see that $X_b^2 \backslash (lb \cup rb)$ is precisely a Lie groupoid, introduced in [**11**].

Let X be a manifold with boundary; it is thus the positive part of a manifold without boundary M endowed with a submanifold V of codimension 1, which is transversally oriented and divides M into two parts. Let us then consider the more general case where one is given M and V a submanifold of codimension 1, with no extra condition. Let $NV = TM/TV$ be the normal bundle of V into M. We define the groupoid $\mathbf{G}(V)$ as follows:

$$\mathbf{G}(V) = M\backslash V \times M\backslash V \cup \widetilde{V{\times}V}$$

where $\widetilde{V{\times}V}$ is the bundle over $V \times V$ of the isomorphisms of N_yV onto N_xV.

The groupoid structure is that of the couples groupoid on $M\backslash V \times M\backslash V$, and on $\widetilde{V{\times}V}$ one has

$$r(x, y, \alpha) = x, s(x, y, \alpha) = y, (x, y, \alpha) \circ (y, z, \alpha') = (x, z, \alpha\alpha').$$

To define its differentiable groupoid structure, let's take two charts Ω_i ($i = 1, 2$) of M which intersect V and such that $V \cap \Omega_i$ is transversally oriented, and $f_i : \Omega_i \to \mathbb{R}$ two defining functions of $V \cap \Omega_i$. By definition, the restriction $\underline{\mathrm{d}f_i}$ of the differential of f_i to $V \cap \Omega_i$ is a non-zero section of the normal bundle of $\overline{V}$. Then

$$W_{f_1,f_2} = \{(x, y, \lambda) \in \Omega_1 \times \Omega_2 \times \mathbb{R}^*, f_1(x) = \lambda f_2(y)\}$$

is a submanifold of $\Omega_1 \times \Omega_2 \times \mathbb{R}^*$, and one defines a chart of $\mathbf{G}(V)$ through $\iota_{\Omega_1,\Omega_2}$: $\mathbf{G}(V)_{\Omega_1}^{\Omega_2} \to W_{\varphi_1,\varphi_2}$:

$$\begin{aligned} \iota_{\Omega_1,\Omega_2}(\gamma) &= (r(\gamma), s(\gamma), \tfrac{f_1(r(\gamma))}{f_2(s(\gamma))}) && \text{si } \gamma \in \Omega_1 \times \Omega_2, \\ &= (r(\gamma), s(\gamma), \lambda) && \text{si } \gamma \in \widetilde{\Omega_1{\times}\Omega_2}. \end{aligned}$$

with $\alpha(\underline{\mathrm{d}f_2}) = \lambda \underline{\mathrm{d}f_1}$.

This gives a Lie groupoid structure to $\mathbf{G}(V)$.

We can now define the groupoid of a manifold with corners. Recall that the s-connected component G^c of a groupoid G is the union of the connected components of the units of the s-fibers G_y; it is the smallest open subgroupoid of G which contains $G^{(0)}$.

DEFINITION 3.1. Let X be a manifold with corners, and let M be a manifold without boundary endowed with a transversally oriented submanifold V of codimension 1 which splits M into two parts, such that X is the positive part of M.

The groupoid of X is the restriction to X of the s-connected component of $\mathbf{G}(V)$:

$$\Gamma(X) = \mathbf{G}^c(V)_X.$$

REMARK 3.2. This definition does not depend on the choice of (M, V).

The definition above can be simplified if one chooses a defining function ρ of ∂X. Under this choice,

$$\Gamma(X) = \{(x, y, \lambda) \in X \times X \times \mathbb{R}_+^*, \rho(x) = \lambda \rho(y)\} = \overset{\circ}{X} \times \overset{\circ}{X} \cup \partial X \times \partial X \times \mathbb{R}_+^*.$$

Thus a pseudodifferential operator on $\Gamma(X)$ is obtained by gluing a pseudodifferential operator on $\overset{\circ}{X}$ and a $\mathbb{R}_+^*$-invariant pseudodifferential on the cylinder made out of ∂X, as in [**1**].

3.3. b-calculus and pseudodifferential calculus on $\Gamma(X)$. To compare these two calculi, we first compare the spaces on which they define their Schwartz kernels. But the map

$$X \times X \times \mathbb{R}_*^+ \supset \Gamma(X) \ni (x, y, \lambda) \mapsto (x, y, \frac{1-\lambda}{1+\lambda}) \in X_b^2 \subset X \times X \times [-1, 1]$$

defines an embedding of $\Gamma(X)$ into X_b^2, whose complementary is precisely $lb \cup rb$; the units of $\Gamma(X)$ coincide with the diagonal Δ_b. .

As $lb \cup rb$ is disjoint from a neighborhood of Δ_b, one has $I_c^\infty(\Gamma(X)) = I_c^\infty(X_b^2)$. It induces the following result:

THEOREM 3.3. *With the notations above,* $\Psi_c^\infty(\Gamma(X)) = \Psi_{b,c}^\infty(X)$.

Obviously, the only difference between the b-calculus and the pseudodifferential calculus on $\Gamma(X)$ lies at the level of their smoothing operators. To further the comparison, we need to introduce a notion of Schwartz space on $\Gamma(X)$.

DEFINITION 3.4. Let G be a Lie groupoid, and μ a Haar system on G. A function $\phi : G \to \mathbb{R}_+$ will be called a length function with polynomial growth if:

- $\phi(\gamma_1\gamma_2) \leq \phi(\gamma_1) + \phi(\gamma_2)$,
- $\phi(\gamma^{-1}) = \phi(\gamma)$,
- ϕ is proper,
- $(\exists c, N)(\forall x \in G^{(0)}, \forall r \in \mathbb{R}_+)(\mu_x(\varphi^{-1}([0, r])) \leq c(r^N + 1))$.

DEFINITION 3.5. The space of rapidly decreasing functions with respect to ϕ is

$$\mathcal{S}^0 = \{f \in C_0(G, \Omega^{\frac{1}{2}}), \forall P \in \mathbb{C}[X], \sup_{\gamma \in G} |P(\varphi(\gamma))f(\gamma)| < \infty\}.$$

The Schwartz space of G with respect to ϕ is the subspace of $\mathcal{S}^0$ of functions f such that

$$\forall K_1, K_2 \in I_c^\infty(G, G^{(0)}; \Omega^{\frac{1}{2}}), K_1 * f * K_2 \in \mathcal{S}^0.$$

To define a Schwartz space on $\Gamma(X)$, one needs to define a length function with polynomial growth.

Let ρ be a defining function of ∂x: then

$$\Gamma(X) = \{(x, y, \lambda) \in X \times X \times \mathbb{R}_+^*, \rho(x) = \lambda \rho(y)\} = \overset{\circ}{X} \times \overset{\circ}{X} \cup \partial X \times \partial X \times \mathbb{R}_+^*.$$

Let's define $\phi(x, y, \lambda) = |\log(\lambda)|$. This is a length function with polynomial growth, and the Schwartz space $\mathcal{S}(\Gamma(X))$ induced by ϕ does not depend on ρ.

Proposition 3.6. *The space of smoothing operators of the b-calculus, Ψ_b^∞ is a subspace of $\mathcal{S}(\Gamma(X))$.*

The C^∞ functions on X_b^2 which are zero in Taylor series at $lb \cup rb$ are indeed rapidly decreasing as well as their derivatives with respect to $\exp(\phi)$. Thus they are rapidly decreasing as well as their derivatives with respect to ϕ.

Remark 3.7. The set of smoothing operators of the *b*-calculus does not have the properties we expect from a Schwartz-type space, especially it is not stable under holomorphic functional calculus.

4. Pseudodifferential calculus on manifolds with corners

4.1. The groupoid of a manifold with embedded corners. In this section we give a few indications to build $\Gamma(X)$ for X a manifold with embedded corners, following [**11**].

Recall that a manifold with corners is a C^∞ manifold modelled on $\mathbb{R}_+^n$. Any point x admits a neighborhood $V \simeq \mathbb{R}_+^k \times U$; k is the codimension of x. An open face $\overset{\circ}{F}$ of codimension k is a connected face of the set of points of codimension k. A face is the closure of an open face. The codimension of X is the maximum of the codimensions of its points.

R. Melrose's definition of manifolds with corners imposes an extra-condition: that the hyperfaces (faces of codimension 1) be embedded, which is equivalent to the existence of a defining function for each hyperface H. For instance, the first manifold below (we call it the two-corners drop) is a manifold with embedded corners, while the second (the one-corner drop) is not, as the boundary meets itself:

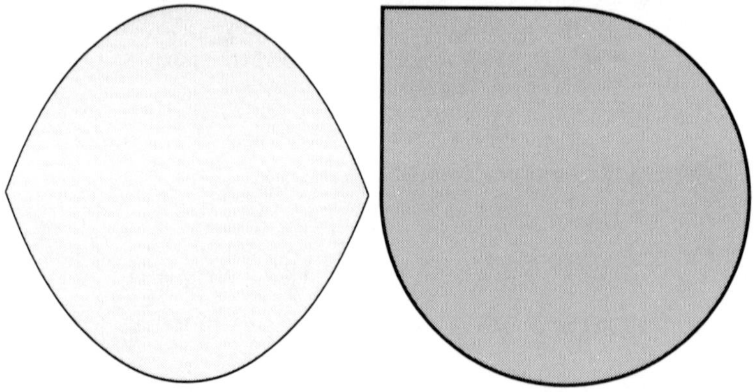

As this condition is not necessary to build a pseudodifferential calculus, we will call these the manifolds with embedded corners.

Let X be a manifold with embedded corners. As in the case of manifolds with boundary, we can embed X in a manifold without boundary, M, endowed with a finite set of submanifolds $(V_i)_{i\in I}$, where each V_i is transversally oriented and splits M in two, and such that X is the positive part of $(M,(V_i)_{i\in I})$. By taking fiber products of the groupoids $\mathbf{G}(V_i)$, and then restricting to X, one gets a Lie groupoid denoted by $\Gamma(X)$.

A useful description of $\Gamma(X)$ is the following:

$$\Gamma(X) = \bigcup_{\overset{\circ}{F}} \left\{(x,y,\alpha), x,y \in \overset{\circ}{F}, \alpha : N_y\overset{\circ}{F} \xrightarrow{\sim} N_x\overset{\circ}{F}\right\},$$

where α sends each hyperface onto itself. Thus, if one is given a defining function for each hyperface, the normal bundle of an open face $\overset{\circ}{F}$ is isomorphic to $\mathbb{R}_+^{codim(\overset{\circ}{F})}$, so that one has the trivialization

$$\Gamma(X) \simeq \bigcup_{\overset{\circ}{F}} \overset{\circ}{F} \times \overset{\circ}{F} \times \mathbb{R}_+^{*\,codim(\overset{\circ}{F})}.$$

4.2. The groupoid of a manifold with corners. In this section, X is a manifold with corners, but not necessarily a manifold with embedded corners: a hyperface can intersect itself, and the normal bundles are no longer trivial. For this reason, the definition above has to be extended to take into account all the isomorphisms of the normal bundle, as it does not have necessarily any meaning to send a hyperface onto itself.

For instance, let X be the Klein bottle whose section is a one-corner drop D: $X = D\times[0,1]/\sim$, with $(d,0) \sim (s(d),1)$, where s is the symetry across the diagonal of the drop. This is a codimension-2 manifold with corners, with one hyperface. On a neighborhood of a point of codimension 2, the normal bundle can be locally trivialized, but it is not possible to put an order on the local hyperfaces: if one follows the corner all around, the local hyperfaces are exchanged one with another.

The groupoid of a manifold with corners X, introduced in [**10**], is

$$\mathcal{G}(X)=\Big\{(x,y,\alpha), x,y\in X, \mathrm{codim}(x)=\mathrm{codim}(y), \alpha : N_yF(y) \xrightarrow{\sim} N_xF(x)\Big\}.$$

In this definition, x and y need not be in the same face, and α is an isomorphism of quadrants; let's further its decription.

Let x and y be two points of same codimension, A and A' be the sets of local hyperfaces at x and y. The normal bundles can be trivialized:

$$N_yF(y) \simeq \mathbb{R}_+^{A'}, N_xF(x) \simeq \mathbb{R}_+^{A},$$

so that an isomorphism $\alpha : N_yF(y) \to N_xF(x)$ is given by a monomial matrix with strictly positive values, *i.e.* by the product of a diagonal matrix with strictly positive values and of the matrix of a bijection $\sigma : A' \to A$.

As before, the groupoid structure is given by

$$s(x,y,\alpha) = y, r(x,y,\alpha) = x, (x,y,\alpha)(y,z,\alpha') = (x,z,\alpha\alpha').$$

To define the C^∞ manifold structure, consider two charts of X, Ω and Ω', of the same codimension. Let A (*resp.* A') be the set of local hyperfaces of Ω (*resp.* Ω'). We have

$$\Omega \xrightarrow{\sim} U \times \mathbb{R}_+^{A}, \Omega' \xrightarrow{\sim} U' \times \mathbb{R}_+^{A'}.$$

Let $\sigma : A' \to A$ be a bijection. One can then define an injective map

$$\begin{array}{rcl} \psi_\sigma : U \times U' \times \mathbb{R}_+^{A'} \times \mathbb{R}_+^{*A'} & \to & \mathcal{G}(X) \\ (u,u',t,\lambda) & \mapsto & (\pi^{-1}(u,\sigma(\lambda t)), \pi'^{-1}(u',t), \alpha_{\sigma,t,\lambda}) \end{array}$$

where $\alpha_{\sigma,t,\lambda}$ is an isomorphism of the normal bundle defined as follows. Let $B'_t = \{i \in A', t_i = 0\}$, and $B_t = \sigma(B)$, be the sets of local hyperfaces of $\pi'^{-1}(u',t)$ and

$\pi^{-1}(u, \sigma(\lambda t)$. Then $\alpha_{\sigma,t,\lambda}$ is defined through the trivializations induced by π and π':

$$\begin{array}{ccc} \alpha_{\sigma,t,\lambda} : N_{\pi'^{-1}(u',t)} F(\pi'^{-1}(u',t)) & \xrightarrow{\simeq} & N_{\pi^{-1}(u,\sigma(\lambda t))} F(\pi^{-1}(u, \sigma(\lambda t))) \\ \simeq \downarrow & & \simeq \downarrow \\ \mathbb{R}_+^{B'} & \xrightarrow{m_{\sigma,t,\lambda}} & \mathbb{R}_+^{B} \end{array}$$

where $m_{\sigma,t,\lambda}$ is the product of the matrix of $\sigma_{|B'}$ by the diagonal matrix composed of the λ_i for $i \in B'$.

This map ψ_σ makes $U \times U' \times \mathbb{R}_+^{A'} \times \mathbb{R}_+^{*A'}$ into a chart of $\mathcal{G}(X)$, and this defines a C^∞ manifold structure for $\mathcal{G}(X)$. This groupoid is amenable.

REMARK 4.1. If X is of codimension greater or equal to 2, then $\mathcal{G}(X)$ is non-Hausdorff.

4.3. Pseudodifferential calculus on manifolds with corners. As $\mathcal{G}(X)$ is a (longitudinally smooth) Lie groupoid, we can define a pseudodifferential calculus on any manifold with corners:

DEFINITION 4.2. Let X be a manifold with corners (where the hyperfaces need not be embedded). The pseudodifferential calculus of X is the pseudodifferential calculus of the Lie groupoid of X, $\mathcal{G}(X)$.

It follows that there is a symbol map, and an analytic index

$$Ind_a : K_*(C_0(T^*X)) \to K_*(C^*(\mathcal{G}(X))).$$

THEOREM 4.3. *The analytic index of a manifold with corners X takes its values in a group which depends only on the codimension of X.*

PROOF. Let X and X' be two manifolds with corners of the same codimension. Let $x \in X$: there exists also $x' \in X'$ such that $\operatorname{codim}(x) = \operatorname{codim}(x')$. This implies that X' intersects any orbit of $\mathcal{G}(X \cup X')$, thus $\mathcal{G}(X \cup X')_X^X = \mathcal{G}(X)$ is an open, full subgroupoid of $\mathcal{G}(X \cup X')$. Equivalently, $\mathcal{G}(X')$ is an open, full subgroupoid of $\mathcal{G}(X \cup X')$, so that $\mathcal{G}(X)$ is equivalent to $\mathcal{G}(X')$. We proved in [**10**] that it implies that $K_*(C^*(\mathcal{G}(X))) = K_*(C^*(\mathcal{G}(X')))$, even if $\mathcal{G}(X)$ is non-Hausdorff. □

In collaboration with P.Y. Le Gall, we computed in [**7**] the K-theory groups of $\mathcal{G}(X)$ for any codimension:

THEOREM 4.4. *For any $n \in \mathbb{N}$, one has the following equality:*

$$0 \to K_*(C^*(\mathcal{G}_n)) \to K_*(C_0(\mathbb{R}_+^{*n}) \rtimes \mathfrak{S}_n) \to K_{1-*}(C^*(\mathcal{G}_{n-1})) \to 0,$$

and

$$K_0(C_0(\mathbb{R}_+^{*n}) \rtimes \mathfrak{S}_n) = \mathbb{Z}^{p_n}, K_1(C_0(\mathbb{R}_+^{*n}) \rtimes \mathfrak{S}_n) = \mathbb{Z}^{i_n},$$

where p_n is the number of partitions of n of the shape $n = \lambda_1 + \cdots + \lambda_{2k}$ with $1 \leq \lambda_1 < \cdots < \lambda_{2k}$, and i_n is the number of partitions of n of the shape $n = \lambda_1 + \cdots + \lambda_{2k+1}$ with $1 \leq \lambda_1 < \cdots < \lambda_{2k+1}$.

4.4. The reduced groupoid of a manifold with corners. The definition of $\mathcal{G}(X)$, as we mentioned before, is very general. In some situations, a smaller groupoid can be built, as we did for manifolds with embedded corners. More generally, it is always possible to consider the s-connected component of $\mathcal{G}(X)$; in the case of manifolds with embedded corners, it coincides with the groupoid defined in the previous section, so we will denote it by $\Gamma(X)$.

To describe $\Gamma(X)$, consider for each open face $\overset{\circ}{F}$ the finite covering $E_{\overset{\circ}{F}}$ whose fiber at a point x is the set of vertices of $N_x\overset{\circ}{F}$. Let $Gal(E_{\overset{\circ}{F}})$ be its Galois covering:

$$Gal(E_{\overset{\circ}{F}}) = \left\{(x, y, \sigma), x, y \in \overset{\circ}{F}, \sigma : E_{\overset{\circ}{F}y} \xrightarrow{\sim} E_{\overset{\circ}{F}x}\right\}.$$

Let also

$$Hol(E_{\overset{\circ}{F}}) = Gal(E_{\overset{\circ}{F}})^c$$

be the s-connected component of $Gal(E_{\overset{\circ}{F}})$.

PROPOSITION 4.5. *Let X be a manifold with corners. Then $\Gamma(X)_x^x$ is the semi-direct product of the holonomy group $Hol(E_{\overset{\circ}{F}})_x^x$ and of $\mathbb{R}_+^{*A}$ where A is the fiber of $E_{\overset{\circ}{F}}$ at x. Moreover,*

$$\Gamma(X) = \bigcup_{\overset{\circ}{F}} \Gamma_{iso}^c(X) \rtimes Hol(E_{\overset{\circ}{F}}),$$

where $\Gamma_{iso}^c(X)$ is the s-connected component of the isotropy subgroupoid of $\Gamma(X)$. The groupoid $\Gamma(X)$ is Hausdorff if and only if for any open face $\overset{\circ}{F}$ the covering $E_{\overset{\circ}{F}}$ is trivial.

This result was proved in [**10**].

5. Applications

By introducing the groupoid of a manifold with corners, we can analyze pseudo-differential and index theory problems with the tools of noncommutative geometry.

5.1. Structure of $\overline{\Psi_0}(\Gamma(X))$ and $\overline{\Psi_0}(\mathcal{G}(X))$. To study the algebra of pseudodifferential operators on a manifold with corners, we first state an easy lemma.

LEMMA 5.1. *Let G be a locally compact topological groupoid, and*

$$F_1 \subset F_2 \subset \ldots \subset F_n \subset G^{(0)},$$

an increasing sequence of saturated, closed subsets of $G^{(0)}$. Then $C^*(G)$ has a composition series

$$C^*(G) \supset I_1 \supset \ldots \supset I_n,$$

where $I_k = C^*(G_{G^{(0)}\setminus F_k})$. Moreover, $I_k/I_{k+1} \simeq C^*(G_{F_{k+1}\setminus F_k})$.

Let X be a manifold with embedded corners. We can apply this lemma to the sequence $F_k = \cup_{dimF'<k}F'$. It gives the following result, which was originally proved by R. Melrose and V. Nistor ([**8**]) with analytic methods:

COROLLARY 5.2. *Let X be a manifold with embedded corners. Then $\overline{\Psi_0}(\Gamma(X))$ admits a composition series*

$$\overline{\Psi_0}(\Gamma(X)) \supset \mathcal{I}_0 \supset \mathcal{I}_1 \supset \ldots \supset \mathcal{I}_n$$

where n is the dimension de X, $\mathcal{I}_0 = C^*(\Gamma(X))$, et $\mathcal{I}_l$ is the closure of the ideal of b-pseudodifferential operators of strictly negative order whose indicial parts are zero on each face $F' \subset F$ of dimension strictement lower than l.

The subquotients are determined by the natural isomorphisms $\sigma : \overline{\Psi_0}(\Gamma(X))/\mathcal{I}_0 \overset{\sim}{\to} C(S^*X)$, and by

$$\mathcal{I}_l/\mathcal{I}_{l+1} \simeq \bigoplus_{\dim F' = l} C_0(\mathbb{R}^{n-l}, \mathcal{K}_{F'}), 0 \leq l \leq n.$$

Now if X is a general manifold with corners, we have a similar result by applying the lemma to the sequence $F_j = \{x \in X, \dim(x) < n + j - p\}$ where n is the dimension of X and p is its codimension.

COROLLARY 5.3. *Let X be a manifold with corners of dimension n and of codimension p. Then there is a composition series*

$$\overline{\Psi_0}(\mathcal{G}(X)) \supset \mathcal{I}_0 \supset \mathcal{I}_1 \supset \ldots \supset \mathcal{I}_p$$

where $\mathcal{I}_0 = C^*(\mathcal{G}(X))$, and $I_k = C^*(\mathcal{G}(X)_{X \setminus F_k})$. Moreover, $\mathcal{I}_l/\mathcal{I}_{l+1}$ is Morita equivalent to $C^*(\mathbb{R}_+^{*n-j} \rtimes \mathfrak{S}_{n-j})$.

5.2. Ellipticity conditions. Another general and easy result gives the ellipticity conditions for pseudodifferential operators on manifolds with corners.

Let G be an amenable, Lie groupoid, and $F \subset G^{(0)}$ a saturated, closed subset. Then the following diagram is commutative:

$$\begin{array}{ccccccccc}
 & & 0 & & & & & & \\
 & & \downarrow & & & & & & \\
 & & C^*(G_{G^{(0)}\setminus F}) & & & & & & \\
 & & \downarrow & & & & & & \\
0 & \to & C^*(G) & \to & \overline{\Psi_0}(G) & \xrightarrow{\sigma} & C_0(S^*G^{(0)}) & \to & 0 \\
 & & \downarrow & & \downarrow{\scriptstyle In_F} & & \downarrow & & \\
0 & \to & C^*(G_F) & \to & \overline{\Psi_0}(G_F) & \longrightarrow & C_0(S^*G^{(0)}_{|F}) & \to & 0 \\
 & & \downarrow & & & & \downarrow & & \\
 & & 0 & & & & 0 & &
\end{array}$$

thus $\overline{\Psi_0}(G) \to \overline{\Psi_0}(G_F)$ is onto, for $C^*(G) \to C^*(G_F)$ and $C^*(S^*G^{(0)}) \to C^*(S^*G^{(0)}_{|F})$ are onto.

PROPOSITION 5.4. *$C^*(G_{G^{(0)}\setminus F}) = Ker\ \sigma \cap Ker\ In_F = Ker\ (\sigma \oplus In_F)$.*

Moreover, the image of $\sigma \oplus In_F$ is the fiber product of $\overline{\Psi_0}(G_F)$ and of $C^*(S^*G^{(0)})$ above $C^*(S^*G^{(0)}_{|F})$.

COROLLARY 5.5. *Let X be a manifold with corners, and G be an open subgroupoid of $\mathcal{G}(X)$ containing $\mathcal{G}(X)^{(0)}$ (for instance, $G = \mathcal{G}(X)$, or $G = \Gamma(X)$). Then the following sequence is exact:*

$$0 \to \mathcal{K} \to \overline{\Psi_0}(G) \to C_0(S^*X) \times_f \overline{\Psi_0}(G_{\partial X}) \to 0,$$

where the last algebra is a fiber product above $C_0(S^*X_{|\partial X})$.

This results implies that a pseudodifferential operator on a manifold with corners is Fredholm if and only if its symbol is invertible, as well as its restriction to the boundary. This generalizes the result of Atiyah-Patodi-Singer, and gives a more general and easier proof to a result of R. Melrose and V. Nistor ([**8**]), obtained in the context of manifolds with embedded corners.

References

1. M. F. Atiyah, V. K. Patodi, and I. M. Singer, *Spectral asymmetry and Riemannian geometry. I*, Math. Proc. Cambridge Philos. Soc. **77** (1975), 43–69.
2. Alain Connes, *A survey of foliations and operator algebras.*, 1982, pp. 521–628 (English).
3. ______, *Noncommutative geometry*, Academic Press, Inc., 1994.
4. Alain Connes and Georges Skandalis, *The longitudinal index theorem for foliations*, Pub. of Research Institute for Math. Sc. **20, N. 6** (1984), 1139–1183.
5. Michel Hilsum and Georges Skandalis, *Morphismes K-orientés d'espaces de feuilles et fonctorialité en théorie de Kasparov*, Ann. Scient. Éc. Norm. Sup. **4è série, t.20** (1987), 325–390.
6. Lars Hörmander, *The analysis of linear partial differential operators III: Pseudo-differential operators.*, Grundlehren der Mathematischen Wissenschaften [Fundamental Principles of Mathematical Sciences], vol. 274, Springer-Verlag, Berlin, 1994.
7. Pierre-Yves Le Gall and Bertrand Monthubert, *K-theory of the indicial algebra of a manifold with corners*, Preprint.
8. R. Melrose and V. Nistor, *K-theory of C^*-algebras of b-pseudodifferential operators*, Geom. Funct. Anal. **8** (1998), no. 1, 88–122.
9. Richard B. Melrose, *The Atiyah-Patodi-Singer index theorem*, A K Peters Ltd., Wellesley, MA, 1993.
10. Bertrand Monthubert, *Groupoïdes et calcul pseudo-différentiel sur les variétés à coins*, Ph.D. thesis, Université Paris 7, 1998.
11. ______, *Pseudodifferential calculus on manifolds with corners and groupoids*, Proc. Amer. Math. Soc. **127** (1999), no. 10, 2871–2881.
12. Bertrand Monthubert and François Pierrot, *Indice analytique et groupoïdes de Lie*, C. R. Acad. Sci. Paris Sér. I Math. **325** (1997), no. 2, 193–198.
13. Victor Nistor, Alan Weinstein, and Ping Xu, *Pseudodifferential operators on differential groupoids*, Pacific J. Math. **189** (1999), no. 1, 117–152.

Université P. Sabatier, UFR MIG, 118 route de Narbonne, 31062 Toulouse cedex 4
E-mail address: monthube@picard.ups-tlse.fr

Contemporary Mathematics
Volume **282**, 2001

Quantization of Poisson algebras associated to Lie algebroids

N.P. Landsman and B. Ramazan

ABSTRACT. We prove the existence of a strict deformation quantization for the canonical Poisson structure on the dual of an integrable Lie algebroid. It follows that any Lie groupoid C^*-algebra may be regarded as a result of a quantization procedure. The C^*-algebra of the tangent groupoid of a given Lie groupoid G (with Lie algebroid $A(G)$) is the C^*-algebra of a continuous field of C^*-algebras over $\mathbb{R}$ with fibers $A_0 = C^*(A(G)) \simeq C_0(A^*(G))$ and $A_\hbar = C^*(G)$ for $\hbar \neq 0$. The same is true for the corresponding reduced C^*-algebras. Our results have applications to, e.g., transformation group C^*-algebras, K-theory, and index theory.

CONTENTS

1. Introduction

The idea of "quantization" has evolved through a number of stages. At the beginning of this century, one meant the fact that at a microscopic scale certain physical quantities (like energy or angular momentum) assume only discrete values. Such discreteness is easily understood within the Hilbert space formalism of quantum mechanics, where self-adjoint operators may or may not have a discrete spectrum, and is no longer seen as the defining property of a quantum theory.

Since about 1925, the idea has referred to the passage from a classical to a "corresponding" quantum theory. There are basically two ways to describe this

1991 *Mathematics Subject Classification.* Primary 46L65, 46L60; Secondary 46L87, 81S99, 22A22.

passage, starting either from the Lagrangian or from the Hamiltonian version of classical mechanics.

As first recognized by Feynman, the Lagrangian approach naturally leads to the path-integral formulation of quantum mechanics, and writing down a path integral may be seen as an act of quantization. All quantum-mechanical observables are constructed from so-called transition amplitudes, which correspond to the integration of certain functions with respect to the path integral. Though intuitively appealing, it is often hard to make this step rigorous. Through the nineties, many ideas of Witten originated from the use of the path integral.

Most mathematically precise work on quantization is based on the Hamiltonian formulation of classical mechanics. It was initially believed that the underlying mathematics consisted of symplectic geometry, but since about 1976 it has been understood that classical mechanics should be based on the concept of a Poisson algebra [**Ki, Li**]. This is a commutative algebra with a Lie bracket that turns each element of the algebra into a derivation with respect to the commutative structure; see Definition 4.1 below. Most Poisson algebras one encounters in physics are of the form $C^\infty(P)$, where P is a so-called Poisson manifold; the commutative algebra structure comes from pointwise multiplication, and the Lie bracket is just the Poisson bracket $\{\, ,\}$. Symplectic manifolds form a special, nondegenerate case.

As first recognized by Heisenberg in 1925, the quantum-mechanical observables of a given physical system should form a noncommutative algebra; the noncommutativity leads to the uncertainty relations that form the physical basis of quantum mechanics. A general, heuristic theory incorporating this idea was given in Dirac's book [**Di**]. The correct mathematical formalism of quantum mechanics, which still stands today, is due to von Neumann [**vN**], who created the abstract theory of Hilbert spaces and self-adjoint operators for this purpose. The only (but crucial) modification to von Neumann's formalism has been to allow other C^*-algebras than $\mathfrak{B}(\mathcal{H})$ (the algebra of all bounded operators on a Hilbert space $\mathcal{H}$) as algebras of observables. This generalization corresponds to admitting superselection rules; in the context of the quantization of finite-dimensional systems this means that one incorporates Poisson manifolds that are not symplectic in the underlying classical theory. The examples in this paper are of such a form.

In the Hamiltonian formalism, quantizing a Poisson algebra A^0 of classical observables amounts to finding a "corresponding" noncommutative associative algebra A of quantum observables, as well as a quantization map $\mathcal{Q} : A^0 \to A$, subject to certain conditions. Initially, practically all of quantization theory was based on a single idea of Dirac [**Di**], which he conceived in 1926 during a Sunday walk near Cambridge. Namely, in quantum mechanics the role of the Poisson bracket of the classical theory should be played by $1/i\hbar$ times the commutator. Here $\hbar$ is a small positive number, which in physics is a constant of nature. Hence if a classical observable f is quantized by a quantum observable $\mathcal{Q}(f)$, one expects that "Dirac's condition"

$$(i\hbar)^{-1}[\mathcal{Q}(f), \mathcal{Q}(g)] = \mathcal{Q}(\{f, g\})$$

holds at some fixed value of $\hbar$. In other words, the quantization map $\mathcal{Q}$ should be a Lie algebra homomorphism with respect to the Poisson bracket and the (rescaled) commutator.

Geometric quantization [**Wo**] was an attempt to make this idea precise. Given a "prequantizable" symplectic manifold S with associated Poisson algebra $C^\infty(S)$, this approach produces a Hilbert space $\mathcal{H}(S)$ (the space of L^2-sections of the prequantization line bundle $L(S)$ over S) and a Lie algebra homomorphism $\mathcal{Q}^{\text{pre}}$ from $C^\infty(S)$ to a certain algebra $\mathcal{A}(S)$ of unbounded operators on $\mathcal{H}(S)$ that are densely defined on the domain of smooth sections of $L(S)$.

The case where S is a coadjoint orbit in the dual $\mathfrak{g}^*$ of the Lie algebra $\mathfrak{g}$ of a Lie group G has been studied in particular detail, for the following reason. Each element of $X \in \mathfrak{g}$ defines a function $\tilde{X}$ on $\mathfrak{g}^*$, and hence on S, by $\tilde{X}(\theta) = \theta(X)$. One has

$$\{\tilde{X}, \tilde{Y}\} = \widetilde{[X, Y]}, \tag{1.1}$$

so that $\mathcal{Q}^{\text{pre}}$ restricts to a Lie algebra homomorphism from $\mathfrak{g}$ to $\mathcal{A}(S)$. Assuming that this (infinitesimal) representation of $\mathfrak{g}$ is integrable to a unitary representation U of G, one finds that U tends to be reducible. In order to obtain an irreducible subrepresentation of U, one must restrict $\mathcal{H}(S)$. In the simplest cases one may simply project onto an irreducible subspace with orthogonal projection P in the commutant of $U(G)$. For general $f \in C^\infty(S)$, one then has to define $\mathcal{Q}(f) = P\mathcal{Q}^{\text{pre}}(f)P$; it is clear that this modification destroys the Lie algebra homomorphism property (except on the linear span of the $\tilde{X}$, since $[\mathcal{Q}^{pre}(\tilde{X}), P] = 0$).

A vast number of other methods has been proposed to achieve irreducibility; see [**Ki**] for a recent overview. The conclusion is that one cannot achieve irreducibility of $U(G)$ while preserving the Lie algebra homomorphism property on all smooth functions. In other examples than coadjoint orbits, one finds that the map $\mathcal{Q}^{\text{pre}}$ is unsatisfactory for other (though related) reasons, the conclusion being the same: a satisfactory quantization map is only a Lie algebra homomorphism in the stated sense on some subspace of the Poisson algebra of classical observables.

The way out of this dilemma emerged in the seventies, mainly as a consequence of the work of Berezin [**Be**], Vey [**Ve**], and Bayen et al. [**BFFLS**]. It is, quite simply, to require Dirac's condition only asymptotically, that is, for $\hbar \to 0$. This, of course, necessitates the dramatic step of defining the quantization data A and $\mathcal{Q}$ for a family $I \subseteq \mathbb{R}$ of values of $\hbar$ that contains 0 as an accumulation point. Hence, given a Poisson algebra A^0, one now needs to find a family of algebras $\{A_\hbar\}_{\hbar \in I}$ and maps $\mathcal{Q}_\hbar : A^0 \to A_\hbar$, with $\mathcal{Q}_0 = \text{id}$, such that $(i\hbar)^{-1}[\mathcal{Q}_\hbar(f), \mathcal{Q}_\hbar(g)] \to \mathcal{Q}_\hbar(\{f, g\})$ in some sense. A method that accomplishes this is generically referred to as a deformation quantization.

There are (at least) two ways to make rigorous sense of the above convergence. The oldest, introduced in the context of quantization theory in [**BFFLS**], is to define $A_\hbar$ and $\mathcal{Q}_\hbar$ as formal power series in $\hbar$. This method, called formal deformation quantization, remains the most popular; see [**Ko**] for a recent highlight.

The second approach, which is based on choosing the $A_\hbar$ to be C^*-algebras, was introduced by Rieffel [**Ri2**]-[**Ri7**]. Here $\hbar$ is simply a real number rather than a formal deformation parameter, and one imposes Dirac's condition asymptotically in norm; see Definition 4.3 below. It is then mathematically natural and physically meaningful to require that the $A_\hbar$ form a continuous field of C^*-algebras over the index set $I \ni \hbar$. In section 4 we give a precise formulation of this approach to deformation quantization, introducing two appropriate generalizations of Rieffel's original definition that lie at the basis of the results in the present paper.

Having discussed the development of modern quantization theory in broad outline, we now turn to the class of classical systems (that is, Poisson manifolds) that are quantized in this paper. Our motivation comes from a number of directions.

The first is the quantization theory of coadjoint orbits discussed above. Although the attempt to quantize individual orbits is only successful in special cases, one could try to quantize all coadjoint orbits of a given Lie group at one stroke by quantizing $\mathfrak{g}^*$ (or $C^\infty(\mathfrak{g}^*)$) as a whole. To do so, one starts from the fact that $\mathfrak{g}^*$ is canonically a Poisson manifold when equipped with the so-called Lie–Poisson structure that may be defined by (1.1). As first shown by Rieffel [**Ri4**] in the special case that G is exponential and nilpotent, one may construct a quantization of $C^\infty(\mathfrak{g}^*)$ in which $A_\hbar = C^*(G)$ for all $\hbar \neq 0$, and the maps $\mathcal{Q}_\hbar$ are essentially given by the pullback of the exponential map from $\mathfrak{g}$ to G, rescaled by $1/\hbar$. We show that this works without any assumption on the Lie group G. The idea that coadjoint orbits in $\mathfrak{g}^*$ are quantized by unitary irreducible representations of G then re-enters through the back door, as follows. The well-known correspondence between nondegenerate representations of $C^*(G)$ and unitary representations of G, preserving irreducibility, is matched by the fact that the irreducible representations of $C^\infty(\mathfrak{g}^*)$ as a Poisson algebra [**La3**] are precisely given by the coadjoint orbits of G (or covering spaces thereof). In general, there is no correspondence through quantization between the irreducible representations of $C^\infty(\mathfrak{g}^*)$ and the irreducible representations of $C^*(G)$; the correspondence through (deformation) quantization is between the algebras in question themselves.

Secondly, one may quantize a particle moving on a Riemannian manifold M (so that the pertinent Poisson manifold is the cotangent bundle T^*M, which is symplectic) by taking $A_\hbar$ to be the C^*-algebra $\mathfrak{B}_0(L^2(M))$ of compact operators on $L^2(M)$ for all $\hbar \neq 0$, and defining a quantization map $\mathcal{Q}_\hbar$ in terms of the geodesics on M. This result may be generalized by coupling the particle to an external Yang–Mills field with gauge group H; one then starts from a principal H-bundle P, so that $M = \mathsf{P}/H$. The Poisson manifold describing the classical theory is then given by $P = (T^*\mathsf{P})/H$, which is not symplectic unless H is discrete, and when H is compact the quantum algebra of observables turns out to be $\mathfrak{B}_0(L^2(\mathsf{P}))^H$; see [**La1, La3**] for details and the generalization to noncompact H. Taking $\mathsf{P} = H = G$ actually reproduces the previous example, since $(T^*G)/G \simeq \mathfrak{g}^*$ as Poisson manifolds, whereas for compact Lie groups one has $\mathfrak{B}_0(L^2(G))^G \simeq C^*(G)$ (and analogously for the noncompact case). The quantization maps in question are also compatible with this specialization.

Thirdly, in the early eighties Lie groupoids and locally compact groupoids started to play a role in C^*-algebras as a result of the work of Connes and of Renault, whereas in the late eighties Lie groupoids entered symplectic and Poisson geometry, as well as quantization theory, through the idea of a symplectic groupoid (cf. [**We2**]). Against this background, it was an obvious idea that the above quantizations ought to be interpretable in terms of Lie groupoids. The key to this possibility lies in the fact the passage from a Lie groupoid to its convolution C^*-algebra has a classical analogue. Namely, like a Lie group, a Lie groupoid G has an associated "infinitesimal" object, its Lie algebroid [**Pr1**] $A(G)$, which is a vector bundle over the base space G_0 of G. The dual bundle $A^*(G)$ of $A(G)$ admits a canonical Poisson structure [**CDW, Cou**], so that eventually one may associate a Poisson algebra $C^\infty(A^*(G))$ to G.

All threads now come together [**La1**], in that in each of the three classes of examples above the classical Poisson algebra is of the form $C^\infty(A^*(G))$ and the quantum C^*-algebra is of the form $C^*(G)$, where G is some Lie groupoid. This is obvious in the first example. In the second, one takes G to be the pair groupoid

$G = M \times M$ on M, whose Lie algebroid is the tangent bundle TM; the Poisson structure on the dual bundle T^*M, which is nothing but the cotangent bundle of M, is just its usual canonical structure (which is symplectic). See also [**CCFGRV**]. Finally, in the third example G is the so-called gauge groupoid of the principal bundle P [**Ma**], with associated Lie algebroid $(T\mathsf{P})/H$.

Based on these examples, it was conjectured in [**La2**] that one should obtain a deformation quantization of this type for any Lie groupoid G. For formal deformation quantization a version of this conjecture was proved in [**NWX**], and a proof in C^*-algebraic deformation quantization above appeared in [**La3, La4, Ra1**].

In this paper we give a precise and complete formulation of all mathematical concepts surrounding this class of C^*-algebraic deformation quantizations, and prove two technically distinct versions of the above conjecture, each of which has its merits. The generalization of Connes's tangent groupoid [**Co2**] $\hat{G}$ to arbitrary Lie groupoids G [**HS, We1**] plays an important role. A significant intermediate result is that $C^*(\hat{G})$ is the C^*-algebra of a continuous field of C^*-algebras over $\mathbb{R}$ with fibers $A_0 = C^*(A(G)) \simeq C_0(A^*(G))$ and $A_\hbar = C^*(G)$ for $\hbar \neq 0$. The same is true for the corresponding reduced C^*-algebras. See section 6.

The sections of the paper are listed in the table of contents following the abstract. Our main results, Theorems 4.4 and 4.6, are stated at the end of section 4. The remainder of the paper develops the proofs of these theorems.

2. Lie groupoids and Lie algebroids

This section is a brief review of our objects of study, mainly intended to establish our notation. We assume that the reader is familiar with the basic theory of locally compact groupoids (cf. [**Re1**]), which we always assume to be second countable. We denote a groupoid as a whole by G, the base is called G_0, which is seen as a subspace of the arrow space (or total space) G_1 under the inclusion map ι. The source and range projections are called s and r, respectively, and for $x \in G_0$ we put

$$G^x = r^{-1}(x) \subset G_1. \tag{2.1}$$

The space of composable arrows is $G_2 = \{(\gamma, \gamma') \in G_1 \times G_1 \mid r(\gamma') = s(\gamma)\}$.

DEFINITION 2.1. *A Lie groupoid is a locally compact groupoid G for which G_1 and G_0 are manifolds, s and r are surjective submersions, and multiplication and inclusion are smooth maps.*

It follows that ι is an immersion, that the inverse is a diffeomorphism, that G_2 is a closed submanifold of $G_1 \times G_1$, that the fibers $r^{-1}(x)$ and $s^{-1}(x)$ are submanifolds of G_1 for all $x \in G_0$, and that all isotropy groups are Lie groups. The basic reference on Lie groupoids is Mackenzie's book [**Ma**]; also see [**Co2, La3**].

A Lie group defines its Lie algebra; one may also study Lie algebras in their own right. The situation for Lie groupoids is similar.

DEFINITION 2.2. *A Lie algebroid on a manifold M is a vector bundle E over M, which apart from the bundle projection $\tau : E \to M$ is equipped with a vector bundle map $\rho : E \to TM$ (called the anchor), as well as with a Lie bracket $[\,,\,]_E$ on the space $C^\infty(M, E)$ of smooth sections of E, satisfying*

$$\rho \circ [X, Y]_E = [\rho \circ X, \rho \circ Y], \tag{2.2}$$

where the right-hand side is the usual commutator of vector fields on $C^\infty(M,TM)$, *and*

$$[X, fY]_E = f[X,Y]_E + ((\rho \circ X)f)Y \tag{2.3}$$

for all $X,Y \in C^\infty(M,E)$ *and* $f \in C^\infty(M)$. *We generally omit the suffix* E *on the Lie bracket.*

It is part of the definition of a bundle map that the anchor is fiber-preserving and linear on each fiber. This concept is due to Pradines [**Pr2**]. The basic reference on Lie algebroids is [**Ma**]; also see [**La3**]. The simplest examples of Lie algebroids are tangent bundles TM, where ρ is the identity, and Lie algebras, for which M is a point.

We now explain how one may associate a Lie algebroid $A(G)$ with a given Lie groupoid G [**Pr1, Ma, La3**]. A left-invariant vector field ξ^L on G_1 is a vector field satisfying $Tr(\xi^L) = 0$ and $TL_\gamma(\xi^L)(\gamma') = \xi^L(\gamma\gamma')$ for all $(\gamma,\gamma') \in G_2$. Here $L_\gamma : G^{s(\gamma)} \to G^{r(\gamma)}$ is defined for each $\gamma \in G_1$ by

$$L_\gamma(\gamma_1) = \gamma\gamma_1. \tag{2.4}$$

Note that the second condition is well-defined because of the first one. The space of all smooth left-invariant vector fields on G_1 is denoted by $C^\infty(G_1,TG_1)^L$, which is easily shown to be a Lie algebra under the usual commutator borrowed from $C^\infty(G_1,TG_1)$. Also, a left-invariant vector field is obviously determined by its values on the unit space G_0. The tangent bundle of G_1 at the unit space has a decomposition

$$T_xG_1 = T_xG_0 \oplus (\ker Tr)_x, \tag{2.5}$$

where $\ker Tr \subset TG_1$ is the vector bundle over G_1 whose fiber $(\ker Tr)_\gamma$ above $\gamma \in G_1$ is the kernel of the derivative $T_\gamma r : T_\gamma G_1 \to T_{r(\gamma)}G_0$ of the range projection $r : G_1 \to G_0$. The special case $\gamma = x \in G_0$ occurs in (2.5).

DEFINITION 2.3. *The Lie algebroid* $A(G)$ *of a Lie groupoid* G *is given by the following:*

1. *The vector bundle* $A(G)$ *over* G_0 *is the normal bundle defined by the embedding* $G_0 \hookrightarrow G_1$; *accordingly, the bundle projection* $\tau : A(G) \to G_0$ *is given by* s *or* r *(which coincide on* G_0*).*
2. *Identifying the normal bundle with* $(\ker Tr)|_{G_0}$ *by (2.5), the anchor is given by* $\rho = Ts : \ker Tr \to TG_0$.
3. *Identifying a section of the normal bundle with an element of* $C^\infty(G_1,TG_1)^L$ *through the previous item, the Lie bracket* $[\, , \,]_{A(G)}$ *is given by the commutator.*

The required equality (2.2) is automatically satisfied (as it holds for all vector fields on G_1). We leave the verification of (2.3) to the reader.

It follows from this definition that a Lie algebra $\mathfrak{g}$ is the Lie algebroid of a Lie group G, and that the tangent bundle TG_0 is the Lie algebroid of the pair groupoid $G_1 = G_0 \times G_0$.

For Lie groups one has an exponential map from the Lie algebra to the group. For manifolds M with connection the exponential map is defined on the tangent bundle TM, which it maps into M. As indicated by Pradines [**Pr3**], these are special cases of a construction that holds for general Lie groupoids G, provided its Lie algebroid $A(G)$ is endowed with a connection. Following [**La3, La4**], we here present a slightly different construction that is more suitable for our application to deformation quantization; also see [**NWX**].

First note that the vector bundles $\ker Tr$ and $s^*A(G)$ (over G_1) are isomorphic via the map TL_γ, applied fiberwise. Hence the connection on $A(G)$, with associated horizontal lift $\ell^{A(G)}$, yields a connection on $\ker Tr$ through pull-back. Going through the definitions, one finds that the associated horizontal lift ℓ of a tangent vector $X = \dot{\gamma} = d\gamma(t)/dt_{t=0}$ in $T_\gamma G_1$ to $Y \in T^r_\gamma G_1$ is

$$\ell_Y(\dot{\gamma}) = \frac{d}{dt}[L_{\gamma(t)*}\ell^{A(G)}_{TL_{\gamma^{-1}}(Y)}(s(\gamma(t)))]_{t=0}, \tag{2.6}$$

which is an element of $T_Y(\ker Tr)$ (here $\ell^{A(G)}(\dots)$ lifts a curve).

Since the bundle $\ker Tr \to G_1$ has a connection, one can define parallel transport $X \to X(t)$ on $\ker Tr$ in precisely the same way as on a tangent bundle with affine connection. That is, the flow $X(t)$ is the solution of

$$\dot{X}(t) = \ell_{X(t)}(X(t)) \tag{2.7}$$

with initial condition $X(0) = X$. The projection of $X(t)$ to G_1 is a "geodesic" $\gamma_X(t)$.

DEFINITION 2.4. *The left exponential map* $\mathrm{Exp}^L : A(G) \to G_1$ *is defined by*

$$\mathrm{Exp}^L(X) = \gamma_X(1) = \tilde{\tau}(X(1)), \tag{2.8}$$

where $\tilde{\tau} : \ker Tr \to G$ *is the restriction of the bundle projection* $TG_1 \to G_1$ *to* $\ker Tr$. *It is assumed that the geodesic flow* $X(t)$ *on* $\ker Tr$ *(defined by the connection on* $\ker Tr$ *pulled back from the one on* $A(G)$*) is defined at* $t = 1$. *If not,* $\mathrm{Exp}^L(X)$ *is undefined.*

There is a symmetrized version of Exp^L that plays a key role in our quantization theory. First note that for all $X \in A(G)$ for which $\mathrm{Exp}^L(X)$ is defined one has

$$r(\mathrm{Exp}^L(X)) = \tau(X); \tag{2.9}$$

recall that τ is the bundle projection of the Lie algebroid. To derive this, note that $r(\gamma_X(0)) = \tau(X)$ by construction, and $dr(\gamma_X(t))/dt = Tr(X(t)) = 0$.

We combine this with the obvious $\tau(\frac{1}{2}X) = \tau(-\frac{1}{2}X)$ to infer that

$$r(\mathrm{Exp}^L(\tfrac{1}{2}X)) = r(\mathrm{Exp}^L(-\tfrac{1}{2}X)) = s(\mathrm{Exp}^L(-\tfrac{1}{2}X)^{-1}).$$

Thus the (groupoid) multiplication in (2.10) below is well-defined.

DEFINITION 2.5. *The Weyl exponential map* $\mathrm{Exp}^W : A(G) \to G_1$ *is defined by*

$$\mathrm{Exp}^W(X) = \mathrm{Exp}^L(-\tfrac{1}{2}X)^{-1}\mathrm{Exp}^L(\tfrac{1}{2}X). \tag{2.10}$$

The following well-known result is a form of the tubular neighbourhood theorem.

PROPOSITION 2.6. *The maps* Exp^L *and* Exp^W *are diffeomorphisms from a neighbourhood* V *of* $G_0 \subset A(G)$ *(as the zero section) to a neighbourhood* W *of* G_0 *in* G_1*, such that* $\mathrm{Exp}^L(x) = \mathrm{Exp}^W(x) = x$ *for all* $x \in G_0$.

Cf. [**La3, La4**] for a proof.

3. The C^*-algebra of a Lie groupoid

In this section we show that to every Lie groupoid G one can associate a C^*-algebra $C^*(G)$. This C^*-algebra was introduced by Bigonnet [**Bg**], but the idea to use half densities for an intrinsic construction of the C^*-algebra of a Lie groupoid first appeared in Connes's work [**Co1**] for the special case of the holonomy groupoid of a foliation. Connes also proposed a general definition in [**Co2**]; this only works if fiber isomorphisms of the type discussed below are understood. The construction we give in this paper is due to Renault [**Re4**]; it is essentially the same as that in [**La2, La3**]. Also see [**CW**]. Further details on densities as used below may be found in [**Tr, La3**].

Let V be an n-dimensional vector space and let α be a fixed real number. A density of weight α on V is a function $\rho : \wedge^n(V) \to \mathbb{C}$ that satisfies

$$\rho(\lambda T) =\mid \lambda \mid^{\alpha} \rho(T)$$

for every $\lambda \in \mathbb{C}$ and $T \in \wedge^n(V)$. The 1-dimensional vector space of the densities of weight α on V is denoted by $|\Omega|^{\alpha}(V)$. If a density is positive at a point, then it is positive everywhere on $\wedge^n(V)\backslash\{0\}$. Hence one may unambiguously speak of positive densities. We have the canonical isomorphisms $|\Omega|^{\alpha+\beta}(V) \simeq |\Omega|^{\alpha}(V)\otimes|\Omega|^{\beta}(V)$ and $|\Omega|^{\alpha}(V \oplus W) \simeq |\Omega|^{\alpha}(V) \otimes |\Omega|^{\alpha}(W)$, where W is another vector space.

Denote the parallelepiped generated by a family $v_1, \dots , v_n$ of vectors in V by $[v_1, \dots , v_n]$. Given a positive 1-density ρ on V, the translation invariant measure μ on V that satisfies $\mu([v_1, \dots , v_n]) = \rho(v_1 \wedge \dots \wedge v_n)$ for every family $\{v_1, \dots , v_n\}$ of vectors is said to be the measure generated by ρ. If $\{e_1, \dots , e_n\}$ is a basis of V and $f \in L^1(V, \mu)$, then

$$\int f\left(\sum_{i=1}^{n} x_i e_i\right) \rho(e_1 \wedge \dots \wedge e_n)dx = \int f d\mu. \tag{3.1}$$

We now generalize this construction to fibers in vector bundles. Let E be a vector bundle on M, with fibers E_x over $x \in M$. For each $x \in M$, consider $|\Omega|^{\alpha}(E_x)$, the 1-dimensional vector space of densities of weight α on the vector space E_x. The vector bundle over M with fibers $|\Omega|^{\alpha}(E_x)$ is denoted by $|\Omega|^{\alpha}(E)$. The sections of the vector bundle $|\Omega|^{\alpha}(E)$ over M are called α-densities. An α-density $\mu \in C^{\infty}(M, |\Omega|^{\alpha}(E))$ is called positive if $\mu(x)$ is a positive density for every $x \in M$. The function associated to μ in the local frame $e = (e_1, \dots , e_p)$ of E on $U \subset M$ is the map $\mu_e \in C^{\infty}(U)$ defined by $\mu_e(x) = \mu(x)\, (e_1(x) \wedge \dots \wedge e_p(x))$.

The set of α-densities on E is a free 1-dimensional module over the ring of functions defined on M. Moreover, $C^{\infty}(M, |\Omega|^{\alpha}(E))$ is a free 1-dimensional module over $C^{\infty}(M)$, and choosing a positive smooth density we can identify densities with functions.

Now let G be a Lie groupoid with Lie algebroid $A(G)$; recall from Definition 2.3 that $A(G)$ is a vector bundle over G_0. For every real number α, one may associate the vector bundle $|\Omega|^{\alpha}(A(G))$ to $A(G)$. In addition, form the vector bundle $|\Omega|^{\alpha}(\ker Tr)$ of α-densities associated to the vector bundle $\ker Tr \subset TG_1$. (defined after (2.5)).

The groupoid G acts (from the left) on the bundle $|\Omega|^{\alpha}(\operatorname{Ker} Tr)$. To define this action we need a further bit of notation. Let V, W be vector spaces with dim V=dim W. For every linear map $f : V \to W$, define $f^{\Omega} : |\Omega|^{\alpha}(F) \to |\Omega|^{\alpha}(E)$ on

$\rho \in |\Omega|^{\alpha}(F)$, $v_1, v_2, \ldots, v_n \in V$ by $(f^{\Omega}\rho)(v_1 \wedge \ldots \wedge v_n) = \rho(f(v_1) \wedge \ldots \wedge f(v_n))$. Obviously, if ρ is a positive density then $f^{\Omega}\rho$ is positive, too.

To the derivative $T_{\gamma\gamma_1}L_{\gamma^{-1}} : T_{\gamma\gamma_1}G^{r(\gamma)} \to T_{\gamma_1}G^{s(\gamma)}$ of L_γ (cf. (2.4)) one associates the corresponding map between the spaces of densities $(T_{\gamma\gamma_1}L_{\gamma^{-1}})^{\Omega}$: $|\Omega|^{\alpha}(T_{\gamma_1}G^{s(\gamma)}) \to |\Omega|^{\alpha}(T_{\gamma\gamma_1}G^{r(\gamma)})$. For simplicity, we denote $(T_{\gamma\gamma_1}L_{\gamma^{-1}})^{\Omega}$ by γ; the action of G on $|\Omega|^{\alpha}(\ker Tr)$ is given by $\gamma : |\Omega|^{\alpha}(T_{\gamma_1}G^{s(\gamma)}) \to |\Omega|^{\alpha}(T_{\gamma\gamma_1}G^{r(\gamma)})$, where $\gamma_1 \in G^{s(\gamma)}$.

Denote the vector bundle $r^*|\Omega|^{1/2}(A(G)) \otimes s^*|\Omega|^{1/2}(A(G))$ over G by $\mathfrak{G}$. The fiber of $\mathfrak{G}$ over $\gamma \in G$ is $\mathfrak{G}_\gamma = |\Omega|^{1/2}(T_{r(\gamma)}G^{r(\gamma)}) \otimes |\Omega|^{1/2}(T_{s(\gamma)}G^{s(\gamma)})$.

DEFINITION 3.1. *The convolution algebra of a Lie groupoid G is the space $C_c^\infty(G_1, \mathfrak{G})$ of smooth compactly supported sections of the vector bundle $\mathfrak{G}$ over G_0, with product $f * g$ of $f, g \in C_c^\infty(G_1, \mathfrak{G})$ defined by*

$$(f * g)(\gamma) = \int_{G^{r(\gamma)}} (\mathrm{id} \otimes \gamma_1) f(\gamma_1)(\gamma_1 \otimes \mathrm{id}) g(\gamma_1^{-1}\gamma), \tag{3.2}$$

where id is the identity map, and involution given by

$$f^*(\gamma) = \widetilde{f(\gamma^{-1})}. \tag{3.3}$$

Here the map

$$\sim : |\Omega|^{1/2}(T_x G^x) \otimes |\Omega|^{1/2}(T_y G^y) \to |\Omega|^{1/2}(T_y G^y) \otimes |\Omega|^{1/2}(T_x G^x)$$

is defined by $a(x) \otimes b(y) \mapsto \overline{b(y) \otimes a(x)}$.

It is not difficult to show that these expressions are well defined. Let us show, for example, that the integral in (3.2) is well defined. We extend the action of G on the bundle $|\Omega|^{1/2}(\ker Tr)$ to the tensor product in an obvious way. For $f, g \in C_c^\infty(G_1, \mathfrak{G})$ we see that $(\mathrm{id} \otimes \gamma_1) f(\gamma_1)$ is an element of $|\Omega|^{1/2}(T_{r(\gamma_1)}G^{r(\gamma_1)}) \otimes |\Omega|^{1/2}(T_{\gamma_1}G^{r(\gamma_1)})$, and that $(\gamma_1 \otimes \mathrm{id}) g(\gamma_1^{-1}\gamma) \in |\Omega|^{1/2}(T_{\gamma_1}G^{r(\gamma_1)}) \otimes |\Omega|^{1/2}(T_{s(\gamma)}G^{s(\gamma)})$. The integrand in (3.2) is then an element of $|\Omega|^{1/2}(T_{r(\gamma)}G^{r(\gamma)}) \otimes |\Omega|^{1}(T_{\gamma_1}G^{r(\gamma_1)}) \otimes |\Omega|^{1/2}(T_{s(\gamma)}G^{s(\gamma)})$, so that it may be integrated as a 1-density on the manifold $G^{r(\gamma)}$. Hence we obtain an element of $|\Omega|^{1/2}(T_{r(\gamma)}G^{r(\gamma)}) \otimes |\Omega|^{1/2}(T_{s(\gamma)}G^{s(\gamma)})$.

PROPOSITION 3.2. *The map $\lambda \mapsto \lambda|_{G_0}$ is a bijection between the left invariant sections of $|\Omega|^{\alpha}(\ker Tr)$ and the sections of $|\Omega|^{\alpha}(A(G))$. This bijection preserves smoothness as well as positivity.*

PROOF. The restriction is well defined, since $|\Omega|^{\alpha}(\ker Tr)$ contains $|\Omega|^{\alpha}(A(G))$. To prove that the map in question is bijective, it is sufficient to check that its inverse is $\rho \mapsto \lambda$, where $\rho \in |\Omega|^{\alpha}(A(G))$ and $\lambda(\gamma) = \gamma \cdot \rho(s(\gamma))$. Ξ

We now define the groupoid C^*-algebra $C^*(G)$ and its reduced counterpart $C_r^*(G)$. Firstly, to endow the convolution *-algebra $C_c^\infty(G_1, \mathfrak{G})$ with a C^*-norm we consider its *-representations on a Hilbert space. Let $\mathcal{R}$ be the set of all *-representations $\pi(C_c^\infty(G_1, \mathfrak{G}))$ for which $f \mapsto \langle \xi, \pi(f)\eta \rangle$ is a Radon measure on $C_c^\infty(G_1, \mathfrak{G})$ for each pair $\xi, \eta \in H$. One then defines a C^*-norm on $C_c^\infty(G_1, \mathfrak{G})$ by

$$\|f\| = \sup_{\pi \in \mathcal{R}} \|\pi(f)\|; \tag{3.4}$$

cf. [**Re1**]. The completion of $C_c^\infty(G_1, \mathfrak{G})$ in this norm is the C^*-algebra $C^*(G)$.

Secondly, for every $x \in G_0$ we define a *-representation π_x of $C_c^\infty(G_1, \mathfrak{G})$ on the Hilbert space $L^2(G^x)$ of half-densities on the manifold G^x by

$$[\pi_x(f)\xi](\gamma) = \int_{G^x} (\mathrm{id} \otimes \gamma_1) f(\gamma_1)(\gamma_1 \otimes \mathrm{id})\xi(\gamma_1^{-1}\gamma), \tag{3.5}$$

for $\gamma \in G^x$, where $\xi \in L^2(G^x)$. Then $\|f\|_r = \sup_{x \in G_0} \|\pi_x(f)\|$ is a norm (cf. [**Re1**]), and the completion of $C_c^\infty(G_1, \mathfrak{G})$ in this norm is the reduced C^*-algebra $C_r^*(G)$.

For an amenable groupoid G we have $C_r^*(G) = C^*(G)$. For the proof of this result and a detailed discussion of amenable groupoids see [**AR**].

Recall the following concept (Def. I.2.2 in [**Re1**]).

DEFINITION 3.3. *A left Haar system on a locally compact groupoid G is a family of measures $(\lambda^x)_{x \in G_0}$ such that the support of λ^x is G^x, the system is left-invariant under the map $L_\gamma : G^{s(\gamma)} \to G^{r(\gamma)}$, and for each $\phi \in C_c(G_1)$, the map*

$$x \mapsto \lambda(\phi)(x) = \int \phi \, d\lambda^x \tag{3.6}$$

defines a continuous function on G_0.

If G is a Lie groupoid, we say that the Haar system is smooth if each such function is smooth for $\phi \in C_c^\infty(G_1)$.

As explained in [**Re1**], one can associate C^*-algebras $C^*(G, \lambda)$ and $C_r^*(G, \lambda)$ to a locally compact groupoid G with Haar system (λ^x). Recall that convolution and involution are given on the dense subalgebra $C_c(G_1)$ by

$$f * g(\gamma) = \int_{G^{s(\gamma)}} f(\gamma\gamma_1) g(\gamma_1^{-1}) d\lambda^{s(\gamma)}(\gamma_1); \tag{3.7}$$

$$f^*(\gamma) = \overline{f(\gamma^{-1})}. \tag{3.8}$$

This turns $C_c(G_1)$ into a *-algebra, which we denote by $C_c(G, \lambda)$. Similarly, if G is smooth one has the *-algebra $C_c^\infty(G, \lambda)$. The C^*-algebras $C^*(G, \lambda)$ and $C_r^*(G, \lambda)$ are completions of $C_c(G, \lambda)$ or $C_c^\infty(G, \lambda)$ in suitable C^*-norms.

Not every locally compact groupoid admits a left Haar system; in the theory of groupoid C^*-algebras one therefore usually postulates its existence. For Lie groupoids the situation is more favourable.

PROPOSITION 3.4. *A Lie groupoid G has a smooth left Haar system $(\lambda^x)_{x \in G_0}$. The associated convolution *-algebra $C_c^\infty(G, \lambda)$ is isomorphic to $C_c^\infty(G_1, \mathfrak{G})$, and the associated C^*-algebras $C^*(G, \lambda)$ and $C_r^*(G, \lambda)$ are isomorphic to $C^*(G)$ and $C_r^*(G)$, respectively.*

PROOF. Let λ be a smooth positive section of $|\Omega|^1(A(G))$. By Proposition 3.2, λ can be extended to a smooth, positive, and left invariant section λ_G of $|\Omega|^1(\ker Tr)$. Then $\lambda_G^{1/2}$ is a smooth left invariant section of $|\Omega|^{1/2}(\ker Tr)$, and $\gamma \mapsto \lambda_G^{1/2}(r(\gamma)) \otimes \lambda_G^{1/2}(s(\gamma))$ is a smooth section of $\mathfrak{G}$. For every section $f \in C_c^\infty(G_1, \mathfrak{G})$ there exists a function $f_\lambda \in C_c^\infty(G_1)$ such that

$$f(\gamma) = f_\lambda(\gamma)\lambda_G^{1/2}(r(\gamma)) \otimes \lambda_G^{1/2}(s(\gamma)). \tag{3.9}$$

An easy calculation shows that for $f, g \in C_c^\infty(G_1, \mathfrak{G})$ we have

$$(f * g)(\gamma) = (f_\lambda * g_\lambda)(\gamma)\lambda_G^{1/2}(r(\gamma)) \otimes \lambda_G^{1/2}(s(\gamma)),$$

where

$$(f_\lambda * g_\lambda)(\gamma) = \int_{G^{r(\gamma)}} f_\lambda(\gamma_1) g_\lambda(\gamma_1^{-1}\gamma)\lambda_G(\gamma_1). \tag{3.10}$$

The right-hand side of the last equality is well defined as the integral of the 1-density $f_\lambda(\gamma_1)g_\lambda(\gamma_1^{-1}\gamma)\lambda_G(\gamma_1)$ over the manifold $G^{r(\gamma)}$. It is also easy to see that

$$f^*(\gamma) = f_\lambda^*(\gamma)\lambda_G^{1/2}(r(\gamma)) \otimes \lambda_G^{1/2}(s(\gamma)), \tag{3.11}$$

where $f_\lambda^*(\gamma) = \overline{f_\lambda(\gamma^{-1})}$.

Let $x \in G_0$. The restriction of λ_G to the submanifold G^x is a 1-density, hence it has an associated measure on G^x. As we identify the 1-densities and their associated measures, we use the same notation λ^x for the restriction of λ_G to G^x and the induced measure. The equality $\gamma \cdot \lambda^{s(\gamma)} = \lambda^{r(\gamma)}$ proves that the λ^x form a left Haar system; its is easy to check that each function (3.6) is smooth.

It is easily checked that, under the correspondence $f \leftrightarrow f_\lambda$ defined in (3.9) and the replacement of λ_G by the left Haar system $(\lambda^x)_{x\in G_0}$, the expressions (3.10) and (3.11) are transformed into (3.7) and (3.8), respectively. We leave the proof of the isomorphisms of the completions of the *-algebras in question to the reader. Ξ

4. Strict deformation quantization

As we have seen, the idea of C^*-algebraic deformation quantization is to relate a given Poisson algebra to a family of C^*-algebras.

DEFINITION 4.1. *A Poisson algebra is a complex commutative associative algebra equipped with a Lie bracket* $\{\,,\,\}$ *for the which the Leibniz rule holds; that is, one has*

$$\{f, gh\} = \{f, g\}h + g\{f, h\}. \tag{4.1}$$

A Poisson manifold is a manifold P *equipped with a Lie bracket* $\{\,,\,\}$ *on* $C^\infty(P)$ *that together with pointwise multiplication turns* $C^\infty(P)$ *into a Poisson algebra.*

This definition is due to Lichnerowicz [**Li**] and Kirillov [**Ki**]. The Lie bracket in a Poisson algebra is usually called the Poisson bracket. For the theory of Poisson manifolds we refer to these papers, and to the textbooks [**MR, Va**]. A Poisson algebra is the classical analogue of a C^*-algebra [**La3**]. This analogy is clear if one sees a C^*-algebra as a non-associative version of a Poisson algebra, in which the commutative product is given by the anti-commutator $xy + yx$ and the Lie bracket is $\{x, y\} = i(xy - yx)$. The Leibniz rule is then satisfied. The analogy is even better when one defines a Poisson algebra as a real algebra, and restricts the above two operations to the self-adjoint part of a C^*-algebra.

In this paper, the key example of a Poisson algebra is provided by the dual bundle E^* of a Lie algebroid E over M (cf. Definition 2.2). This Poisson structure was discovered in [**CDW, Cou**]. The corresponding bracket is most easily defined by listing special cases by which it is uniquely determined; these are

$$\{f, g\} = 0; \tag{4.2}$$

$$\{\tilde{X}, f\} = (\rho \circ X)f; \tag{4.3}$$

$$\{\tilde{X}, \tilde{Y}\} = \widetilde{[X, Y]}_{A(G)}. \tag{4.4}$$

Here $f, g \in C^\infty(M)$ are regarded as functions on E^* in the obvious way, and $\tilde{X} \in C^\infty(E^*)$ is defined by a section $X \in C^\infty(M, E)$ through $\tilde{X}(\theta) = \theta(X(\tau^*(\theta)))$. See [**CDW**] for an intrinsic definition. In particular, a Lie groupoid G canonically determines a Poisson algebra $C^\infty(A^*(G))$. This is the generic Poisson algebra that we are going to quantize.

Our first definition of C^*-algebraic deformation quantization uses the concept of a continuous field of C^*-algebras. We here state a reformulation of Dixmier's familiar definition [**Di**] due to Kirchberg–S. Wassermann [**KW**], which is tailor made for our applications.

DEFINITION 4.2. *A continuous field of C^*-algebras $(C, \{A_t, \varphi_t\}_{t\in T})$ over a locally compact Hausdorff space T consists of a C^*-algebra C, a collection of C^*-algebras $\{A_t\}_{t\in T}$, and a set $\{\varphi_t : C \to A_t\}_{t\in T}$ of surjective *-homomorphisms, such that for all $c \in C$*

1. *the function $t \to \|\varphi_t(c)\|$ is in $C_0(T)$;*
2. *one has $\|c\| = \sup_{t\in T} \|\varphi_t(c)\|$;*
3. *there is an element $fc \in C$ for any $f \in C_0(T)$ for which $\varphi_t(fc) = f(t)\varphi_t(c)$ for all $t \in T$.*

The continuous cross-sections of the field in the sense of [**Di**] consist of those elements $\{a_t\}_{t\in T}$ of $\prod_{t\in T} A_t$ for which there is a (necessarily unique) $a \in C$ such that $a_t = \varphi_t(a)$ for all $t \in T$.

Our first definition of C^*-algebraic deformation quantization is now as follows [**La3, La4**].

DEFINITION 4.3. *A strict deformation quantization of a Poisson manifold P consists of*

1. *a dense Poisson algebra $A^0 \subset C_0(P)$ under the given Poisson bracket $\{\, ,\}$ on P;*
2. *A subset $I \subseteq \mathbb{R}$ containing 0 as an accumulation point;*
3. *a continuous field of C^*-algebras $(C, \{A_\hbar, \varphi_\hbar\}_{\hbar\in I})$, with $A_0 = C_0(P)$;*
4. *a linear map $\mathcal{Q} : A^0 \to C$ that satisfies (with $\mathcal{Q}_\hbar(f) \equiv \varphi_\hbar(\mathcal{Q}(f))$)*

$$\mathcal{Q}_0(f) = f; \tag{4.5}$$

$$\mathcal{Q}_\hbar(f^*) = \mathcal{Q}_\hbar(f)^*, \tag{4.6}$$

for all $f \in A^0$ and $\hbar \in I$, and for all $f, g \in A^0$ satisfies Dirac's condition

$$\lim_{\hbar\to 0} \|(i\hbar)^{-1}[\mathcal{Q}_\hbar(f), \mathcal{Q}_\hbar(g)] - \mathcal{Q}_\hbar(\{f, g\})\| = 0. \tag{4.7}$$

In view of the comment after Definition 4.2, for fixed $f \in A^0$ each family $\{\mathcal{Q}_\hbar(f)\}_{\hbar\in I}$ is a continuous cross-section of the continuous field in question. In view of (4.5) this implies, in particular, that

$$\lim_{\hbar\to 0} \|\mathcal{Q}_\hbar(f)\mathcal{Q}_\hbar(g) - \mathcal{Q}_\hbar(fg)\| = 0. \tag{4.8}$$

This shows that a strict deformation quantization yields asymptotic morphisms in the sense of E-theory [**Co2**]. See [**La3**] for an extensive discussion of quantization theory from the above perspective.

Every good definition is the hypothesis of a theorem. Indeed, our first main result [**La3, La4**] is as follows.

THEOREM 4.4. *Let G be a Lie groupoid, with associated C^*-algebras $C^*(G)$ and $C^*_r(G)$, and Poisson manifold $A^*(G)$. There exists a strict deformation quantization of $A^*(G)$ for which $I = \mathbb{R}$, $A_0 = C_0(A^*(G))$, and $A_\hbar = C^*(G)$ for $\hbar \neq 0$. In particular, there exists a continuous field of C^*-algebras over $\mathbb{R}$ with these fibers. The same claim holds with $C^*(G)$ replaced by $C^*_r(G)$.*

A different definition of C^*-algebraic deformation quantization, which generalizes Rieffel's original definition [**Ri2**], was introduced in [**Ra1**]. This definition is closer to the notion of formal deformation quantization introduced in [**BFFLS**] than Definition 4.3.

DEFINITION 4.5. *Let P be a Poisson manifold. A semi-strict deformation quantization of P consists of*

1. *a dense Poisson algebra $A^0 \subset C_0(P)$ under the given Poisson bracket $\{\, ,\}$ on P;*
2. *A subset $I \subseteq \mathbb{R}$ containing 0 as an accumulation point;*
3. *For each $h \in I$, an associative product $\times_\hbar$, an involution $^{*_\hbar}$, and a C^*-seminorm $\|\cdot\|_\hbar$ on A^0, such that*
   (a) *For $\hbar = 0$ we recover the usual product, involution and norm of $C_0(P)$;*
   (b) *For each $f \in A^0$, the map $h \mapsto \|f\|_\hbar$ is continuous;*
   (c) *For all $f, g \in A^0$ one has*

$$\lim_{\hbar \to 0} \|(i\hbar)^{-1}(f \times_\hbar g - g \times_\hbar f) - \{f, g\}\|_\hbar = 0. \tag{4.9}$$

A strict deformation quantization cannot necessarily be turned into a semi-strict one, since $\mathcal{Q}(A^0)\mathcal{Q}(A^0)$ is not necessarily contained in $\mathcal{Q}(A^0)$, so that one may not be able to transfer the products on the $A_\hbar$ to an $\hbar$-dependent product on A^0. Conversely, a semi-strict deformation quantization is not necessarily strict, for the seminorms can fail to be norms. However, let $f \in A^0$ be a non-null function. Then $\|f\|_0 \neq 0$ and by the condition 3.2 in Definition 4.5 one has $\|f\|_\hbar \neq 0$ for h in a neighborhood of 0.

For each $\hbar \in I$, $N_\hbar = \{f \in A^0 | \; \|f\|_\hbar = 0\}$ is a two-sided closed ideal in $(A^0, \times_\hbar, {}^{*_\hbar}, \|\cdot\|_\hbar)$, and $A^0/N_\hbar$ is a pre-C^*-algebra, with completion $A^-_\hbar$. Defining $\mathcal{Q}_\hbar$ as the canonical projection of A^0 onto $A^0/N_\hbar$, we almost obtain a strict deformation quantization of P; the only difficulty is that the $A^-_\hbar$ may not form a continuous field of C^*-algebras.

The counterpart of Theorem 4.4 for strict deformation quantization is as follows [**Ra1**].

THEOREM 4.6. *Let $A(G)$ be the Lie algebroid of a Lie groupoid, with associated Poisson manifold $A^*(G)$. There exists a semi-strict deformation quantization of $A^*(G)$ over $I = \mathbb{R}$.*

As explained above, this theorem does not follow from Theorem 4.4, though the proofs have many elements in common. We now briefly outline the organization of the proofs. Sections 5 and 6 address the continuous field of C^*-algebras claimed to exist in Theorem 4.4, and in similar vein contain the heart of the proof of condition (b) of Definition 4.5 in Theorem 4.6. Section 7 develops the theory of the Fourier transform on vector bundles. In section 8 this theory is used to define the map $\mathcal{Q}$ of Theorem 4.4 as well as the product $\times_\hbar$, the involution $*_\hbar$, and the seminorm $\|\cdot\|_\hbar$ of Theorem 4.6. The proofs of both theorems are then complete up to Dirac's condition (4.7) and (4.9). Section 9 develops local techniques that give detailed information on the Poisson structure on $A^*(G)$. Finally, (4.7) and (4.9) are proved in section 10.

5. Continuous fields of groupoids

Rieffel has developed useful techniques for proving continuity of fields of C^*-algebras occurring in examples of strict deformation quantization in the context of groups [**Ri1**]. The main result of this section, Theorem 5.5, generalizes these results to the context of groupoids. This theorem, which is an application of results of Blanchard [**Bl**], gives information on the field of C^*-algebras associated to a continuous field of locally compact groupoids. Although the examples studied in this paper concern Lie groupoids, we prove Theorem 5.5 for the general case of locally compact groupoids.

We first mention a known result. Recall that, when $E \subset G_0$, G_E is the subgroupoid of G consisting of all $\gamma \in G_1$ for which $r(\gamma) \in E$ and $s(\gamma) \in E$.

PROPOSITION 5.1. *Let G be a locally compact groupoid with left Haar system $(\lambda^x)_{x\in G_0}$, and let U be an open invariant subset of G_0. Write $F = G_0 \backslash U$.*

1. *The following sequence is exact:*

$$0 \longrightarrow C^*(G_U) \xrightarrow{e} C^*(G) \xrightarrow{i^*} C^*(G_F) \longrightarrow 0. \tag{5.1}$$

Here e is firstly defined as map from $C_c(G_U)$ to $C_c(G_1)$ by extending a function on G_U to one on G_1 by making it zero on the complement of G_U, and secondly extended to a map from $C^*(G_U)$ to $C^*(G)$ by continuity. Similarly, i^* is firstly defined from $C_c(G_1)$ to $C_c(G_F)$ as the pullback of the inclusion $G_F \hookrightarrow G_1$, and then extended by continuity.

2. *If the groupoid G_F is amenable, then the following sequence is exact:*

$$0 \longrightarrow C^*_r(G_U) \xrightarrow{e} C^*_r(G) \xrightarrow{i^*} C^*_r(G_F) \longrightarrow 0. \tag{5.2}$$

This result was given by Torpe [**To**] in the case of C^*-algebras associated to foliations, and by Hilsum and Skandalis [**HS**] in the general case; also see [**Re1**], p.102. A complete proof, based on Renault's theorem of disintegration of representations [**Re2**], may be found in [**Ra1**]. A counterexample that shows that the second sequence of reduced C^*-algebras is not necessarily exact for non-amenable G_F is given in [**Re3**].

DEFINITION 5.2. *A field of groupoids is a triple (G, T, p), with G a groupoid, T a set, and $p : G \to T$ a surjective map such that $p = p_0 \circ r = p_0 \circ s$, where $p_0 = p_{|G_0}$. If G is locally compact, T Hausdorff, and p continuous and open, we say that (G, T, p) is a continuous field of locally compact groupoids. If G is a Lie groupoid, T a manifold, and p a submersion, (G, T, p) is called a smooth field of Lie groupoids.*

If (G, T, p) is a field of groupoids and $Y \subset T$, then $A = p_0^{-1}(Y)$ is an invariant subset of G_0, and $G_A = p^{-1}(Y)$ is a subgroupoid of G. In the case of a continuous field of locally compact groupoids,

$$G(t) = p^{-1}(\{t\}) \tag{5.3}$$

is a closed locally compact subgroupoid of G for every $t \in T$. In the case of a smooth field of Lie groupoids, $G(t)$ is a Lie groupoid for every $t \in T$.

In this section, (G, T, p) is a continuous field of locally compact groupoids over a locally compact Hausdorff space T, for which there exists a left Haar system $(\lambda^x)_{x \in G_0}$ on G.

Let $C_0(T)$ be the C^*-algebra of continuous functions on T that vanish at infinity. We define a structure of $C_0(T)$ left module on the space of continuous compactly supported functions $C_c(G_1)$ by $(fa)(\gamma) = f(p(\gamma))a(\gamma)$, where $f \in C_0(T)$ and $a \in C_c(G_1)$. For every $f \in C_0(T)$ and $a, b \in C_c(G_1)$, the following equalities hold:

$$\begin{aligned} f(a*b) &= (fa)*b = a*(fb); \\ (fa)^* &= f^*a^*. \end{aligned} \tag{5.4}$$

We also have $C_0(T)C_c(G_1) = C_c(G_1)$. Indeed, for every $a \in C_c(G_1)$, there is a function $f \in C_c(T)$ such that $f = 1$ on $p(\operatorname{supp} a)$, and then $a = fa \in C_0(T)C_c(G_1)$.

An easy rewriting of Lemma 1.13 from [**Re1**] proves

LEMMA 5.3. *For every representation $\pi : C_c(G_1) \to \mathcal{L}(\mathcal{H})$ that is continuous for the inductive limit topology there is a unique continuous representation $\phi : C_0(T) \to \mathcal{L}(\mathcal{H})$ such that $\pi(fa) = \phi(f)\pi(a)$.*

Using the previous lemma for $f \in C_0(T)$, $a \in C_c(G_1)$, and π continuous representation of $C_c(G_1)$, we have $\|\pi(fa)\| \le \|\phi(f)\|\,\|\pi(a)\| \le \|f\|\|a\|$, hence $\|fa\| \le \|f\|\|a\|$. For every $f \in C_0(T)$, the map $C_c(G_1) \ni a \mapsto fa \in C_c(G_1)$ has a continuous extension on $C^*(G)$. This provides a $C_0(T)$ Banach module structure on $C^*(G)$.

Corollary 1.9 in [**Bl**] shows that $C_0(T)C^*(G)$ is closed in $C^*(G)$. But one has $C_0(T)C^*(G) \supset C_0(T)C_c(G_1) = C_c(G_1)$, and we obtain that $C^*(G)$ is a nondegenerate module. For every $a, b \in C^*(G)$ and every $f \in C_0(T)$, the conditions $f(a*b) = (fa)*b = a*(fb)$ and $(fa)^* = f^*a^*$ are easily verified, hence $C^*(G)$ is a $C_0(T)$ *-algebra (cf. [**Bl**]). Similar arguments prove that $C_r^*(G)$ is a $C_0(T)$ *-algebra.

LEMMA 5.4. *Let T, Y be topological spaces, $p : Y \to T$ an open onto map, and $\varphi : Y \to \mathbb{R}$ an upper semicontinuous function such that for every $t \in T$ one has $sup\{\varphi(y) | p(y) = t\} < \infty$. Then the function $\psi : T \to \mathbb{R}$ given by $\psi(t) = \sup_{p(y)=t} \varphi(y)$ is lower semi-continuous.*

PROOF. Let t_i be a net converging in T to t, and pick $a \in \mathbb{R}$. We show that $\psi(t_i) \le a$ implies $\psi(t) \le a$. Let $y \in Y$ be such that $p(y) = t$. A lemma on open maps (cf. [**FD**], p.126) proves the existence of a net $y_i \in Y$, for which $p(y_i) = t_i$ and $y_i \to y$. For every i, $\varphi(y_i) \le \psi(t_i) \le a$. But φ is lower semicontinuous, so $\varphi(y) \le a$ and $\psi(t) \le a$. Ξ

We now come to the main result of this section.

THEOREM 5.5. *Let (G, T, p) be a continuous field of locally compact groupoids, take $a \in C_c(G_1)$, and, for each $t \in T$, write $a_t = a_{|G(t)}$. Then*

1. *The map $t \mapsto \|a_t\|_{C^*(G(t))}$ is upper semicontinuous.*
2. *The map $t \mapsto \|a_t\|_{C_r^*(G(t))}$ is lower semicontinuous.*

PROOF. 1. For $t \in T$, $C_t(T) = \{f \in C_0(T) | f(t) = 0\}$ is a closed two sided ideal, hence $C_t(T)C^*(G)$ is a closed two sided ideal in $C^*(G)$. Corollary 1.9 in [**Bl**] applied to the $C_t(T)$ Banach module $C^*(G)$ shows that $C_t(T)C^*(G)$ is closed in $C^*(G)$. Let $\pi_t : C^*(G) \to C^*(G)/C_t(T)C^*(G)$ be the quotient map. By Lemma 1.10 from

[**Bl**] one has $\|\pi_t(a)\| = \inf_{f\in C_0(T)} \|(1-f+f(t))a\|$; hence the map $t \mapsto \|\pi_t(a)\|$ is upper semicontinuous as the infimum of a family of continuous functions.

Denote the closed invariant subset $p_0^{-1}(\{t\})$ of G_0 by F, and let $U = G_0\backslash F$. Obviously, $G(t) = G_F$. We claim that

$$i\,(C_c(G_U)) = C_t(T)C_c(G), \tag{5.5}$$

where $i : C^*(G_U) \to C^*(G)$ is as in Proposition 5.1. Let $a \in i\,(C_c(G_U))$. In T there is an open neighborhood V of the compact set $p(\operatorname{supp} a)$ such that $t \notin V$. We take $f \in C_c(T)$ with $f = 1$ on $p(\operatorname{supp} a)$ and $\operatorname{supp} f \subset V$, and then $a = fa \in C_t(T)C_c(G_1)$. The other inclusion is obvious.

Taking the norm closure in (5.5), we obtain $i(C^*(G_U)) = C_t(T)C^*(G)$. With Proposition 5.1 this shows that $C^*(G(t)) = C^*(G)/C_t(T)C^*(G)$, so $\|\pi_t(a)\| = \|a_t\|$, which ends the proof.

2. The left regular representation $\pi_x^L(f) : L^2(G_1, \lambda^x) \to L^2(G_1, \lambda^x)$ of $C_c(G_1)$ associated to $x \in G_0$ is given by

$$\pi_x^L(f)\xi(\gamma) = \int_{G^x} f(\gamma\gamma_1)\xi(\gamma_1^{-1})d\lambda^x(\gamma_1). \tag{5.6}$$

Set $\|\xi\|_\infty = \sup_{x\in G_0} \|\xi\|_{L^2(G_1,\lambda^x)}$; then

$$\left\|\pi_x^L(f)\right\| = \sup\left\{\left\langle\pi_x^L(f)\xi,\eta\right\rangle \,|\, \xi,\eta \in C_c(G_1), \|\xi\|_\infty < \infty, \|\eta\|_\infty < \infty\right\}.$$

The map $x \mapsto \left\langle\pi_x^L(f)\xi,\eta\right\rangle$ is continuous for each $f \in C_c(G_1)$ and $\xi,\eta \in C_c(G_1)$, hence $x \mapsto \left\|\pi_x^L(f)\right\|$ is lower semicontinuous. But $\|a_t\| = \sup_{t\in p^{-1}(t)} \|\pi_x^L(a_t)\|$, and Lemma 5.4 ends the proof. ∎

COROLLARY 5.6. *If (G, T, p) is a continuous field of locally compact groupoids and $a \in C_c(G_1)$, then*

1. *The maps $t \mapsto \|a_t\|$ and $t \mapsto \|a_t\|_r$ are continuous at every $t \in T$ for which the groupoid $G(t)$ is amenable.*
2. *If G is amenable, then the maps $t \mapsto \|a_t\|$ and $t \mapsto \|a_t\|_r$ are continuous.*

PROOF. 1. Straightforward.

2. By Proposition 5.2.3 from [**AR**], G is amenable if and only if $G(t)$ is amenable for every $t \in T$. ∎

6. The tangent groupoid

The tangent groupoid of a manifold was introduced by Connes [**Co2**]. This idea was generalized to arbitrary Lie groupoids by Hilsum and Skandalis [**HS**] under the name normal groupoid, and by Weinstein [**We1**] under the name blowup. Connes's construction corresponds to the pair groupoid of a manifold. We here use the name "tangent groupoid" also for the general case.

DEFINITION 6.1. *Let G be a Lie groupoid with Lie algebroid $A(G)$. The tangent groupoid $\hat{G}$ is a Lie groupoid with base $\hat{G}_0 = [0,1] \times G_0$, defined by the following structures.*

- *As a set, $\hat{G}_1 = A(G) \cup \{(0,1] \times G_1\}$. We write elements of $\hat{G}$ as pairs $(\hbar, u)$, where $u \in A(G)$ for $\hbar = 0$ and $u \in G$ for $\hbar \neq 0$. Thus $A(G)$ is identified with $\{0\} \times A(G)$.*

- *As a groupoid,* $\hat{G} = \{0 \times A(G)\} \cup \{(0,1] \times G\}$*. Here* $A(G)$ *is regarded as a Lie groupoid over* G_0 *with* $s = r = \tau$ *(the bundle projection of* $A(G)$*), and addition in the fibers as the groupoid multiplication. The groupoid operations in* $(0,1] \times G$ *are those in* G*. For example, for* $\hbar \neq 0$ *one has*

$$\hat{r}(\hbar, u) = (\hbar, r(u)). \tag{6.1}$$

- *The smooth structure on* $\hat{G}_1$*, making it a manifold with boundary, is as follows. To start, the open subset* $\mathcal{O}_1 = (0,1] \times G_1 \subset \hat{G}_1$ *inherits the product manifold structure. Let* $G_0 \subset V \subset A(G)$ *and* $\iota(G_0) \subset W \subset G_1$*, as in Theorem 2.6. Let* $\mathcal{O}$ *be the open subset of* $[0,1] \times A(G)$ *(equipped with the product manifold structure; this is a manifold with boundary, since* $[0,1]$ *is), defined as* $\mathcal{O} = \{(\hbar, X) \mid \hbar X \in V\}$*. Note that* $\{0\} \times A(G) \subset \mathcal{O}$*. The map* $\psi : \mathcal{O} \to \hat{G}_1$ *is defined by*

$$\begin{aligned} \psi(0, X) &= (0, X); \\ \psi(\hbar, X) &= (\hbar, \mathrm{Exp}^W(\hbar X)). \end{aligned} \tag{6.2}$$

Since $\mathrm{Exp}^W : V \to W$ *is a diffeomorphism (cf. Proposition 2.6) we see that* ψ *is a bijection from* $\mathcal{O}$ *to* $\mathcal{O}_2 = \{0 \times A(G)\} \cup \{(0,1] \times W\}$*. This defines the smooth structure on* $\mathcal{O}_2$ *in terms of the smooth structure on* $\mathcal{O}$*. Since* $\mathcal{O}_1$ *and* $\mathcal{O}_2$ *cover* $\hat{G}_1$*, this specifies a smooth structure on* $\hat{G}_1$*.*

Using Exp^L rather than Exp^W in this definition, one obtains the smooth structure defined in [**HS**], which is equivalent to the one constructed above. A proof that the change of coordinates from $\mathcal{O}_1$ to $\mathcal{O}_2$ on their intersection is smooth may be found in [**HS**].

We omit the proof that $\hat{G}$ is a Lie groupoid; see [**HS**], and, for full details, [**Ra1**]. It follows that $\hat{G}$ is a smooth field of Lie groupoids over $\mathbb{R}$.

PROPOSITION 6.2. *Let* G *be a Lie groupoid with tangent groupoid* $\hat{G}$*. Writing* $\mathbb{R}^* = \mathbb{R}\backslash\{0\}$*, the following sequences are exact:*

$$0 \to C_0(\mathbb{R}^*) \otimes C^*(G) \to C^*(\hat{G}) \to C_0(A^*(G)) \to 0; \tag{6.3}$$

$$0 \to C_0(\mathbb{R}^*) \otimes C^*_r(G) \to C^*_r(\hat{G}) \to C_0(A^*(G)) \to 0. \tag{6.4}$$

PROOF. Consider the open invariant subset $U = G_0 \times \mathbb{R}^*$ of $\hat{G}_0$, with complement $F = G_0 \backslash U$. Then $\hat{G}_U = G \times \mathbb{R}^*$ and $\hat{G}_F = A(G) \times \{0\}$. By Proposition 5.1 we have the exact sequence

$$0 \to C^*(G \times \mathbb{R}^*) \to C^*(\hat{G}) \to C^*(A(G)) \to 0.$$

Gelfand's theorem implies $C^*(A(G)) = C_0(A^*(G))$ (also cf. section 7), and since $C^*(G \times \mathbb{R}^*)$ is the C^*-algebra of the trivial field of the C^*-algebras $C^*(G)$ on $\mathbb{R}^*$, we have $C^*(G \times \mathbb{R}^*) = C_0(\mathbb{R}^*) \otimes C^*(G)$.

The second exact sequence is a consequence of Proposition 5.1 as well, since $A(G)$ is commutative as a groupoid (cf. Definition 6.1), and therefore amenable [**AR**]. Ξ

COROLLARY 6.3. *(cf. [**Co2**]) Let* M *be a manifold. Then the tangent groupoid* $\widehat{M \times M} = (M \times M \times \mathbb{R}^*) \bigcup (TM \times \{0\})$ *is amenable, and the sequence*

$$0 \to C_0(\mathbb{R}^*) \otimes \mathcal{K}(L^2(M)) \to C^*(\widehat{M \times M}) \to C_0(T^*M) \to 0 \tag{6.5}$$

is exact.

PROOF. By Proposition 5.2.3 in [**AR**], a groupoid bundle is amenable if and only if all its fibers are amenable. Since the Lie algebroid $A(G)$ is amenable as a groupoid, the amenability of $\hat{G}$ is reduced to the amenability of the groupoid G. In this case, $C^*(\hat{G}) = C^*_r(\hat{G})$, cf. [**Re1, AR**]. Now use Proposition 6.2 for the amenable groupoid $G = M \times M$ to finish the proof. Ξ

Recall Definition 4.2, whose notation we adopt.

THEOREM 6.4. *Let G be a Lie groupoid with tangent groupoid $\hat{G}$. For $\hbar \in \mathbb{R}$, define*

$$\begin{aligned} G(0) &= A(G); \\ G(\hbar) &= G \ \forall \hbar \neq 0. \end{aligned} \tag{6.6}$$

Define $\hat{\varphi}_\hbar : C^\infty_c(\hat{G}) \to C^\infty_c(G(\hbar))$ as the pullback of the inclusion $G(\hbar) \hookrightarrow \hat{G}$ (cf. Definition 6.1); in other words, $\hat{\varphi}_\hbar(\hat{f})$ is the restriction of $\hat{f} \in C^\infty_c(\hat{G})$ to $G(\hbar)$.

1. *Each $\hat{\varphi}_\hbar$ may be extended by continuity to a surjective *-homomorphism $\varphi_\hbar : C^*(\hat{G}) \to C^*(G(\hbar))$, and also to a surjective *-homomorphism $\varphi_{(r)\hbar} : C^*_r(\hat{G}) \to C^*_r(G(\hbar))$.*
2. *The C^*-algebras $C = C^*(\hat{G})$ and $A_\hbar = C^*(G(\hbar))$, and the maps $\varphi_\hbar$ form a continuous field of C^*-algebras over $I = \mathbb{R}$.*
3. *The C^*-algebras $C = C^*_r(\hat{G})$ and $A_\hbar = C^*_r(G(\hbar))$, and the maps $\varphi_{(r)\hbar}$ form a continuous field of C^*-algebras over $I = \mathbb{R}$.*

PROOF.
1. Using the definition of the norms in question, one checks that each map $\hat{\varphi}_\hbar$ is contractive. The *-homomorphism property is obvious on $C^\infty_c(\hat{G})$ from the definition of $\hat{G}$ and $C^*(G)$, and extends by continuity.
2. Condition 1 in Definition 4.2 follows at $\hbar \neq 0$ since the field of C^*-algebras is trivial away from 0. At $\hbar = 0$ continuity follows from Corollary 5.6, since $A(G)$, being a commutative groupoid, is amenable [**AR**].
3. The reduced case is proved in the same way. Ξ

For a different proof of this theorem, note that each family $\hat{\varphi}_\hbar(\hat{f})$, $\hat{f} \in C^\infty_c(\hat{G})$, is stable under convolution and involution, and dense in $C^*_{(r)}(G)$. One then uses Prop. 10.3.2 in [**Di**] and the equivalence between Kirchberg–S. Wassermann's Definition 4.2 and Dixmier's definition Def. 10.3.1 in [**Di**] of a continuous field of C^*-algebras. Condition 10.1.2 (ii) in [**Di**] is then proved in the same way as above.

A different proof of Theorem 6.4, which makes no use of the results in section 5 is possible as well [**La3, La4**]. This proof is based on the following lemma, due Lee; cf. Theorem 4 in [**Le**].

LEMMA 6.5. *Let C be a C^*-algebra, and let $\psi : \mathrm{Prim}(C) \to T$ be a continuous and open map from the primitive spectrum $\mathrm{Prim}(C)$ (equipped with the Jacobson topology [**Di**]) to a locally compact Hausdorff space T. Define $I_t = \cap\psi^{-1}(t)$; i.e., $c \in I_t$ iff $\pi_I(c) = 0$ for all $I \in \psi^{-1}(t)$ (here $\pi_I(C)$ is the irreducible representation whose kernel is I). Note that I_t is a closed two-sided ideal in C.*

Taking $A_t = C/I_t$ and $\varphi_t : C \to A_t$ to be the canonical projection, the collection $(C, \{A_t, \varphi_t\}_{t\in T})$ is a continuous field of C^*-algebras.

For a proof see [**Le**] or [**ENN1**]. We apply this lemma with $C = C^*(\hat{G})$ and $T = I = \mathbb{R}$. In order to verify the assumption in the lemma, we first note that $I_0 \simeq C_0((0,1]) \otimes C^*(G)$, as follows from a glance at the topology of $\hat{G}$. Hence

$\text{Prim}(I_0) = (0,1] \times \text{Prim}(C^*(G))$, with the product topology. Furthermore, one has $C^*(\hat{G})/I_0 \simeq C^*(A(G)) \simeq C_0(A^*(G))$. Hence $\text{Prim}(C^*(\hat{G})/I_0) \simeq A^*(G)$. Using this in Prop. 3.2.1 in [**Di**], with $A = C^*(\hat{G})$ and I the ideal I_0 generated by those $f \in C_c^\infty(\hat{G})$ that vanish at $\hbar = 0$, yields the decomposition

$$\text{Prim}(C^*(\hat{G})) \simeq A^*(G) \cup \{(0,1] \times \text{Prim}(C^*(G))\}, \tag{6.7}$$

in which $A^*(G)$ is closed. This does not provide the full topology on $\text{Prim}(C^*(\hat{G}))$, but it is sufficient to know that $A^*(G)$ is not open. If it were, $(0,1] \times \text{Prim}(C^*(G))$ would be closed in $\text{Prim}(C^*(\hat{G}))$, and this possibility can be excluded by looking at the topology of $\hat{G}$ and the definition of the Jacobson topology.

Using (6.7), we can define a map $\psi : \text{Prim}(C^*(\hat{G})) \to [0,1]$ by $\psi(I) = 0$ for all $I \in A^*(G)$ and $\psi(\hbar, I) = \hbar$ for $\hbar \neq 0$ and $I \in \text{Prim}(C^*(G))$. It is clear from the preceding considerations that ψ is continuous and open. Using this in Lemma 6.5, one sees that $I_\hbar$ is the ideal in $C^*(\hat{G})$ generated by those $\hat{f} \in C_c^\infty(\hat{G})$ vanish at $\hbar$. Hence $A_0 \simeq C_0(A^*(G))$, as above, and $A_\hbar \simeq C^*(G)$ for $\hbar \neq 0$. Theorem 6.4 then follows from Lemma 6.5.

We now give a left Haar system for the tangent groupoid $\hat{G}$. Firstly, pick a positive section $\mu \in C^\infty(G_0, |\Omega|^1(A(G)))$, and let λ be the associated smooth, positive, left invariant section of $|\Omega|^1(\ker Tr)$, as in Proposition 3.2; hence $\mu = \lambda|_{G_0}$. Now define a section $\hat{\lambda} \in C^\infty(\hat{G}_1, |\Omega|^1(\ker T\hat{r})$ by

$$\begin{aligned} \hat{\lambda}(x, X, 0) &= \lambda(x); \\ \hat{\lambda}(\gamma, \hbar) &= |\hbar|^{-p}\lambda(\gamma). \end{aligned} \tag{6.8}$$

Here p is the dimension of the typical fiber of $A(G)$; the factor $|\hbar|^{-p}$ is necessary in order to have a smooth system also at $\hbar = 0$, as is easily verified using the manifold structure on $\hat{G}$. This section is smooth, positive and left invariant, so that it defines a left Haar system $(\hat{\lambda}^{(\hbar,x)})_{(\hbar,x)\in\hat{G}_0}$ for $\hat{G}$ by Proposition 3.4. The *-algebraic structure on $C_c^\infty(\hat{G}, \hat{\lambda})$ defined by (3.7) and (3.8) with (6.8) becomes

$$\hat{f} * \hat{g}(0, \xi_x) = \int_{\tau^{-1}(x)} \hat{f}(0, \xi - \eta_x)\hat{g}(0, \eta_x)d\mu^x(\eta_x); \tag{6.9}$$

$$\hat{f} * \hat{g}(\hbar, \gamma) = |\hbar|^{-p} \int_{G^{s(\gamma)}} \hat{f}(\hbar, \gamma\gamma_1)\hat{g}(\hbar, \gamma_1^{-1})d\lambda^{s(\gamma)}(\gamma_1); \tag{6.10}$$

$$\hat{f}^*(0, \xi) = \overline{\hat{f}(0, -\xi)}; \tag{6.11}$$

$$\hat{f}^*(\hbar, \gamma) = \overline{\hat{f}(\hbar, \gamma^{-1})}. \tag{6.12}$$

Here $\xi_x \in \tau^{-1}(x)$, and μ^x is the measure on $\tau^{-1}(x) \subset A(G)$ generated by the density $\mu(x)$ (cf. (3.1)). Finally, (λ^x) is the left Haar system on G defined by λ according to Proposition 3.4. These formulae should be compared with (3.7) and (3.8).

7. The Fourier transform on vector bundles

In this section we extend the notion of a Fourier transform from $\mathbb{R}^n$ to vector bundles, and show that it remains an isomorphism of a suitably defined Schwartz space of rapidly decreasing functions.

Let M be a manifold of dimension n, and let E be a vector bundle over M with fiber dimension p. Now E is a commutative Lie groupoid if the source and range projections are both equal to the bundle projection, and groupoid multiplication is addition in each fiber. The C^*-algebra $C^*(E)$ is then commutative, too. Using the proposition on page 582 of [**FD**], it can be shown that the maximal ideal space of $C^*(E)$ is E^*. The Gelfand transform is an isomorphism between the C^*-algebras $C^*(E)$ and $C_0(E^*)$. Its explicit form is a fiberwise Fourier transform $\mathcal{F}: C^*(E) \to C_0(E^*)$. On the convolution algebra $C_c^\infty(E, \mathfrak{E})$ the Fourier transform is given, for $\theta_x \in E_x^* \subset E^*$, by

$$(\mathcal{F}\rho)(\theta_x) = \int_{E_x} e^{-i\langle\theta_x, \xi_x\rangle} \rho(\xi_x).$$

It is easier to write formulas for functions rather than densities. To do so, fix a positive 1-density $\mu \in C^\infty(M, |\Omega|^1(E))$; cf. the proof of Proposition 3.4. For every $x \in M$, the density $\mu(x) \in |\Omega|^1(E_x)$ defines a translation invariant measure μ^x on E_x, and the family $(\mu^x)_{x\in M}$ is a smooth Haar system for E. The Fourier transform is denoted by $\mathcal{F}_\mu : C^*(E, \mu) \to C_0(E^*)$ in order to emphasize its dependence on μ. For $f \in L^1(E, d\mu^x)$, $\theta_x \in E_x^*$, we then have

$$(\mathcal{F}_\mu f)(\theta_x) = \int f(\xi_x) e^{-i\langle\theta_x, \xi_x\rangle} d\mu^x(\xi_x). \tag{7.1}$$

We now generalize the definition given by Rieffel in the case of the trivial bundle $M \times \mathbb{R}^p$ (cf. [**Ri3**]) to define a Schwartz space on E. But first we fix some notations.

The variables in $\mathbb{R}^n \times \mathbb{R}^p$ are denoted (u, v), and for $\beta \in \mathbb{N}^p$ we put $\partial_v^\beta F = \dfrac{\partial^{\beta_1+\cdots+\beta_p} F}{\partial v_1^{\beta_1} \cdots \partial v_p^{\beta_p}}$. The Fourier transform $\mathcal{F}_u$ on $\mathbb{R}^n \times \mathbb{R}^p$ with u constant is given by $(\mathcal{F}_u F)(u, w) = \int F(u, v) e^{-i\langle v, w\rangle} dv$.

Let $(e_1, \dots, e_p)$ be a local frame of E, with dual frame $(e_1^*, \dots, e_p^*)$ of E^*, let (q, λ) be local coordinates on E, and finally let (q, ϵ) be local coordinates of E^*. The expression of a function f on E in local coordinates is denoted by the corresponding capital letter F. Then (3.1) for $f \in L^1(E, \mu^x)$ becomes

$$\int f\left(\sum_{i=1}^n v_i e_i(x)\right) \mu_e(x) dv = \int f d\mu^x. \tag{7.2}$$

DEFINITION 7.1. *We say that a function $f : E \to \mathbb{C}$ is M-compactly supported when $\pi(\operatorname{supp} f)$ is relatively compact in M. The set of continuous M-compactly supported functions is called $C_{c,M}(E)$.*

A function $f \in C_{c,M}(E)$ is said to be rapidly decreasing if $\lambda^\alpha f(q, \lambda)$ is bounded for every $\alpha \in \mathbb{N}^p$, where $\lambda^\alpha = \lambda_1^{\alpha_1} \dots \lambda_p^{\alpha_p}$, with $\lambda_1, \dots, \lambda_p$ the coordinates of λ in the frame $(e_1, \dots, e_p)$.

This definition is independent of the frame. Indeed, let $(e_1, \dots, e_p)$ be a frame of E on an open subset U of M, and suppose that $\lambda^\alpha f(q, \lambda)$ is bounded for every $\alpha \in \mathbb{N}^p$. If $(e_1', \dots, e_p')$ is another frame of E on U, let $g : U \to GL(p, \mathbb{R})$ be the function given by $e' = eg$. If we write the same point of E in the two frames, its coordinates are related by $\lambda_i' = \sum_j \lambda_j g_{ji}$. To show that $\lambda'^\alpha f(q, \lambda')$ is bounded, remark that expanding $\lambda_1'^{\alpha_1} \dots \lambda_p'^{\alpha_p}$ we obtain a polynomial in $\lambda_1, \dots, \lambda_p$ with continuous functions on M that vanishes outside the compact closure of $\pi(\operatorname{supp} f)$ as coefficients.

DEFINITION 7.2. *We say that a function* $f \in C^\infty(E)$ *is of Schwartz type on* E *if* f *is* M*-compactly supported and* $\partial^\beta_\lambda f = \dfrac{\partial^{\beta_1+\cdots+\beta_p} f}{\partial\lambda_1^{\beta_1}\cdots\lambda_p^{\beta_p}}$ *is rapidly decreasing for every* $\beta = (\beta_1, \dots, \beta_p) \in \mathbb{N}^p$. *The set of Schwartz functions on* E *is denoted by* $S(E)$.

PROPOSITION 7.3. $S(E)$ *is a dense* **-subalgebra of* $C^*(E,\mu)$ *and* $S(E^*)$ *is a dense* **-subalgebra of* $C_0(E^*)$.

PROOF. We show that $S(E)$ is closed under convolution. For f, g in $S(E)$ it is obvious that $f * g$ is M-compactly supported. Using a partition of unity argument, it easily follows that $f * g \in S(E)$. It is trivial that $S(E)$ is closed under involution. Now $S(E)$ contains the smooth compactly supported functions on E, and by a result from [**FD**], page 140, $C^\infty_c(E)$ is dense in $C^*(E,\mu)$; it follows that $S(E)$ is dense in $C^*(E,\mu)$.

For the second part of the proposition, $S(E^*)$ is obviously a *-subalgebra of $C_0(E^*)$ containing the dense subalgebra $C^\infty_c(E^*)$ of $C_0(E^*)$. Ξ

LEMMA 7.4. *If* $f \in S(E)$, *then* $\mathcal{F}_\mu f \in S(E^*)$

PROOF. The restriction of f to E_x is integrable because it is rapidly decreasing, so (7.1) makes sense in this case. It is then easy to show that $\mathcal{F}_\mu f$ is smooth and M-compactly supported.

Choosing a suitable partition of unity, we may assume that the projection onto M of the support of f is a subset of the domain of a chart $\alpha : U \to \mathbb{R}^n$ of M. We can also suppose that there exists a frame $(e_1, \dots, e_p)$ of E on U.

The expression of f in the corresponding local coordinates of E is given by $F : \alpha(U) \times \mathbb{R}^p \to \mathbb{C}$, $F(u,v) = f(\sum v_i e_i(\alpha^{-1}(u)))$. It is straightforward to see that the local expression of $\mathcal{F}_\mu f$, denoted by $\hat{F}$, satisfies $\hat{F}(u,w) = \mu_e(\alpha^{-1}(u))(\mathcal{F}_u F)(u,w)$. The relation $w^\alpha \partial^\beta_w \hat{F}(u,w) = \mu_e(\alpha^{-1}(u))(-i)^{|\alpha|+|\beta|}\mathcal{F}_u\left(\partial^\alpha_v(v^\beta F)\right)(u,w)$ reduces the problem to the fact that $\mathcal{F}_u G$ is bounded for every G rapidly decreasing on the trivial bundle $\mathbb{R}^n \times \mathbb{R}^p$ of base $\mathbb{R}^n$.

But $|\mathcal{F}_u G(u,w)| \le \displaystyle\int |F(u,v)|\,dv$, and the proof is finished by the remark that on the right side we have a continuous compactly supported function in u. Ξ

For a vector space V and a nowhere vanishing element ρ of $|\Omega|^\alpha(V)$, let ρ^* be the element of $|\Omega|^\alpha(V^*)$ that satisfies $\rho^*(v_1^* \wedge \cdots \wedge v_n^*) = \dfrac{1}{\rho(v_1 \wedge \cdots \wedge v_n)}$ for every pair of dual bases $\{v_1, \dots, v_n\}$ on V and $\{v_1^*, \dots, v_n^*\}$ on V^*. Then $\mu_*(x) = (\mu(x))^*$ defines a 1-density $\mu_* \in C^\infty(M, |\Omega|^1(E^*))$, and hence a left Haar system $(\mu_*^x)_{x\in M}$ on E^* (seen as a Lie groupoid in the same way as E).

LEMMA 7.5. *The restriction of the Fourier transform* $\mathcal{F}_\mu$ *to* $S(E)$ *is a bijection onto* $S(E^*)$, *with inverse*

$$(\mathcal{F}_\mu^{-1} g)(\xi_x) = (2\pi)^{-p} \int g(\theta_x) e^{i\langle\theta_x,\xi_x\rangle} d\mu_*^x(\theta_x). \tag{7.3}$$

PROOF. Compute in local coordinates. Ξ

Using these two lemmas we have

PROPOSITION 7.6. $\mathcal{F}_\mu$ *is an algebra isomorphism between* $S(E)$ *with the convolution* **-algebra structure inherited from* $C^*(E,\mu)$ *and* $S(E^*)$ *with the* **-algebra structure borrowed from* $C_0(E^*)$.

In the following lemma, which is used in the proof of Proposition 9.1, we list a series of properties of the Fourier transform on $S(E)$.

LEMMA 7.7. *If $f, g \in S(E)$, $\phi \in S(E^*)$ and $a \in C^\infty(M)$, then*

1. $[(a \circ \pi)f]\hat{} = (a \circ \pi)\hat{f}$;
2. $\frac{\partial \hat{f}}{\partial q_j}(\theta_x) = \widehat{\frac{\partial f}{\partial q_j}}(\theta_x) + \frac{\partial ln\mu_e}{\partial q_j}(x)\hat{f}(\theta_x)$;
3. $\frac{\partial \hat{f}}{\partial \epsilon_i}(\theta_x) = -i(\xi_i f)\hat{}(\theta_x)$;
4. $\frac{\partial \check{\phi}}{\partial \lambda_i}(\xi_x) = i(w_i\phi)\check{}(\xi_x)$;
5. $i\theta_k \hat{f}(\theta_x) = \left(\frac{\partial f}{\partial \lambda_k}\right)\hat{}(\theta_x)$.

Here $(\cdot)\hat{}$ is the Fourier transform $\mathcal{F}_\mu$ of $(\cdot)$ and $(\cdot)\check{}$ is the inverse Fourier transform $\mathcal{F}_\mu^{-1}$ of $(\cdot)$.

The Schwartz functions do not form the only class of interest to us.

DEFINITION 7.8. *The Paley–Wiener functions on E^*, denoted by $C^\infty_{PW}(E^*)$, consist all functions in $S(E^*)$ whose (inverse) Fourier transform is in $C^\infty_c(E)$.*

We use this definition for $E = A(G)$; the Poisson algebra A^0 in Theorems 4.4 and 4.6 is $C^\infty_{\rm PW}(A^*(G))$. Writing the Poisson bracket and the pointwise product in terms of the Fourier transform, one quickly establishes that A^0 is indeed a Poisson algebra (cf. section 9).

8. Weyl quantization

Having constructed the continuous field of C^*-algebras called for in Theorem 4.4 in section 6, it remains to define a Poisson algebra $A^0 \subset C_0(A^*(G))$ and a map $\mathcal{Q} : A^0 \to C^*(\hat{G})$, or equivalently, a family of involutive maps $\mathcal{Q}_\hbar : A^0 \to C^*(G)$ satisfying (4.7). These maps will, in addition, provide the data of Theorem 4.6. We do so by a generalization of Weyl quantization on $T^*\mathbb{R}^n$; cf. [**La3**].

We pick a positive 1-density $\mu \in C^\infty(G_0, |\Omega|^1(A(G)))$, with associated left Haar system $(\lambda^x)_{x\in G_0}$ on G (see section 3), left Haar system $(\hat{\lambda}^{(\hbar,x)})_{(\hbar,x)\in \hat{G}_0}$ on $\hat{G}$ (see section 6), and left Haar system $(\mu^x)_{x\in G_0}$ on $A(G)$ (see section 7). We shall work with the concrete C^*-algebras $C^*(G, \lambda)$ and $C^*(\hat{G}, \hat{\lambda})$ rather than with their intrinsically defined versions $C^*(G)$ and $C^*(\hat{G})$; see section 3. Moreover, since the argument below is the same for the reduced C^*-algebras, we will not consider that case explicitly.

Now choose some function $\kappa \in C^\infty(A(G), \mathbb{R})$ with support in V (cf. Proposition 2.6), equaling unity in some smaller tubular neighbourhood of G_0, as well as satisfying $\kappa(-\xi) = \kappa(\xi)$ for all $\xi \in A(G)$.

DEFINITION 8.1. *Let G be a Lie groupoid with Lie algebroid $A(G)$. We put*

$$A^0 = C^\infty_{PW}(A^*(G)) \subset A_0 = C_0(A^*(G)). \tag{8.1}$$

For $\hbar \neq 0$, the Weyl quantization of $f \in A^0$ is the element $\mathcal{Q}^W_\hbar(f) \in C^\infty_c(G_1)$, regarded as a dense subalgebra of $C^*(G, \lambda)$, defined by

$$\begin{aligned} \mathcal{Q}^W_\hbar(f)(\mathrm{Exp}^W(\xi)) &= |\hbar|^{-p}\kappa(\xi)\mathcal{F}_\mu^{-1} f(\xi/\hbar) \ \forall \xi \in V; \\ \mathcal{Q}^W_\hbar(f)(\gamma) &= 0 \ \forall \gamma \notin W. \end{aligned} \tag{8.2}$$

Here the Weyl exponential $\mathrm{Exp}^W : A(G) \to G$ *is defined in (2.10).*

This definition is possible by virtue of Proposition 2.6. (An analogous definition using Exp^L would not satisfy (4.6); such a definition would generalize the Kohn–Nirenberg calculus of pseudodifferential operators rather than the Weyl calculus.) By our choice of A^0, the operator $\mathcal{Q}_\hbar^W(f)$ is independent of κ for small enough $\hbar$ (depending on f).

PROPOSITION 8.2. *For each* $f \in A^0$, *the operator* $\mathcal{Q}_\hbar^W(f)$ *of Definition 8.1 satisfies* $\mathcal{Q}_\hbar^W(f)^* = \mathcal{Q}_\hbar^W(f^*)$, *and the family* $\{\mathcal{Q}_\hbar^W(f)\}_{\hbar\in\mathbb{R}}$, *with* $\mathcal{Q}_0^W(f) = f$, *is a continuous cross-section of both continuous fields of* C^**-algebras in Theorem 6.4.*

PROOF. It is immediate from (3.8) and (2.10) that for real-valued $f \in A^0$ the operator $\mathcal{Q}_\hbar^W(f)$ is self-adjoint in $C^*(G, \lambda)$; this implies the first claim.

Take $f \in C^\infty_{\mathrm{PW}}(A^*(G))$. The function $\mathcal{Q}^W(f)$ on $\hat{G}_1$ that is defined by

$$
\begin{aligned}
\mathcal{Q}^W(f)(0,\xi) &= \mathcal{F}_\mu^{-1} f(\xi);\\
\mathcal{Q}^W(f)(\hbar, \mathrm{Exp}^W(\xi)) &= \kappa(\xi)\mathcal{F}_\mu^{-1} f(\xi/\hbar) \ \forall \xi \in V;\\
\mathcal{Q}^W(f)(\gamma) &= 0 \ \forall \gamma \notin W
\end{aligned}
\tag{8.3}
$$

and is in $C_c^\infty(\hat{G}_1)$; cf. Definition 6.1. In other words, $\mathcal{Q}^W(f)$ is an element of $C^*(\hat{G}, \hat{\lambda})$.

For $\hat{f} \in C_c^\infty(\hat{G})$, define the restriction maps

$$
\begin{aligned}
\hat{\varphi}_0(\hat{f}) : \theta &\mapsto \mathcal{F}_\mu \hat{f}(0, \theta);\\
\hat{\varphi}_\hbar(\hat{f}) : \gamma &\mapsto \hat{f}(\hbar, \gamma) \ (\hbar \neq 0).
\end{aligned}
\tag{8.4}
$$

Then $\hat{\varphi}_0$ is a surjective *-homomorphisms from $C_c^\infty(\hat{G}_1, \hat{\lambda})$ to $C^\infty_{\mathrm{PW}}(A^*(G), \mu)$, and each $\hat{\varphi}_\hbar$ is a surjective *-homomorphisms from $C_c^\infty(\hat{G}_1, \hat{\lambda})$ to $C_c^\infty(G, |\hbar|^{-p}\lambda)$. These maps are contractive, and extend by continuity to surjective *-homomorphisms from $C^*(\hat{G}, \hat{\lambda})$ to $C_0(A^*(G))$ and $C^*(G, |\hbar|^{-p}\lambda)$, respectively. However, in Theorem 4.4 we have $A_0 = C_0(A^*(G))$ and $A_\hbar = C^*(G, \lambda)$ for $\hbar \neq 0$. Hence the maps $\varphi_\hbar$ of Definition 4.2 should be taken as

$$
\begin{aligned}
\varphi_0(\hat{f}) : \theta &\mapsto \mathcal{F}_\mu \hat{f}(0, \theta);\\
\varphi_\hbar(\hat{f}) : \gamma &\mapsto |\hbar|^{-p} \hat{f}(\hbar, \gamma) \ (\hbar \neq 0).
\end{aligned}
\tag{8.5}
$$

These maps extend to surjective *-homomorphisms from $C^*(\hat{G}, \hat{\lambda})$ to $C_0(A^*(G))$ and $C^*(G, \lambda)$, respectively. It follows that $A_\hbar \simeq C^*(\hat{G}, \hat{\lambda})/\ker \varphi_\hbar$ for all $\hbar \in \mathbb{R}$, and that $\mathcal{Q}_\hbar^W(f) = \varphi_\hbar(\mathcal{Q}^W(f))$. The proposition follows. Ξ

We have now proved Theorem 4.4 up to Dirac's condition, and turn to Theorem 4.6. We in addition need to choose the tubular neighbourhood W of $G_0 \subset G_1$ so that $\gamma\gamma' \subset W$ whenever $(\gamma, \gamma') \in G_2 \cap (W \times W)$.

We define a semi-strict deformation quantization of $A^*(G)$ in the following way. For given $f \in A^0 = C^\infty_{\mathrm{PW}}(A^*(G))$, we choose a cutoff function χ_f on $\mathbb{R}$ that is 1 in a neighbourhood of 0 and has support well inside the set of values of $\hbar$ for which $\mathcal{Q}^W(f)$ as defined in (8.3) is independent of κ. It is clear from the choice of A^0 that this can be done. For each $\hbar \in \mathbb{R}^*$, we then have:

- an involution $f^{*_\hbar} = \overline{f}$ (independent of $\hbar$);
- a semi-norm $\|f\|_\hbar = \chi_f(\hbar)\|\mathcal{Q}_\hbar^W(f)\|_{C^*(G,\lambda)}$;

- a product $\times_\hbar$ defined by the condition

$$\mathcal{Q}_\hbar^W(f \times_\hbar g) = \chi_f \chi_g(\hbar) \mathcal{Q}_\hbar^W(f) * \mathcal{Q}_\hbar^W(g),$$

where $*$ is the product in $C^*(G, \lambda)$.

The existence of the product follows from a detailed but straightforward analysis of the support properties of $\mathcal{Q}_\hbar^W(f) * \mathcal{Q}_\hbar^W(g)$, leading to the conclusion that, seen as a function on G_1 for fixed $\hbar$, under the stated assumptions its support is contained inside W; see [**Ra1**]. Hence it can be pulled back using Exp^W and can subsequently be Fourier-transformed so as to yield the desired function $f \times_\hbar g$.

The fact that $*_\hbar$ is indeed an involution is a consequence of the self-adjointness property (4.6) of $\mathcal{Q}_\hbar^W$. The same comment applies to the C^*-property of the semi-norm. In general, these are not norms, since for a given $f \neq 0$ there may well exist values of $\hbar$ such that $\chi(\hbar)\mathcal{Q}_\hbar^W(f) = 0$, seen as a function on $G(\hbar) \subset \hat{G}$.

The conditions defining a semi-strict deformation quantization, except (4.9), are now trivially satisfied as a consequence of Theorem 4.4 as proved so far. A direct proof of condition (b) in Definition 4.5 is also immediate, based on Theorem 6.4.

9. The local structure of the Poisson bracket

In this section, taken from [**Ra1**], we express the structure of Lie algebroids E, as well as the Poisson bracket on $C^\infty(E^*)$, in local coordinates. This is of interest in itself, but in the context of quantization it is a key tool for proving Dirac's condition (4.7) or (4.9).

Recall Definition 2.2 of a Lie algebroid. Let $\{e_1, e_2, \dots, e_p\}$ be a local frame on $U \subset M$ for E, and let $(q_1, \dots, q_n, \lambda_1, \dots, \lambda_p)$ be local coordinates of E, with the q_i's local coordinates for M and the λ_j's local coordinates for the fibers associated to the frame $\{e_1, e_2, \dots, e_p\}$. Then, locally, the fact that E is a Lie algebroid implies the existence of structure functions c_{ijk} and a_{ij} in $C^\infty(U)$ such that

$$[e_i, e_j] = \sum_k c_{ijk} e_k; \tag{9.1}$$

$$\rho(e_i) = \sum_j a_{ij} \frac{\partial}{\partial q_j}. \tag{9.2}$$

The Poisson bracket (4.2) - (4.4) of the functions $\phi, \psi \in C^\infty(E^*)$ is then locally given by

$$\{\phi, \psi\} = \sum_{i,j} \rho(e_i)(q_j) \left(\frac{\partial \phi}{\partial \epsilon_i} \frac{\partial \psi}{\partial q_j} - \frac{\partial \phi}{\partial q_j} \frac{\partial \psi}{\partial \epsilon_i} \right) + \sum_{i,j} \widetilde{[e_i, e_j]} \frac{\partial \phi}{\partial \epsilon_i} \frac{\partial \psi}{\partial \epsilon_j};$$

recall the notation explained below (4.4). Taking the value at the point $(x, \alpha) \in E_x^*$ and using the structure functions of the Lie algebroid, we obtain

$$\begin{aligned} \{\phi, \psi\}(x, \alpha) &= \sum_{i,j} a_{ij}(x) \left(\frac{\partial \phi}{\partial \epsilon_i}(x, \alpha) \frac{\partial \psi}{\partial q_j}(x, \alpha) - \frac{\partial \phi}{\partial q_j}(x, \alpha) \frac{\partial \psi}{\partial \epsilon_i}(x, \alpha) \right) \\ &+ \sum_{i,j,k} c_{ijk}(x) \epsilon_k(\alpha) \frac{\partial \phi}{\partial \epsilon_i}(x, \alpha) \frac{\partial \psi}{\partial \epsilon_j}(x, \alpha). \end{aligned} \tag{9.3}$$

Fix a positive 1-density $\mu \in C^\infty(M, |\Omega|^1(E))$. We define a Poisson bracket on the dense subalgebra $S(E)$ of $C^*(E, \mu)$, by transporting the Poisson bracket of

$C^\infty(E^*)$ using the Fourier transform $\mathcal{F}_\mu$. That is,

$$\{f,g\}_\mu = \mathcal{F}_\mu^{-1}\left(\{\mathcal{F}_\mu f, \mathcal{F}_\mu g\}\right). \tag{9.4}$$

PROPOSITION 9.1. *Eq. (9.4) defines a Poisson bracket on the convolution algebra $S(E)$. The Poisson bracket of $f, g \in S(E)$ is explicitly given by*

$$\begin{aligned} \{f,g\}_\mu(\xi_x) &= -i\sum_{i,j} a_{ij}(x)\left(\xi_i f * \frac{\partial g}{\partial q_j} - \xi_i g * \frac{\partial f}{\partial q_j}\right)(\xi_x) \\ &- i\sum_{i,j} a_{ij}(x)\frac{\partial \ln \mu_e}{\partial q_j}(x)\left(\xi_i f * g - \xi_i g * f\right)(\xi_x) \\ &+ i\sum_{i,j,k} c_{ijk}(x)\frac{\partial}{\partial \lambda_k}\left(\xi_i f * \xi_j g\right)(\xi_x). \end{aligned} \tag{9.5}$$

PROOF. Eq. (9.3) proves that $S(E^*)$ is stable under the Poisson bracket $\{\cdot,\cdot\}$, hence (9.4) is well defined. Since the Fourier transform is an algebra isomorphism between $S(E)$ and $S(E^*)$, it can easily be shown that we have a Poisson algebra structure on $S(E)$. Eq. (9.5) is a consequence of Lemma 7.7 and (9.3). ∎

We now specialize to the case of relevance to us, where $E = A(G)$ is the Lie algebroid of a Lie groupoid G. We analyze the local structure of G in the neighborhood of a fixed point $x_0 \in G_0 \subset G$ in a convenient parametrization; also cf. [**NWX**].

The map r is a submersion at $x_0 \in G_0 \subset G_1$, hence there exists an open neighborhood U of 0 in $\mathbb{R}^n$, an open neighborhood V of 0 $\mathbb{R}^m$, and parametrizations $\psi : U \times V \to G_1$ and $\varphi : U \to G_0$ such that

$$\begin{aligned} &1.\ \psi(0,0) = x_0; \\ &2.\ r(\psi(u,v)) = \varphi(u); \\ &3.\ \psi(U \times \{0\}) = \psi(U \times V) \textstyle\bigcap G_0. \end{aligned} \tag{9.6}$$

It follows from the first two conditions that $\varphi(u) = \psi(u,0)$. To the parametrization ψ of the Lie groupoid G one can associate a parametrization $\theta : U \times \mathbb{R}^m \to A(G)$ of the neighborhood $A(G)_{\varphi(U)}$ of the fiber $A(G)_{x_0}$ in $A(G)$, given by $\theta(u,v) = (\varphi(u), \frac{\partial \psi}{\partial v}(u,0)v)$.

For every $x_0 \in G_0$ there exists a neighborhood $\psi(U \times V)$ of x_0 in G that is diffeomorphic to the neighborhood $\theta(U \times V)$ of $(x_0, 0)$ in $A(G)$ by $\alpha = \psi \circ \theta^{-1}$. Moreover, $\alpha(A(G)_x) \subset G^x$ for each $x \in \varphi(U)$. This result can be formulated in a stronger form, based on the existence of an exponential map for a Lie groupoid; see section 2. Namely, taking α equal to the restriction of Exp^L (cf. Definition 2.4) to V, and α_x as the restriction of α to $A(G)_x \bigcap V$, one achieves that $\alpha(A(G)_x \bigcap V) = G^x \bigcap W$ and that $\alpha'_x(0)$ is the identity of $A(G)_x$.

The submersion $\sigma = \varphi^{-1} \circ s \circ \psi : U \times V \to U$ is the local expression of the source map s in the parametrization ψ. One has $\sigma(u,0) = u$, since $\varphi(\sigma(u,0)) = s(\psi(u,0)) = \psi(u,0) = \varphi(u)$. Similar expression may be given for the multiplication and the inversion.

The following theorem gives a version of the Baker-Campbell-Hausdorff formula for Lie groupoids. This will enter the proof of Dirac's condition.

THEOREM 9.2. *Let G be a Lie groupoid. Then*

1. $(\psi(u,v),\psi(u_1,w)) \in G_2$ *if and only if* $u_1 = \sigma(u,v)$. *In that case, their product is of the form* $\psi(u,v)\psi(\sigma(u,v),w) = \psi(u,p(u,v,w))$, *where* $p : U \times V \times V \to V$ *is a smooth map which has the expansion*

$$p(u,v,w) = v + w + B(u,v,w) + O_3(u,v,w), \tag{9.7}$$

with B *bilinear in* (v,w), *and* $O_3(u,v,w)$ *of the order of* $\|(v,w)\|^3$.
2. *Let* $(u,v) \in U \times V$ *be such that* $\psi(u,v)^{-1} \in \psi(U \times V)$. *Then* $\psi(u,v)^{-1} = \psi(\sigma(u,v),w)$, *where* w *satisfies* $p(u,v,w) = 0$. *Moreover,* $w = -v + B(u,v,v) + O_3(u,v)$, *with* $O_3(u,v)$ *of the order of* $\|v\|^3$.

PROOF. We merely sketch the proof. For details cf. [**Ra4**].

1. Set $g = \psi(u,v)$ and $h = \psi(u_1,w)$. With $s(g) = \varphi(\sigma(u,v))$ and $r(h) = \varphi(u_1)$, we have $(g,h) \in G_2$ if and only if $u_1 = \sigma(u,v)$. Also, $r(gh) = \varphi(u)$ implies the existence of a unique $p(u,v,w) \in V$ such that $\psi(u,v)\psi(\sigma(u,v),w) = \psi(u,p(u,v,w))$. Hence we obtain a map $p : U \times V \times V \to V$ that satisfies $p(u,0,w) = w$ and $p(u,v,0) = v$. Using these equations in a Taylor expansion of p yields (9.7).

2. Similar to 1. Ξ

We can now give explicit formulae for the structure functions of the Lie algebroid $A(G)$. First, remark that the family $\{e_1, e_2, \dots, e_m\}$ defined by $e_i(\varphi(u)) = \theta(u, f_i)$, $i = \overline{1,m}$, where $\{f_1, f_2, \dots, f_m\}$ is the canonical basis of $\mathbb{R}^m$, is a frame of $A(G)$ on $\varphi(U)$. Also, recall that the anchor of $A(G)$ is given by $\rho = Ts$ (cf. Definition 2.3.2), that the local coordinate functions of G_0 are $q_j = pr_j \circ \varphi^{-1}$, and that $B_1, \dots, B_m$ are the coordinates of $B : U \times V \times V \to V$ in the base $\{f_1, f_2, \dots, f_m\}$ of $\mathbb{R}^m$. Finally, recall that $a_{ij} = \rho(e_i)(q_j)$, and that the c_{ijk} are given by $[e_i, e_j] = \sum c_{ijk} e_k$.

PROPOSITION 9.3. *For each* $u \in U$, *the structure functions of the Lie algebroid* $A(G)$ *are given by* $a_{ij}(\varphi(u)) = \dfrac{\partial \sigma_j}{\partial v_i}(u,0)$ *and* $c_{ijk}(\varphi(u)) = B_k(u, f_i, f_j) - B_k(u, f_j, f_i)$.

PROOF. Direct calculations. See [**Ra1**], [**Ra4**] for details. Ξ

We now return to the Poisson bracket on $A(G)$. Recall Proposition 9.1, and put $E = A(G)$. The next proposition gives a local expression of the Poisson bracket on $S(A(G))$.

Let $\lambda \in C^\infty(G_0, |\Omega|^1(A(G)))$ be a positive 1-density and Λ the local expression in the parametrization θ of the function associated to λ in the local frame $\{e_1, e_2 \dots, e_m\}$.

PROPOSITION 9.4. *For* $f, g \in S(A(G))$, *let* $h = \{f,g\}_\lambda$ *be the Poisson bracket of* f *and* g *given by (9.5). Denote the local expressions of* f, g, h *in the parametrization* θ *by* F, G, H, *respectively. Then*

$$H(u,v) = i \int F(u,w) G(u, v-w) \sum_j \left[B_j(u,w,f_j) - B_j(u,f_j,w) \right] \Lambda(u) dw$$

$$+ i \int \sum_k \left[B_k(u,w,v) - B_k(u,v,w) \right] F(u,w) \frac{\partial G}{\partial v_k}(u, v-w) \Lambda(u) dw$$

$$-i\sum_{i,j}\frac{\partial\sigma_j}{\partial v_i}(u,0)\frac{\partial\Lambda}{\partial u_j}(u)\int w_i\left[F(u,w)G(u,v-w)-G(u,w)F(u,v-w)\right]dw$$
$$-i\sum_{i,j}\frac{\partial\sigma_j}{\partial v_i}(u,0)\int w_i\left[F(u,w)\frac{\partial G}{\partial u_j}(u,v-w)-G(u,w)\frac{\partial F}{\partial u_j}(u,v-w)\right]\Lambda(u)dw.$$

PROOF. Replacing the structure functions in (9.5) by their expressions calculated in Proposition 9.3, we can directly identify the first two lines of (9.5) with the last two in the formula of $H(u,v)$. Denote the expression in the last line of (9.5) by l. We then have

$$\begin{aligned}
l(\theta(u,v)) &= -\sum_{i,j,k}c_{ijk}(\varphi(u))\frac{\partial}{\partial v_k}\left(\int w_iF(u,w)(v_j-w_j)G(u,v-w)\Lambda(u)dw\right)\\
&= \sum_{i,j}\int(B_j(u,f_j,w_if_i)-B_j(u,w_if_i,f_j))F(u,w)G(u,v-w)\Lambda(u)dw\\
&+ \sum_{i,j,k}\int[B_k(u,(v_j-w_j)f_j,w_if_i)-B_k(u,w_if_i,(v_j-w_j)f_j)]\\
&\cdot\quad F(u,w)\frac{\partial G}{\partial v_k}(u,v-w)\Lambda(u)dw\\
&= \sum_{j}\int(B_j(u,f_j,w)-B_j(u,w,f_j))F(u,w)G(u,v-w)\Lambda(u)dw\\
&+ \sum_{k}\int(B_k(u,v-w,w)-B_k(u,w,v-w))\\
&\cdot\quad F(u,w)\frac{\partial G}{\partial v_k}(u,v-w)\Lambda(u)dw.
\end{aligned}$$

By bilinearity one has $B_k(u,w,v-w)-B_k(u,v-w,w)=B_k(u,w,v)-B_k(u,v,w)$; this finishes the proof. Ξ

10. Proof of Dirac's condition

We now use the results of the preceding section to prove (4.7) and (4.9) in the case at hand. We start with two lemmas.

Let ψ be a parametrization of G as in section 9, and denote the function associated to λ in the local frame of TG generated by ψ by λ_0. Put $\Lambda=\lambda_0\circ\psi$.

LEMMA 10.1. *One has* $\lambda_0(\gamma)=\dfrac{\lambda_0(s(\gamma))}{|detJ_{L_\gamma}(s(\gamma))|}$ *for every* γ *in the image of the parametrization* ψ.

PROOF. The left invariance of λ means $\lambda(\gamma)=\gamma\cdot\lambda(s(\gamma))$. It only remains to write the action of G on $|\Omega|^1(\ker Tr)$ in the local parametrization ψ. We leave this to the reader. Ξ

LEMMA 10.2. *Let* $\mu_{u,v}(t)=\lambda_0(\psi(u,tv))$. *Then*
$$\mu'_{u,v}(0)=\sum_{i,j}\frac{\partial\Lambda}{\partial u_i}(u,0)\frac{\partial\sigma_i}{\partial v_j}(u,0)v_j-[B_1(u,v,f_1)++\cdots+B_m(u,v,f_m)]\Lambda(u,0).$$

PROOF. By Lemma 10.1, $\mu_{u,v}(t) = \dfrac{\lambda_0(\psi(\sigma(u,tv),0))}{|det J_{L_{\psi(u,tv)}}(\psi(\sigma(u,tv),0))|}$. Write

$$\begin{aligned} a_{u,v}(t) &= \lambda_0(\psi(\sigma(u,tv),0)); \\ b_{u,v}(t) &= |det J_{L_{\psi(u,tv)}}(\psi(\sigma(u,tv),0))|. \end{aligned}$$

We have $a_{u,v} = \Lambda(\sigma(u,tv),0)$ and $a'_{u,v}(0) = \sum_{i,j} \dfrac{\partial \Lambda}{\partial u_i}(u,0)\dfrac{\partial \sigma_i}{\partial v_j}(u,0)v_j$.

The local expression of the multiplication map $L_{\psi(u,tv)} : G^{\varphi(\sigma(u,tv))} \to G^{\varphi(u)}$ is given by $p(u,tv,\cdot)$ and this shows, after some calculation,

$$b_{u,v}(t) = \begin{vmatrix} 1+tB_1(u,v,f_1) & tB_1(u,v,f_2) & \cdots & tB_1(u,v,f_m) \\ tB_2(u,v,f_1) & 1+tB_2(u,v,f_2) & \cdots & tB_2(u,v,f_m) \\ \cdots & \cdots & \cdots & \cdots \\ tB_m(u,v,f_1) & tB_m(u,v,f_2) & \cdots & 1+tB_m(u,v,f_m) \end{vmatrix}.$$

Taking the derivative we obtain $b'_{u,v}(0) = B_1(u,v,f_1) + B_2(u,v,f_2) + \cdots + B_m(u,v,f_m)$, and to finish the proof remark that $a_{u,v}(0) = \Lambda(u,0)$, $b_{u,v}(0) = 1$. Ξ

Now let $(\gamma,\hbar) \in \hat{G}$, with $\hbar \neq 0$. The smooth structure in a neighborhood of $(\gamma,\hbar)$ is given by a parametrization of the form $\hat{\psi} = \psi \times id : U \times \mathbb{R} \to \hat{G}$, where $\psi : U \to G$ is a parametrization of G in a neighborhood of γ and U is an open subset of $\mathbb{R}^{n+m}$. Write $\alpha = \hat{\psi}^{-1} : \hat{\psi}(U \times \mathbb{R}) \to U \times \mathbb{R}$, and define an element of $T_{(\gamma,\hbar)}\hat{G}$ by

$$\frac{\partial}{\partial \hbar}\hat{f}_{|(\gamma,\hbar)} = \frac{\partial \hat{f}}{\partial \alpha_{n+m+1}}(\gamma,\hbar).$$

Here $\hat{f} \in C^\infty(\hat{G})$. This definition is independent of the parametrization ψ of G. Thus we obtain a vector field $\dfrac{\partial}{\partial \hbar} \in C^\infty(\hat{G}, T\hat{G})$.

The following result is the key lemma towards the proof of Dirac's condition.

LEMMA 10.3. *Let $\hat{f},\hat{g} \in C_c^\infty(\hat{G},\hat{\lambda})$ (seen as a convolution algebra), and write $\hat{f}_0 = \hat{f}_{|\hat{G}(0)}$, $\hat{g}_0 = \hat{g}_{|\hat{G}(0)}$ for their the restrictions to the Lie algebroid $A(G)$. Then*

$$\frac{\partial}{\partial \hbar}[\hat{f},\hat{g}]_{|\hbar=0} = i\{\hat{f}_0,\hat{g}_0\}_\mu.$$

PROOF. The proof is only sketched; for details cf. [**Ra1, Ra3**]. Using a partition of unity, one may assume that $\hat{f}$ and $\hat{g}$ have their support contained in the image of the parametrization $\hat{\psi}$. Let $\hat{k}$ be the convolution of $\hat{f}$ and $\hat{g}$, and let K be the expression of $\hat{k}$ in the parametrization $\hat{\psi}$. After some calculation and the use of (9.7) we obtain

$$\begin{aligned} K(u,v,\hbar) &= \int F(u,w,\hbar)G(\sigma(u,\hbar w), v - w + \hbar B(u,w,w) \\ &- \hbar B(\sigma(u,\hbar w),w,v) + O(\hbar^2),\hbar)\mu_{u,w}(\hbar)dw. \end{aligned}$$

This leads to an expression for the commutator $\hat{f} * \hat{g} - \hat{g} * \hat{f}$. We then differentiate K with respect to $\hbar$ at $\hbar = 0$, use Lemma 10.2 and some changes of variables, and eventually recover the local expression of the Poisson bracket as given in Proposition 9.4. Ξ

We are now in a position to prove Dirac's condition (4.9) and (4.7). We only deal with the former; the latter is proved in exactly the same way. The essence of the proof is that Dirac's condition follows from the continuity condition (b) in Definition 4.5; in the context of Definition 4.3 this condition is satisfied as a consequence of the definition of a continuous field of C^*-algebras.

Define $\hat{V} \subset A(\hat{G})$ as V in Proposition 2.6, but now for the tangent groupoid $\hat{G}$ rather than G. For $f, g \in C_c^\infty(A(G))$, let $\varphi : \hat{V} \to \mathbb{C}$ be defined by

$$\begin{aligned} \varphi(x, X, 0) &= 0; \\ \varphi(x, X, \hbar) &= \frac{(f \times_\hbar g)(x, X) - (g \times_\hbar f)(x, X)}{\hbar} - i\{f, g\}(x, X) \forall \hbar \neq 0. \end{aligned}$$

Lemma 10.3 shows that $\varphi \in C^\infty(\hat{V})$, and arguments like those in the discussion of the product $f \times_\hbar g$ in section 8 show that $\operatorname{supp}\varphi$ is compact in $\hat{V}$. Write $\hat{\varphi} = \varphi \circ \hat{\alpha}^{-1} \in C_c^\infty(\hat{\alpha}(\hat{V}))$. One may regard φ as an element of $C_c^\infty(A(G) \times \mathbb{R})$ and $\hat{\varphi}$ as an element of $C_c^\infty(\hat{G})$ by extending these functions by 0 outside $\hat{V}$ and $\hat{\alpha}(\hat{V})$, respectively. Remark that for every $\hbar$ one has function $\varphi(\cdot, \hbar) \in A^0$. Applying Theorem 6.4 and recalling the notation (6.6), where we look at the $G(\hbar)$ as subgroupoids of $\hat{G}$, one then has

$$\lim_{\hbar \to 0} \|\varphi(\cdot, \hbar)\|_\hbar = \lim_{\hbar \to 0} \left\| \hat{\varphi}_{|G(\hbar)} \right\|_{C^*(G(\hbar), \lambda)} = \left\| \hat{\varphi}_{|G(0)} \right\|_{C^*(G(0), \mu)} = 0.$$

This finishes the proof of Dirac's condition. Having giving the other parts of the proof in section 8, this finishes the proof of Theorems 4.4 and 4.6.

11. Examples and comments

Since a vast number of interesting C^*-algebras are defined by some Lie groupoid, one obtains a large reservoir of examples of our theorems, of which in this section we merely scratch the surface.

EXAMPLE 11.1. *Quantization of the Lie-Poisson structure on a dual Lie algebra*

This example was already introduced in the Introduction. A Lie group is a Lie groupoid with $G_0 = e$. A left-invariant Haar measure on G provides a left Haar system; the ensuing convolution algebra $C^*(G)$ is the usual group C^*-algebra. The Poisson structure on the dual Lie algebra $A^*(G)$ is the well-known Lie-Poisson structure [**MR, Va**]. No connection is needed to define the exponential map, and one has

$$\operatorname{Exp}^L(X) = \operatorname{Exp}^W(X) = \operatorname{Exp}(X), \tag{11.1}$$

where $X \in \mathfrak{g}$ and $\operatorname{Exp} : \mathfrak{g} \to G$ is the usual exponential map. We obtain a semi-strict or strict deformation quantization of $A^*(G)$ for any Lie group G, but the situation is particularly favourable when G is exponential (in that Exp is a diffeomorphism). One may then omit the cutoff functions κ and χ in section 8, and $\mathcal{Q}_\hbar^W(f)$ in (8.2) is simply given by

$$\mathcal{Q}_\hbar^W(f) : \operatorname{Exp}(X) \to (2\pi\hbar)^{-n} \int_{\mathfrak{g}^*} e^{i\langle\theta, X\rangle/\hbar} f(\theta) d^n\theta. \tag{11.2}$$

This is precisely Rieffel's prescription [**Ri4**]; his assumption that G be nilpotent is now seen to be unnecessary in order to obtain a strict deformation quantization.

Moreover, in the exponential case our semi-strict quantization is easily shown to be strict, since the semi-norms $\|\cdot\|_\hbar$ are now norms. More generally, if a Lie groupoid is diffeomorphic to its Lie algebroid then the semi-strict deformation quantization we have given is always strict; see [**Ra1**].

EXAMPLE 11.2. *Transformation group C^*-algebras*

Let a Lie group H act smoothly on a manifold M. The transformation groupoid (or action groupoid) $G = H \times M$ is defined by the operations $s(x,m) = x^{-1}m$ and $r(x,m) = m$, so that $((x,m),(y,m')) \in G_2$ when $m' = x^{-1}m$. In that case, $(x,m)\cdot(y,x^{-1}m) = (xy,m)$. The inclusion is $m \mapsto (e,m)$, and the inverse is $(x,m)^{-1} = (x^{-1},x^{-1}m)$.

Each left-invariant Haar measure dx on H leads to a left Haar system on G. The corresponding groupoid C^*-algebra is the usual transformation group C^*-algebra $C^*(G) = C^*(H,M)$, cf. [**Re1**].

The Lie algebroid $\mathcal{H} \times M$ (where $\mathcal{H}$ is the Lie algebra of H) is a trivial bundle over M, with anchor $\rho(X,m) = -\xi_X(m)$ (the fundamental vector field on M defined by $X \in \mathcal{H}$). Identifying sections of $\mathcal{H}\times M$ with $\mathcal{H}$-valued functions $X(\cdot)$ on M, the Lie bracket on $C^\infty(M,\mathcal{H}\times M)$ is

$$[X,Y]_{\mathcal{H}\times M}(m) = [X(m),Y(m)]_{\mathcal{H}} + \xi_Y X(m) - \xi_X Y(m). \tag{11.3}$$

The associated Poisson bracket coincides with the semi-direct product bracket defined in [**KM**].

The trivial connection on $\mathcal{H}\times M \to M$ yields the exponential maps

$$\mathrm{Exp}^L(X,m) = (\mathrm{Exp}(X),m); \tag{11.4}$$

$$\mathrm{Exp}^W(X,m) = (\mathrm{Exp}(X),\mathrm{Exp}(\tfrac{1}{2}X)m). \tag{11.5}$$

The cutoff function κ in (8.2) is independent of m, and coincides with the function appearing in Example 11.1. For small enough $\hbar$, a function $f \in C^\infty_{\mathrm{PW}}(\mathcal{H}^*\times M)$ is then quantized by

$$\mathcal{Q}^W_\hbar(f) : (\mathrm{Exp}(X),m) \to (2\pi\hbar)^{-n}\int_{\mathcal{H}^*} e^{i\langle\theta,X\rangle/\hbar} f(\theta,\mathrm{Exp}(-\tfrac{1}{2}X)m)d^n\theta. \tag{11.6}$$

When $H = \mathbb{R}^n$ and M has a H-invariant measure, the map $f \to \mathcal{Q}^W_\hbar(f)$ is equivalent to the deformation quantization considered by Rieffel [**Ri2**], who already proved that it is strict.

Finally, when H is exponential we are in the situation discussed at the end of the previous example, so that our semi-strict deformation quantization is strict.

EXAMPLE 11.3. *Weyl quantization on $T^*\mathbb{R}^n$*

As already remarked in the Introduction, the tangent bundle TM of a manifold M is the Lie algebroid of the pair groupoid $G = M \times M$. Hence our formalism produces a (semi) strict deformation quantization of the cotangent bundle T^*M of any manifold with linear connection.

In order to understand Weyl quantization in the light of our formalism, we must change some signs: we add minus signs in front of the right-hand sides of (4.3) and (4.4), and change $f(\xi/\hbar)$ in (8.2) to $f(-\xi/\hbar)$ (cf. [**La3**]) for the rationale behind these signs). For $M = \mathbb{R}^n$ with flat connection (8.2) then simply becomes

$$\mathcal{Q}^W_\hbar(f)\Psi(x) = (2\pi\hbar)^{-n}\int_{T^*\mathbb{R}^n} e^{ip(x-y)/\hbar} f(p,\tfrac{1}{2}(x+y))\Psi(y)d^npd^ny, \tag{11.7}$$

where $\Psi \in L^2(\mathbb{R}^n)$; we here use the fact that $C^*(\mathbb{R}^n \times \mathbb{R}^n) = \mathfrak{B}_0(L^2(\mathbb{R}^n))$. This is precisely Weyl's original prescription, now written in a form that emphasizes its geometric origin. For $\frac{1}{2}(x+y)$ is the midpoint of the geodesic from x to y, and $x-y$ is a tangent vector to the geodesic at this midpoint. Looked at in this way, Weyl quantization may easily be generalized to manifolds with connection [**La1, La3**], and forms a special case of what we have called Weyl quantization for general Lie groupoids in this paper. This answers Rieffel's Question 20 in [**Ri5**].

The fact that this quantization is strict, and in particular satisfies (4.7), had already been proved by Rieffel [**Ri6**]. The associated continuous field of C^*-algebras has fibers $A_0 = C_0(T^*\mathbb{R}^n)$ and $A_\hbar = \mathfrak{B}_0(L^2(\mathbb{R}^n))$ for $\hbar \neq 0$. The C^*-algebra C in Definition 4.2 is $C^*(H_n)$, the group algebra of the simply connected Heisenberg group on $\mathbb{R}^n$; also see [**ENN1**]. This is indeed the C^*-algebra of the tangent groupoid of $\mathbb{R}^n$.

When M be Riemannian and exponential (in that the exponential map is a diffeomorphism between the tangent bundle TM and the pair groupoid $M \times M$), our semi-strict deformation quantization is strict [**Ra1**]. As is well known, for $M = \mathbb{R}^n$ with flat metric the product $\times_\hbar$ is then equal to the Moyal product [**Mo, BFFLS**]. For a detailed analysis of this situation in the setting of the present paper see [**CCFGRV**].

We close with two remarks.

REMARK 11.4. There are clear connections between index theory (in the sense of Atiyah–Singer), C^*-algebraic K-theory, and quantization; see, e.g., [**Fe, Hi, ENN2, Ro**]. Moreover, Connes [**Co2**] discovered a beautiful proof of the index theorem that is based on the tangent groupoid of a manifold.

Let M be a compact Riemannian manifold, with associated cosphere bundle S^*M and C^*-algebra of bounded classical pseudodifferential operators $\Psi^0(M)$. The interpretation of index theory in terms of the K-theory of C^*-algebras comes from the short exact sequence of C^*-algebras

$$0 \to \mathfrak{B}_0(L^2(M)) \to \Psi^0(M) \xrightarrow{\sigma} C(S^*M) \to 0; \tag{11.8}$$

here σ is the symbol map. Any short exact sequence $0 \to J \to A \to B \to 0$ leads to a connecting map $\partial : K^1(B) \to K_0(J)$, so that for (11.8) one obtains

$$\partial : K_1(C(S^*M)) = K^1(S^*(M)) \to K_0(\mathfrak{B}_0) = \mathbb{Z}.$$

The analytic index of an elliptic pseudodifferential operator $P \in \Psi^0(M)$ is given by $\partial([\sigma(P)]) \in \mathbb{Z}$, where, for $f \in C(X)$, $[f]$ is the class in $K^1(X)$ defined by f. One may prove the Atiyah–Singer index theorem from the existence of a strict deformation quantization $\mathcal{Q}_\hbar : C_0(T^*M) \to \mathfrak{B}_0(L^2(M))$; cf. [**Hi**].

Seen in the light of Lie groupoids and quantization, its is obvious that the above data may be generalized [**MP**] by replacing the pair groupoid on M by an arbitrary Lie groupoid G.Then $\mathfrak{B}_0(L^2(M))$ is replaced by $C^*(G)$, the cosphere bundle S^*M is replaced by a cosphere bundle $S^*A(G)$ in $A^*(G)$, and the appropriate generalization $\Psi^0(G)$ (with some abuse of notation) of $\Psi^0(M)$ has recently been defined as well [**NWX, MP**]. Thus (11.8) is generalized to

$$0 \to C^*(G) \to \Psi^0(M) \xrightarrow{\sigma} C_0(S^*A(G)) \to 0; \tag{11.9}$$

see [**MP**].

On may therefore expect that our strict deformation quantization of $A^*(G)$ eventually leads to an index theorem, where the index takes values in $K_0(C^*(G))$.

REMARK 11.5. There is no analogue of Lie's Third Theorem for Lie algebroids. Examples given in [**AM**] show the existence of Lie algebroids that are not associated to any Lie groupoid. For an arbitrary Lie algebroid $A(G)$, Pradines proved (cf. [**Pr4**]) that one can merely construct a local Lie groupoid G which has $A(G)$ as its Lie algebroid. In this paper, we always assume that we have a given Lie groupoid G with Lie algebroid $A(G)$. In order to use the ideas from this paper to quantize all Lie algebroids, we would need to extend two constructions, the C^*-algebra and the tangent groupoid which can be associated to every Lie groupoid, to the context of local Lie groupoids. The construction of the tangent groupoid has been extended to local Lie groupoids in the context of pseudo-differential operators by Nistor, Weinstein and Xu (cf. [**NWX**]). The missing ingredient in order to extend our results to arbitrary Lie algebroids is therefore the construction of the C^*-algebra of a local Lie groupoid.

Acknowledgement. B.R. is greatly indebted to Jean Renault for many helpful discussions.

References

[AM] R. Almeida, P. Molino, Suites d'Atiyah et feuilletages transversalement complets, *C. R. Acad. Sci. Paris* **A300** (1985) 13–15.

[AR] C. Anantharaman Delaroche, J. Renault, Amenable groupoids, preprint 98-05, Univ. Orleans, 1998.

[BFFLS] F. Bayen, M. Flato, C. Fronsdal, A. Lichnerowicz, and D. Sternheimer, Deformation theory and quantization I, II, *Ann. Phys. (N.Y.)* **110** (1978) 61–110, 111–151.

[Be] F.A. Berezin, General concept of quantization, *Commun. Math. Phys.* **40** (1975) 153–174.

[Bg] B. Bigonnet, Construction d'une C^*-algèbre associée à certains feuilletages de Stefan, *C. R. Acad. Sci. Paris* **307** (1988) 307–310.

[Bl] E. Blanchard, Déformations de C*-algèbres de Hopf, *Bull. Sc. math.* **124**(1) (1996) 141–215.

[CW] A. Cannas da Silva, A. Weinstein, *Geometric Models for Noncommutative Algebras*, University of California, Berkeley - AMS, Providence, 1999.

[CCFGRV] J.F. Cariñena, J. Clemente-Gallardo, E. Follana, J.M. Gracia-Bonda, A. Rivero, and J.C. Várilly, Connes' tangent groupoid and strict quantization, *J. Geom. Phys.* **32** (1999) 79–96.

[Co1] A. Connes, Sur la théorie noncommutative de l'intégration, *Lecture Notes in Mathematics* **725** (1980) 19–143.

[Co2] A. Connes, Noncommutative Geometry, Academic Press, San Diego, 1994.

[CDW] A. Coste, P. Dazord, A. Weinstein, Groupoïdes symplectiques, *Publ. Dept. Math. Univ. Lyon1* **2A** (1987) 1–62.

[Cou] T.J. Courant, Dirac Manifolds. *Trans. Amer. Math. Soc.* **319** (1990) 631–661.

[Di] P.A.M. Dirac, *The Principles of Quantum Mechanics.* Clarendon Press, Oxford, 1930.

[Di] J. Dixmier, *C^*-Algebras*, North–Holland, Amsterdam, 1977.

[ENN1] G.A. Elliott, T. Natsume, and R. Nest, The Heisenberg group and K-theory, *K-Theory* **7** (1993) 409–428.

[ENN2] G.A. Elliott, T. Natsume, and R. Nest, The Atiyah–Singer index theorem as passage to the classical limit in quantum mechanics, *Commun. Math. Phys.* **182** (1996) 505–533.

[Fe] B.V. Fedosov, *Deformation Quantization and Index Theory*, Akademie-Verlag, Berlin, 1997.

[FD] J.M.G. Fell, R.S. Doran, *Representations of *-Algebras, Locally Compact Groups, and Banach *-Algebraic Bundles, Vol. 1*, Academic Press, Boston, 1988.

[HS] M. Hilsum, G. Skandalis, Morphismes K-orientés d'espaces de feuilles et fonctorialité en théorie de Kasparov, *Ann. Scient. Ec. Norm. Sup.* **20** (1987) 325–390.

[Hi] N. Higson, On the K-theory proof of the index theorem, *Contemp. Math.* **148** (1993) 67–86.

[KW] E. Kirchberg and S. Wassermann, Operations on continuous bundles of C^*-algebras, *Math. Ann.* **303** (1995) 677–697.

[Ki] A.A. Kirillov, Merits and demerits of the orbit method, *Bull. A.M.S.* **36** (1999) 433–488.

[Ko] M. Kontsevich, M. Deformation quantization of Poisson manifolds I, e-print q-alg/9709040 (1997).

[KM] P.S. Krishnaprasad and J.E. Marsden, Hamiltonian structure and stability for rigid bodies with flexible attachments. *Arch. Rat. Mech. An.* **98** (1987) 137–158.

[La1] N.P. Landsman, Strict deformation quantization of a particle in external gravitational and Yang-Mills fields, *J. Geom. Phys.* **12** (1993) 93–132.

[La2] N.P. Landsman, Classical and quantum representation theory, in *Proc. Seminarium mathematische structuren in de veldentheorie*, eds. E.A. de Kerf and H.G.J. Pijls, *CWI syllabus* **39** (1996) 135–163.

[La3] N.P. Landsman, Mathematical Topics Between Classical and Quantum Mechanics, Springer, New York, 1988.

[La4] N.P. Landsman, Lie groupoid C^*-algebras and Weyl quantization, *Comm. Math. Phys.* **206** (1999) 367–381.

[Le] R.-Y. Lee, On the C^*-algebras of operator fields. *Indiana Univ. Math. J.* **25** (1976) 303–314.

[Li] A. Lichnerowicz, Les variétés de Poisson et leurs algèbres de Lie associées, *J. Diff. Geom.* **12** (1977) 253–300.

[Ma] K. Mackenzie, *Lie Groupoids and Lie Algebroids in Differential Geometry*, Cambridge University Press, Cambridge, 1987.

[MR] J.E. Marsden, and T.S. Ratiu, *Introduction to Mechanics and Symmetry*, Springer, New York, 1994.

[MP] B. Monthubert, and F. Pierrot, Indice analytique et groupoïdes de Lie, *C. R. Acad. Sci. Paris Sér. I Math.* **325** (1997) 193–198.

[Mo] J.E. Moyal, Quantum mechanics as a statistical theory, *Proc. Cambridge Phil. Soc.* **45** (1949), 99–124.

[vN] J. von Neumann, *Mathematische Grundlagen der Quantenmechanik*, Springer, Heidelberg, 1932.

[NWX] V. Nistor, A. Weinstein, P. Xu, Pseudodifferential operators on differential groupoids, *Pac. J. Math.* **189** (1999) 117–152.

[Pr1] J. Pradines, Théorie de Lie pour les groupoïdes différentiables. Relations entre propriétés locales et globales, *C. R. Acad. Sci. Paris* **A263** (1966) 907–910.

[Pr2] J. Pradines, Théorie de Lie pour les groupoïdes différentiables. Calcul différentiel dans la catégorie des groupoïdes infinitésimaux, *C. R. Acad. Sci. Paris* **A264** (1967) 245–248.

[Pr3] J. Pradines, Géométrie différentielle au-dessus d'un groupoïde, *C. R. Acad. Sci. Paris* **A266** (1968) 1194–1196.

[Pr4] J. Pradines, Troisième théorème de Lie pour les groupoïdes différentiables, *C. R. Acad. Sci. Paris* **A267** (1968) 21–23.

[Ra1] B. Ramazan, *Quantification par Déformation des Variétés de Lie-Poisson*, Ph.D. Thesis, University of Orléans, 1998.

[Ra2] B. Ramazan, Limite classique de C^*-algèbres de groupoïdes de Lie, *C. R. Acad. Sci. Paris* **I 329** (1999) 603–606.

[Ra3] B. Ramazan, The tangent groupoid and deformation quantization, preprint IMAR, 1999.

[Ra4] B. Ramazan, Une formule du type Baker-Campbell-Hausdorff type pour les groupoïdes de Lie, preprint, 1999, submitted to *Revue Roumaine Math. Pures Appl.*

[Re1] J. Renault, A groupoid Approach to C*-algebras, *Lecture Notes in Mathematics* **793** (1980).

[Re2] J. Renault, Représentation des produits croisés d'algèbres de groupoïdes, *J. Operator Theory* **18** (1987) 67–97.

[Re3] J. Renault, The ideal structure of groupoid crossed product C*-algebras, *J. Operator Theory* **25** (1991) 3–36.

[Re4] J. Renault, Introduction aux C^*-algèbres. Eléments de KK-Théorie, Notes de cours, Ecole d'été CIMPA, Sfax septembre 1995.

[Ri1] M.A. Rieffel, Continuous fields of C^*-algebras coming from group cocycles and actions, *Math. Ann.* **283** (1989) 631–643.

[Ri2] M.A. Rieffel, Deformation quantization of Heisenberg manifolds, *Comm. Math. Phys.* **122** (1989) 531–562.

[Ri3] M.A. Rieffel, Deformation quantization for operator algebras, *Proc. Symp. Pure Math.* **51** (1990) 411–423.

[Ri4] M.A. Rieffel, Lie group convolution algebras as deformation quantization of linear Poisson structures, *Amer. J. Math.* **112** (1990) 657–686.

[Ri5] M.A. Rieffel, Deformation quantization for actions of R^d, *Mem. A.M.S.* **106**, no.506 (1993).

[Ri6] M.A. Rieffel, Quantization and C^*-algebras, in *C^*-algebras: 1943–1993, Contemp. Math.* **167** (1994) 66–97.

[Ri7] M.A. Rieffel, Questions on quantization, in *Operator algebras and operator theory, Contemp. Math.* **228** (1998) 315–326.
e-print quant-ph/9712009.

[Ro] J. Rosenberg, Behavior of K-theory under quantization, in: *Operator Algebras and Quantum Field Theory*, eds. Doplicher, S., R. Longo, J.E. Roberts, and L. Szido, pp. 404–415, International Press, Cambridge (MA), 1996.

[To] A.M. Torpe, K-theory for the leaf space of foliations by Reeb components, *J. Funct. Anal.* **61** (1985) 15–71.

[Tr] F. Treves, Introduction to Pseudodifferential and Fourier Integral Operators, *Vol. 2*, Plenum Press, New York, 1980.

[Va] I. Vaisman, *Lectures on the Geometry of Poisson Manifolds*, Birkhäuser, Basel, 1994.

[Ve] J. Vey, Déformation du crochet de Poisson sur une variété symplectique, *Commun. Math. Helv.* **50** (1975) 421–454.

[We1] A.Weinstein, Blowing up realizations of Heisenberg-Poisson manifolds, *Bull. Sc. math.* **(2) 113** (1989) 381-406.

[We2] A.Weinstein, Poisson geometry. *Diff. Geom. Appl.* **9** (1998) 213–238.

[Wo] N.M.J. Woodhouse, *Geometric Quantization*, 2nd ed. Clarendon Press, Oxford, 1992.

Korteweg-de Vries Institute for Mathematics, University of Amsterdam, Plantage Muidergracht 24, 1018 TV Amsterdam,The Netherlands
E-mail address: `npl@wins.uva.nl`

Institute of Mathematics of the Romanian Academy, Calea Griviţei 21, P.O. Box 1-764, 70700 Bucharest, Romania
E-mail address: `ramazan@pompeiu.imar.ro`

Selected Titles in This Series

(Continued from the front of this publication)

257 **Peter A. Cholak, Steffen Lempp, Manuel Lerman, and Richard A. Shore, Editors,** Computability theory and its applications: Current trends and open problems, 2000

256 **Irwin Kra and Bernard Maskit, Editors,** In the tradition of Ahlfors and Bers: Proceedings of the first Ahlfors-Bers colloquium, 2000

255 **Jerry Bona, Katarzyna Saxton, and Ralph Saxton, Editors,** Nonlinear PDE's, dynamics and continuum physics, 2000

254 **Mourad E. H. Ismail and Dennis W. Stanton, Editors,** q-series from a contemporary perspective, 2000

253 **Charles N. Delzell and James J. Madden, Editors,** Real algebraic geometry and ordered structures, 2000

252 **Nathaniel Dean, Cassandra M. McZeal, and Pamela J. Williams, Editors,** African Americans in Mathematics II, 1999

251 **Eric L. Grinberg, Shiferaw Berhanu, Marvin I. Knopp, Gerardo A. Mendoza, and Eric Todd Quinto, Editors,** Analysis, geometry, number theory: The Mathematics of Leon Ehrenpreis, 2000

250 **Robert H. Gilman, Editor,** Groups, languages and geometry, 1999

249 **Myung-Hwan Kim, John S. Hsia, Yoshiyuki Kitaoka, and Rainer Schulze-Pillot, Editors,** Integral quadratic forms and lattices, 1999

248 **Naihuan Jing and Kailash C. Misra, Editors,** Recent developments in quantum affine algebras and related topics, 1999

247 **Lawrence Wasson Baggett and David Royal Larson, Editors,** The functional and harmonic analysis of wavelets and frames, 1999

246 **Marcy Barge and Krystyna Kuperberg, Editors,** Geometry and topology in dynamics, 1999

245 **Michael D. Fried, Editor,** Applications of curves over finite fields, 1999

244 **Leovigildo Alonso Tarrío, Ana Jeremías López, and Joseph Lipman,** Studies in duality on noetherian formal schemes and non-noetherian ordinary schemes, 1999

243 **Tsit Yuan Lam and Andy R. Magid, Editors,** Algebra, K-theory, groups, and education, 1999

242 **Bernhelm Booss-Bavnbek and Krzysztof Wojciechowski, Editors,** Geometric aspects of partial differential equations, 1999

241 **Piotr Pragacz, Michał Szurek, and Jarosław Wiśniewski, Editors,** Algebraic geometry: Hirzebruch 70, 1999

240 **Angel Carocca, Víctor González-Aguilera, and Rubí E. Rodríguez, Editors,** Complex geometry of groups, 1999

239 **Jean-Pierre Meyer, Jack Morava, and W. Stephen Wilson, Editors,** Homotopy invariant algebraic structures, 1999

238 **Gui-Qiang Chen and Emmanuele DiBenedetto, Editors,** Nonlinear partial differential equations, 1999

237 **Thomas Branson, Editor,** Spectral problems in geometry and arithmetic, 1999

236 **Bruce C. Berndt and Fritz Gesztesy, Editors,** Continued fractions: From analytic number theory to constructive approximation, 1999

235 **Walter A. Carnielli and Itala M. L. D'Ottaviano, Editors,** Advances in contemporary logic and computer science, 1999

234 **Theodore P. Hill and Christian Houdré, Editors,** Advances in stochastic inequalities, 1999

For a complete list of titles in this series, visit the AMS Bookstore at **www.ams.org/bookstore/**.